P9-CFQ-698

BELTSVILLE SYMPOSIA in AGRICULTURAL RESEARCH

A Series of Annual Symposia Sponsored by
THE BELTSVILLE AGRICULTURAL RESEARCH CENTER
Northeastern Region, Science and Education Administration
United States Department of Agriculture

[6] Strategies of Plant Reproduction

PREVIOUS SYMPOSIA IN THIS SERIES:

[1] Virology in Agriculture
May 10–12, 1976
Published, 1977

[2] Biosystematics in Agriculture
May 8–11, 1977
Published, 1978

[3] Animal Reproduction
May 14–17, 1978
Published, 1979

[4] Human Nutrition Research
May 6–9, 1979
Published, 1981

[5] Biological Control in Crop Production
May 18–21, 1980
Published, 1981

Beltsville Symposia in Agricultural Research

[6] Strategies of Plant Reproduction

Werner J. Meudt, Editor

Invited papers presented at a symposium held
May 17–20, 1981, at the Beltsville Agricultural
Research Center (BARC), Beltsville, Maryland 20705

Organized by THE BARC SYMPOSIUM VI COMMITTEE
Albert A. Piringer, Chairman

Sponsored by
THE BELTSVILLE AGRICULTURAL RESEARCH CENTER
Northeastern Region, Agricultural Research Service
United States Department of Agriculture

ALLANHELD, OSMUN Publishers

GRANADA London Toronto Sydney

Dedicated to the Memory
of
Dr. Sterling B. Hendricks
1902–1981

ALLANHELD, OSMUN & CO. PUBLISHER, INC.

Published in the United States of America in 1983
by Allanheld, Osmun & Co. Publishers, Inc.
(A Division of Littlefield, Adams & Company)
81 Adams Drive, Totowa, New Jersey 07512

First published in Great Britain 1983 by Granada Publishing
Granada Publishing Limited – Technical Books Division
Frogmore, St Albans, Herts AL2 2NF
and
36 Golden Square, London W1R 4AH
117 York Street, Sydney, NSW 2000 Australia
61 Beach Road, Auckland, New Zealand
ISBN 0 246 11946 2

Library of Congress Cataloging in Publication Data
Main entry under title:

Strategies of plant reproduction.

(Beltsville symposia in agricultural research; 6)
Includes indexes.
1. Plants—Reproduction—Congresses. I. Meudt, Werner J. II. BARC Symposium VI Committee. III. Beltsville Agricultural Research Center. IV. Series.
QK825.S87 1982 582′.016 82-11594
ISBN 0-86598-054-3

83 84 85 / 10 9 8 7 6 5 4 3 2 1
Printed in the United States of America

Granada®
Granada Publishing®

Contributors and Their Affiliations

Yosef Ben-Tal
Department of Olive and Viticulture
Institute of Horticulture
The Volcani Center
Bet-Dagan, Israel

Udo Blum
Botany Department
North Carolina State University
Raleigh, North Carolina 27650

Lowell E. Campbell
Agricultural Engineering Laboratory
USDA-ARS
Beltsville Agr. Research Center
Beltsville, Maryland 20705

H. Marc Cathey
National Arboretum
USDA-ARS
Washington, D.C. 20002

Meryl N. Christiansen
Plant Physiology Institute
USDA-ARS
Beltsville Agr. Research Center
Beltsville, Maryland 20705

Charles F. Cleland
Smithsonian Institution
Radiation Biology Laboratory
12441 Parklawn Drive
Rockville, Maryland 20852

Perry B. Cregan
Cell Culture & Nitrogen Fixation Lab
USDA-ARS
Beltsville Agr. Research Center
Beltsville, Maryland 20705

Wayne L. Decker
Department of Atmospheric Science
University of Missouri-Columbia
Columbia, Missouri 65211

Gerald F. Deitzer
Smithsonian Institution
Radiation Biology Laboratory
12441 Parklawn Drive
Rockville, Maryland 20852

Robert J. Downs
Southern Plant Envir. Laboratories
North Carolina State University
2003 Gardner Hall
Raleigh, North Carolina 27650

D. A. Evans
DNA Plant Technology Corporation
2611 Branch Pike
Cinnaminson, New Jersey 08077

C. E. Flick
DNA Plant Technology Corporation
2611 Branch Pike
Cinnaminson, New Jersey 08077

W. P. Hackett
Department of Environmental Horticulture
University of California
Davis, California 95616

J. R. Harlan
Crop Evolution Laboratory
Agronomy Department
University of Illinois
Urbana, Illinois 61801

Allen S. Heagle
USDA-ARS
Plant Pathology Department
North Carolina State University
Raleigh, North Carolina 27650

Walter W. Heck
USDA-ARS
Botany Department
North Carolina State University
Raleigh, North Carolina 27650

J. Heslop-Harrison
Welsh Plant Breeding Station
Plas Gogerddan, Aberystwyth
SY 23 3EB, UK

Hans Kende
MSU-DOE Plant Research Laboratory
Michigan State University
East Lansing, Michigan 48824

H. F. Linskens
Department of Botany
Molecular Biology Department
University Nijemgen
The Netherlands

Robert Ornduff
Department of Botany
University of California
Berkeley, California 94720

Lee H. Pratt
Botany Department
University of Georgia
Athens, Georgia 30602

Richard A. Reinert
USDA-ARS
Plant Pathology Department
North Carolina State University
Raleigh, North Carolina 27650

Roy M. Sachs
Department of Environmental Horticulture
University of California
Davis, California 95616

W. R. Sharp
DNA Plant Technology Corporation
2611 Branch Pike
Cinnaminson, New Jersey 08077

H. E. Sommer
School of Forest Resources
University of Georgia
Athens, Georgia 30602

George L. Steffens
Fruit Laboratory
USDA-ARS
Beltsville Agr. Research Center
Beltsville, Maryland 20705

R. W. Thimijan
Agri. Engineering Laboratory
USDA-ARS
Beltsville Agr. Research Center
Beltsville, Maryland 20705

William J. VanDerWoude
Seed Research Laboratory
USDA-ARS
Beltsville Agr. Research Center
Beltsville, Maryland 20705

Daphne Vince-Prue
Glasshouse Crops Research Institute
Littlehampton BN 16 3PU
Sussex, UK

P. E. Wareing
Department of Botany & Microbiology
The University College of Wales
Aberystwyth, Dyfed SY23 3DA, UK

Max W. Williams
USDA-ARS
The Fruit Research Laboratory
Wenatchee, Washington 98801

Harold W. Woolhouse
John Innes Institute
Norwich NR4 7UH
Norfolk, UK

Symposium Organization

Beltsville Agricultural Research Center (BARC)

P. A. PUTNAM, DIRECTOR

BARC SCIENCE SEMINAR COMMITTEE

A. DeMilo, Chairman
M. Bakst
S. Batra
G. Carpenter
R. Jasper
W. Meudt
R. Romanowski
R. Sayre
H. Schoene
H. Skoog
W. Wergin
C. Tabor
R. Zimmerman

BARC SYMPOSIUM VI SUBCOMMITTEE

A. A. Piringer, Chairman

ORGANIZING COMMITTEE

A. A. Piringer, Chairman
H. M. Cathey
M. N. Christiansen
M. Faust
W. J. Meudt
J. Romberger
G. Schaeffer
W. VanDerWoude

POSTER SESSION

T. Devine

LOGO

H. M. Cathey
W. J. Meudt

FINANCE COMMITTEE

H. Moline, Chairman
C. Tabor

LOCAL ARRANGEMENT

J. Anderson, Chairman
L. Campbell
G. Carpenter
W. Conway
J. Maas
R. Yaklich

PUBLICITY

W. J. Meudt
W. VanDerWoude

Contents

[one] Introduction

1 THE REPRODUCTIVE VERSATILITY OF FLOWERING PLANTS: AN OVERVIEW by J. Heslop-Harrison 3

[two] Transfer of Information and Genetic Interplay

2 INTERPRETATIONS OF SEX IN HIGHER PLANTS by Robert Ornduff 21

3 POLLINATION PROCESSES: UNDERSTANDING FERTILIZATION AND LIMITS TO HYBRIDIZATION by H. F. Linskens 35

4 DIRECTING THE ACCELERATED EVOLUTION OF CROP PLANTS by Jack R. Harlan 51

[three] Photocontrol Systems

5 PHOTOPERIODIC CONTROL OF PLANT REPRODUCTION by Daphne Vince-Prue 73

6 EFFECT OF FAR-RED ENERGY ON THE PHOTOPERIODIC CONTROL OF FLOWERING IN WINTEX BARLEY (*HORDEUM VULGARE* L.) by Gerald F. Deitzer 99

7 MOLECULAR PROPERTIES OF PHYTOCHROME AND THEIR RELATIONSHIP TO PHYTOCHROME FUNCTION by Lee H. Pratt 117

8 MECHANISMS OF PHOTOTHERMAL INTERACTIONS IN THE PHYTOCHROME CONTROL OF SEED GERMINATION by William J. VanDerWoude 135

[four] Hormonal Control Systems

9 SOME CONCEPTS CONCERNING THE MODE OF ACTION OF PLANT HORMONES by Hans Kende 147

10 HORMONAL REGULATION OF FLOWERING AND SEX EXPRESSION by Charles F. Cleland and Yosef Ben-Tal 157

11 HORMONAL CONTROL OF STOLEN AND TUBER DEVELOPMENT, ESPECIALLY IN THE POTATO PLANT by P. F. Wareing 181

12 HORMONAL CONTROL OF PLANT EMBRYOGENY AND SYNTHESIS OF EMBRYO-SPECIFIC PROTEINS by Ian Sussex 197

13 HORMONAL CONTROL OF SENESCENCE ALLIED TO REPRODUCTION IN PLANTS by Harold W. Woolhouse 201

[five] Management and Control of Plant Reproduction

14 OPPORTUNITIES AND NEEDS TO CONTROL PLANT REPRODUCTION by M. N. Christiansen and George L. Steffens 237

15 GENETIC CONTROL OF NITROGEN METABOLISM IN PLANT REPRODUCTION by P. B. Cregan 243

16 SOURCE-SINK RELATIONSHIPS AND FLOWERING by R. M. Sachs and W. P. Hackett 263

17 MANAGING FLOWERING, FRUIT SET, AND SEED DEVELOPMENT IN APPLE WITH CHEMICAL GROWTH REGULATORS by Max W. Williams 273

18 STRATEGIES AND SPECIFICATIONS FOR MANAGEMENT OF IN VITRO PLANT PROPAGATION by W. R. Sharp, D. A. Evans, C. E. Flick and H. E. Sommer 287

[six] Environmental and Stress Factors

19 SIMULATION OF ENVIRONMENTAL IMPACTS ON PRODUCTIVITY by Wayne L. Decker and Clarence M. Sakamoto 307

20 RADIATION AND PLANT RESPONSE: A NEW VIEW by H. M. Cathey, L. E. Campbell and R. W. Thimijan 323

21 EFFECTS OF AIR POLLUTION ON CROP PRODUCTION
by Walter W. Heck, Udo Blum, Richard A. Reinert
and Allen S. Heagle 333

22 CLIMATE SIMULATIONS
by Robert J. Downs 351

INDEXES 369

Foreword

This is the sixth Beltsville Symposium in Agricultural Research sponsored by the Beltsville Agricultural Research Center, U.S. Department of Agriculture. The previous symposia in this series were (I) Virology in Agriculture (1976); (II) Biosystematics (1977): (III) Animal Reproduction (1978); (IV) Human Nutrition Research: Questions and Answers (1979); and (V) Biological Control in Crop Production (1980). Symposium (VII) for 1982 will be entitled "Genetic Engineering as a Tool in Agricultural Research."

Symposium VI was planned as a modern synopsis of plant reproduction with emphasis on both recent research and possible applications in crop reproduction.

When scientists select and develop crop plants, they play a vital role in changing the normal course of events that mother nature has imposed on plants. Perhaps one of the most drastic of these changes is the limiting of the genetic variability that enables these plants to survive stress. Survival of crop plants depends on our constant concern during cultivation. We are engaged in a continued battle between discovering new crops that can survive the constant pressures imposed by climate pests and the need to improve crop plant production. This requires our good understanding of these natural processes if we are to nationally apply our knowledge in manipulating and guiding the plants' own strategies of reproduction and development.

Paul A. Putnam, Director
Beltsville Agricultural Research Center

one
INTRODUCTION

1] The Reproductive Versatility of Flowering Plants: An Overview

by J. HESLOP-HARRISON*

In contemplating the wide range of reproductive mechanisms found among the flowering plants, clearly the first polarization to note is between those systems that involve a sexual process and those that do not. Most angiosperms show some form of sexual behavior at some period of their existence; many combine with it a capacity for asexual reproduction, and a minority today reproduce primarily by such asexual means, having abandoned sexuality. Sexuality is of course the engine for generating variation, and there seems no reason for departing from the proposition inherent in the writings of Weismann of almost a century ago that (in the words of Darlington, 1939) "Sexual reproduction is . . . a mechanism which secures the greatest possibilities of recombination of genetic differences. This is its one primary and universal function. All others derive from it." Asexual reproduction, on the other hand, by short-circuiting the twin processes of meiosis and syngamy, eludes their consequences—gene segregation and recombination—and so freezes the flow of variation.

Historically, our understanding of the way genetical variation may be generated, recombined, exposed, conserved, concealed, and lost in the course of sexual reproduction dates from the 1930s and 1940s, notably from Darlington's *The Evolution of Genetic Systems* (1939), from which the above quotation is taken, and Mather's paper on polygenic inheritance and natural selection of 1943. In more recent decades, wide-ranging discussions of the functioning of genetic systems in plant evolution have been given in Stebbins's *Variation and Evolution in Plants* (1950) and *Chromosomal Evolution in Higher Plants* (1971), Mather's *Genetical Structure of Populations* (1973) and in Grant's paper on the regulation of recombination in plants (1958) and his *Plant Speciation* (1971). Notwithstanding the occasional polemics, I know of nothing in the current literature of substance sufficient to require radical revision of the broadly agreed basic positions set out in these publications.

So how, then, might we define the genetic system? Darlington (1939) identified the three principal components as (slightly paraphrased) (a) the

*Welsh Plant Breeding Station, Plas Gogerddan, Aberystwyth SY23 3EB, UK.

STRATEGIES OF PLANT REPRODUCTION (BARC Symposium number 6—Werner J. Meudt, ed.)
Allanheld, Osmun, Totowa

chromosomal system, which subsumes the molecular mechanism of gene segregation and recombination; (b) the breeding system, which governs the level of hybridity in a population; and (c) the intrinsic and extrinsic factors determining the interaction in sizes of breeding groups. In what follows I shall not be concerned at all with (a) and (c), although much with aspects of (b). But we must first consider some of the features that surround, support, and facilitate the primary sexual system in the flowering plants.

ADJUNCTS OF SEXUALITY

The cardinal points of the sexual cycle are meiosis and syngamy, and in the angiosperms, as in all vascular plants and many lower plants, these events are separated in time by the intercalation of a haploid phase, the gametophyte generation. The angiosperm gametophytes are unisexual and much reduced structurally, in the male to two cell generations in the developing pollen grain and pollen tube, and in the female, in general, to three in the development of the embryo sac. Fertilization is siphonogamous, the gametes being delivered into the embryo sac via the pollen tube that enters one of the cells of the sac, a synergid, where two male gametes are discharged. Emerging from this cell, the gametes—or at least the nuclei they convey, for as yet there is some uncertainty about what actually is transferred—pass toward and fuse with the egg and the endosperm nucleus, the latter itself a product of the fusion of two other nuclei of the sac. This phenomenon of double fertilization is a strikingly unique characteristic of the group—not universal among all members, but one of the unmistakable stigmata of angiospermy, nevertheless.

Commonly, the properties and behavior of the gametophytes are ignored in consideration of angiosperm genetics, and for that matter they also attract little attention physiologically (Heslop-Harrison 1979a). Yet they are not unimportant factors in the reproductive system. The invention of siphonogamy, present also in the higher gymnosperms, conferred a benefit of inestimable value for the further evolution of land plants, namely independence from water as a medium of fertilization. Undoubtedly this facilitated the wholesale colonization of drier habitats, but it produced at the same time a new situation without direct parallel in any other organisms. At each step in the prelude to fertilization—pollen capture, germination, tube penetration and growth—there are opportunities for interaction between a haploid male gametophyte and tissues of the diploid, sporophytic parent of the prospective female gametophytic partner; and the special circumstance is that the male gametophyte, unlike an animal gamete, possesses some residual capacity for gene transcription. Evidently the angiosperms early seized upon this as a means of regulating mating behavior both within and between species, involving the haploid genome in the process.

On the female side, we are only now beginning to make some sense of the functions of the embryo sac, and in doing so to acknowledge its remarkable role in the reproductive economy. That it should turn out to have significant functions is perhaps not too surprising, since the embryo sac is, after all, a conserved element of the angiosperm state which, judging from its taxonomic distribution, must have been present from the time of the earliest ancestors of the group (Heslop-Harrison 1958). The treasuring of a structure of such strange character, and the equally strange double fertilization associated with it, over perhaps 100+ million years of evolution, can only mean that it confers unique advantages, and advantages of a kind that can be foregone only at some peril. The 8-nucleate type is the modal one for the group, and it is a striking fact that while the commonest origin is monosporic, the same 8-nucleate phenotype may be reached through bisporic and tetrasporic pathways, and even through pathways involving nuclear fusions (Maheshwari 1948; review, Heslop-Harrison 1972). The form, and presumably the activities associated with it, are conserved; but the means of attaining that form have been free to vary. One is reminded of that fundamental element of cell metabolism, the ribosome, the spatial structure of which has evidently been conserved during evolution even though its components have varied. As to what the special functions of the embryo sac actually are, one might contemplate first its state of isolation in the nucellus of the ovule, surrounded by parental tissues, but invested by its own membranes and without plasmodesmatal connections outside its boundaries (Fig.1.1). The circumstance strongly suggests a prophylactic role (Heslop-Harrison 1972), and this surmise is now supported by direct observation of the exclusion of viruses (Carroll and Mayhew 1976). As for the internal organization of the embryo sac, the modal 8-nucleate state is certainly associated with functional differentiation. It seems assured that autonomous action of the embryo sac is required for the guidance of the pollen tube (Schwemmle 1968; review, Heslop-Harrison 1979b), and circumstantial evidence suggests that this guidance is supplied by the secretory activity of the synergids. One of the pair of synergids is concerned also in receiving the pollen tube, and in doing so it performs another sanitary function, that of ridding the male gametes of tube cytoplasm before the fertilizations. The antipodals, at least in some families, appear to be specialized for synthesis and secretion as attested by their cytoplasmic organization and the polytene state of their nuclei, and presumably they play a part in the early nutrition of the zygote and embryo before the development of the endosperm. The significance of the double fertilization itself and the consequent triploidy of the endosperm remain obscure, but the fact that its function, too, is rarely surrendered—and then only in a few advanced and specialized families such as the Orchidaceae—leaves little doubt that it also is a closely guarded and valuable part of the general angiospermic inheritance.

The alternation of generations is normally locked to the cardinal points of the sexual cycle, the sporophyte-gametophyte transition to meiosis, and the gametophyte-sporophyte transition to syngamy. One of the concomitants of meiosis is the severance of the meiocytes from the parental tissues and the development of ensheathing membranes and walls. We have shown for the male meiocyte that this is part of a remarkable sequence of changes in the pattern of compartmentation in the anther, and that this radically affects the movement of certain metabolites (Heslop-Harrison and Mackenzie 1967). So far as the sporogenous tissue and its products are concerned, the events

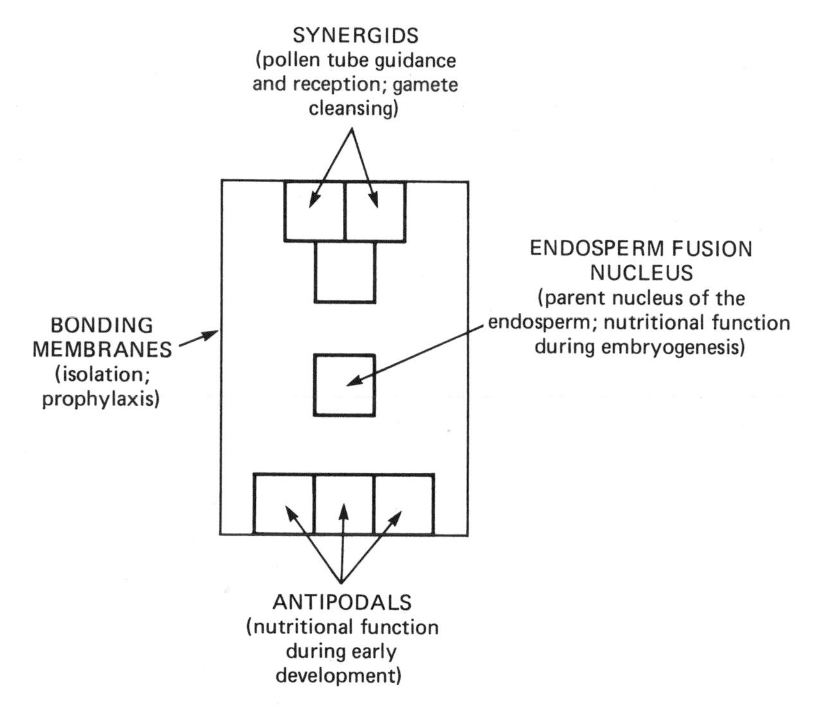

Figure 1.1. An 8-nucleate embryo sac, showing the known or surmised functions of the various components.

presumably provide the isolation necessary for the expression of the genetic independence of the gametophytes during their early development (Heslop—Harrison 1964, 1979b). At the same time, the cytoplasm and the organelles of the meiocytes undergo a radical cycle of dedifferentiation and redifferentiation. This involves the elimination of the cytoplasmic ribosomes and concomitant changes in the organelles (Mackenzie, Heslop-Harrison, and Dickinson 1967; Dickinson and Heslop-Harrison 1970). This event, also, can be interpreted in relation to the sporophyte-gametophyte transition, and we believe that it is actually the process whereby the cytoplasm is restandardized and cleansed of its residual diplophase information before the activation of the gametophytic genome (for general reviews, see Heslop-Harrison 1971, 1979b; Dickinson and Heslop-Harrison 1977). It seems an acceptable extension to this idea that the cytoplasmic clean-up might also impinge upon infective entities present in the mother cell, removing them before the sealing of the walls that provide protection within which the gametophytes—and on the female side, also the embryo—will later accomplish their differentiation. For plants, anyway, we probably need look no further than this for an explanation of the long-postulated "lineage-rejuvenating" properties of sexual reproduction.

My use of the term "adjunct" in the title of this section was well considered. The processes outlined in the immediately foregoing paragraphs are associated with sexuality in the angiosperm and normally locked in time to the sexual cycle; but the link is not indissoluble, for they are not parts of the primary sexual process. The gametophyte-sporophyte alternation can proceed without chromosome reduction and increase, and evidently the extraneous benefits arising from the elaborate compartmentations, dedifferentiations and redifferentiations can be taken without any change whatever in the nuclear constitution.

The same may be said of the seed habit, similarly associated with the sexual cycle. As with siphonogamy, the invention of the seed—made likewise by pre-angiosperms—was a major factor in the invasion of the dry land. The protection and cossetting of the embryo and the furnishing of means for its dispersal marked an evolutionary advance comparable with internal gestation in the mammals, and it evidently yielded similar advantages in opening up new ecological opportunities. But here again we have an adaptation that however closely linked with sexuality is not part of it; and some of the benefits of seminifery may be taken without any alternation of generations whatever, and possibly also without the reduction of the diploid soma to a single isolated cell, which true alternation necessarily involves.

THE SEXUAL SYSTEM AND THE CONTROL OF BREEDING BEHAVIOR

Nothwithstanding various deviations, it is evident enough that certain basic elements of the sexual reproductive systems and its adjuncts—notably among the latter the alternation of generations itself, and various organizational and functional characteristics of the gametophytes—are virtually universal among the flowering plants. Overlying these strata of broadly uniform structure and behavior there are countless variations in the devices by which the consummation of the sexual union—initiated by the actual physical contact of the male and female gametophytes—is eventually attained.

The devices are, of course, the elements of the breeding system. Their overall functions are (a) the general one of ensuring that the opportunity for sexual union actually occurs and (b) to regulate which unions will ultimately be permitted. The devices under (a) are those that harness animate or inanimate agencies in the environment to convey the male gametophytes to the vicinity of the females, essential for bisexual reproduction in sessile organisms; under (b) are those concerned with manipulating the agencies under (a) to enhance the probability of certain unions against others, or which achieve the same end by imposing physiological controls upon the male gametophyte after its transfer.

These may perhaps seem very cumbersome ways of referring to pollination mechanisms and incompatibility systems, but I use this form of statement to emphasize that achieving gamete transfer and imposing choice on what is transferred—or permitted to do the job after transfer—are often closely associated or even inseparable functions. The point is easily illustrated from many examples of insect pollination. As Sprengel and, later, Darwin demonstrated more than a century ago, the angiosperms show exceptional virtuosity in exploiting the sentience of animals to achieve their reproductive ends. Presumably this arose through a long history of co-evolution, the most dramatic culmination being in the relationship of insect and flower. The foraging behavior of sentient pollen vectors is not random, but is determined by the characteristics of the flower; and these may be concerned not simply with ensuring pollination but with ensuring particular *kinds* of pollination. Accordingly, flower character can be a critical factor in achieving selective gamete transfer, and so it can indeed form a very flexible component of the breeding system.

I have listed this and other aspects of flower structure and physiology that seem to have impact on the breeding system in Table 1.1. A further word seems necessary concerning diploid (sporophytic) sexuality as expressed in flower type. The data tabulated by Yampolsky and Yampolsky (1922) provide an interesting basis for comparison with other groups. They show that the dioecy, overwhelmingly the most common situation in animals, occurs in less than 4% of angiosperms, and that sexual polymorphism of any kind is found in only some 7% of dicotyledonous genera and 6% of monocotyledons. Of the monomorphic genera, some 78% are hermaphrodites, taking dicotyledons and monocotyledons together, the remainder showing various forms of monoecy.

Whatever the other consequences of dioecy may be, the primary one is that it must enforce cross-pollination and so promote outbreeding. The same must be true in varying degree of androdioecy, gynodioecy, and other kinds of sexual polymorphism; and one may surmise that monoecy can have a similar consequence. Yet it is clear enough that the separation of the sexes, whether on the same or different individuals, is scarcely the preferred way of achieving outbreeding among the flowering plants; and as various authors have noted, the cost of attaining it this way may be high in overall reproduction competence. The other devices listed in Table 1.1 evidently fit the general angiosperm economy more effectively. The prevalence of self-incompatibility, now recorded on reasonably secure experimental grounds from several thousand species widely distributed among the 330 or so families, indicates clearly that this, in its various forms, is the principal mechanism in the group.

This chapter is scarcely the appropriate context for a general discussion of outbreeding mechanisms, but some comment is required when they constitute so clearly a central feature of angiosperm reproductive systems. Darwin (1876, 1877) perceived this, and addressed himself also to the question, what benefit accrues from outbreeding? His answer is well known: it was that inbreeding, the converse of outbreeding, is deleterious (as he proved in several genera by experiment); therefore natural selection will favor those races or species that avoid it, since they will leave more vigorous progeny. The later unraveling of the genetical significance of sex led to a very different view, namely that outbreeding is favored because it raises the level of heterozygosity in the population and leads to a freer release of variation and so greater evolutionary flexibility. Of the various earlier statements of this proposition and its implications, that of Mather (1943) was perhaps the most clearly argued.

Yet many still find Darwin's explanation—that inbreeding is to be avoided because it is deleterious—difficult to abandon, and some have attempted to

Table 1.1 Structural and Functional Features Associated with the Control of the Breeding System

Dependent on structural adaptations
- Polyoecy and monoecy
- Heterostyly, and other structural features of the flower influencing the distribution of pollen
- Flower characteristics affecting the foraging behavior of pollen vectors

Dependent on developmental timing
- Dichogamy, including protandry and protogyny

Dependent on control of the pollen tube
- Self-incompatibility systems, including certation effects

elevate inbreeding depression to the status of the prime factor in bringing about the evolution of outbreeding systems. Selfing does often, but not of course invariably, entail inbreeding depression. Various reasons have been advanced to explain this (Wright 1977), including the propositions—each well enough supported in some context or the other—that inbreeding may be deleterious because it exposes recessive lethals (e.g., Darlington 1939), and that outbreeding may be favorable because it generates heterozygosity, which is desirable *per se* because it provides biochemical diversity in the individual (e.g., Haldane 1954; Lerner 1954). As Mather (1973) has shown, the latter proposition is difficult to support as a generalization for plants. Indeed, it is to Mather that we owe the most convincing explanation of the situation, based on the idea of functional genic balance—*within* the haploid genomes in the inbreeder, and *within* and *between* the genomes in the heterozygous outbreeder. A change in the breeding habit in either inbreeder or outbreeder leads to the loss of balance and so of fitness, and this is only to be made good by further adjustment under selection. Accordingly, balance and breeding system "must change together and in step with one another, and the need for continuing co-ordination must require the changes to be suitably gradual, step by step, rather than abrupt" (Mather 1973). Such a view leaves unscathed the fundamental proposition that the long-term advantage of outbreeding system lies in the adaptive flexibility they offer through facilitating the flow of variability.

ASEXUAL AND SUB-SEXUAL SYSTEMS

The flowering plants practice a wide variety of asexual reproductive methods, either complementary to the sexual process or in replacement for it. A simplified classification is given in Table 1.2, which is based upon the principles set out by Gustafson in his monograph of 1947 (Heslop-Harrison 1972).

The easiest processes to understand are those coming under the general heading of vegetative reproduction. These depend upon the severance and

Table 1.2 Classification of Apomictic Phenomena

A. Vegetative reproduction. A propagule other than the seed is employed; seminifery is abandoned.

B. Agamospermy. The seed is retained, but the sexual process is abandoned
   1. Adventitious embryony. The embryo forms from somatic tissue; the gametophytes are abandoned.
   2. Gametophytic apomixis. The embryo sac is diploid, and the egg develops parthenogenetically.
      i. Apospory. The embryo sac is formed from a diploid somatic cell.
      ii. Diplospory. The embryo sac is formed from an archesporial cell, by the elimination of meiosis, or meiotic failure followed the restitution of a diploid nucleus.

separate establishment as physiologically independent individuals of segments of the parent; the propagules may be little modified for a reproductive function, or they may be almost as well adapted as a seed for dispersal and propagation, being furnished with the appropriate reserves and protected by resistant ensheathments. It is a trivial semantic point whether we refer to reproduction by these means as apomixis, although if the word is to be used at all in this connection it is as well to follow Stebbins (1950) and apply it only when the vegetative process displaces the sexual more or less completely. The essential feature is that vegetative propagules carry the parental nuclear genotype unchanged, and—perhaps almost as significant—that they also retain the parental organellar genotypes passed on through the cytoplasm. Further: by the very nature of their origin, they inevitably risk taking as part of their inheritance such systemically infective pathogens as may be present in the tissues of the parent at the time of their formation.

The agamospermic systems use the vehicle of the seed, and the propagules gain thereby the advantages of seminifery in respect to protection and dispersal, and may share similar physiological properties in resistance to environmental extremes and control of dormancy. The extent to which the special advantages associated with the alternation of generations outlined above are gained depends on the precise mechanism. The principal pathways in the different systems are set out in Figure 1.2. Adventitious embryony is vegetative propagation on a micro scale: the embryo is produced from the nucellus without an intervening gametophytic stage and then usurps the site of the sexual sac, or competes directly with it. There are indications from *Citrus* that the parental cells may pass through a phase of cytoplasmic "restandardization" comparable with that undergone by the embryo sac mother cell, and if so they do indeed receive such benefits as the process may have to donate in the elimination of pervasive infection. Both apospory and diplospory preserve the alternation of generations and seemingly also the cytoplasmic and other concomitants, including most of the paraphernalia of membranes and walls, pollen-tube guidance systems and synergid functions.

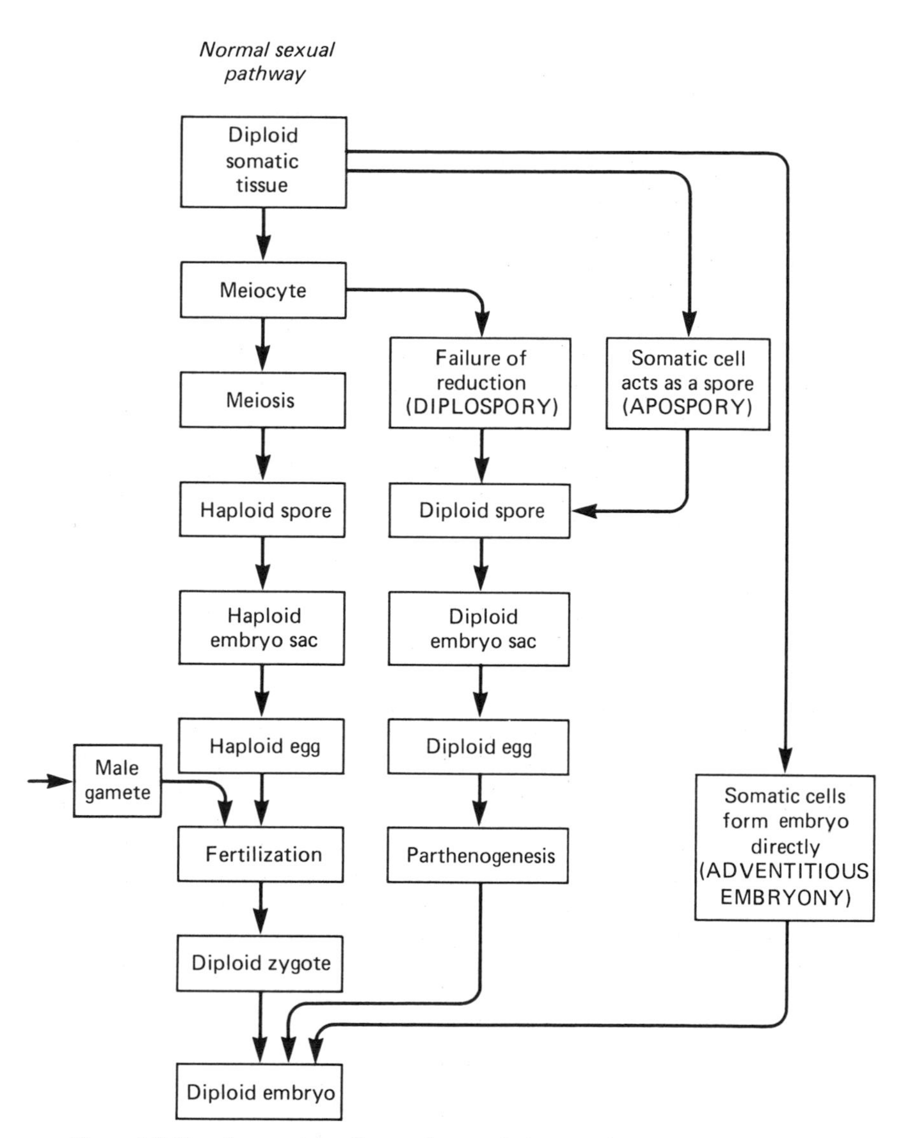

Figure 1.2. Developmental pathways in apomictic systems, set out so as to show the relationship with the normal sexual pathway.

This last fact points again to the importance of the female gametophyte in angiosperm reproduction. Many apomicts in families as remote as the Gramineae and the Rosaceae are pseudogamous, requiring pollination for the formation of viable seeds. Strangely, the second fertilization, that providing the primary endosperm nucleus, may be retained; and where this is so, the apomictic sacs have functional synergids and an appropriately placed central cell nucleus, even when the total nuclear complement is less than eight, as in many diplosporous grasses (Knox and Heslop-Harrison, 1963).

A full apomictic cycle comprising replacements for both meiosis and syngamy (Table 1.2) should provide no opportunity at all for gene segregation and recombination, but some of the systems clearly do entail a degree of recombination. Indeed, the diplosporous mechanism allows for just this, since the diploid embryo sac mother cell nucleus is derived by restitution following a failed meiosis (Darlington 1937). The well-documented variation arising in presumed apomictic lineages of dandelions, poas, and other diplosporous genera no doubt arises from this. Darlington applied the term "sub-sexual" to such processes, and this description is applicable also to other devices in the angiosperm armory, such as the partial apomictic system in canine roses, where some of the chromosomes of an odd-polyploid set pair and recombine while, in the female lineage, others are transmitted through meiosis without change.

REPRODUCTIVE VERSATILITY AND ENVIRONMENTAL INFLUENCE

A striking feature of the reproductive systems of many flowering plants—although one much neglected in theoretical treatments—is the capacity to combine simultaneously two or more reproductive methods, or to switch between different modes in the lifetime of the individual. This reproductive versatility is extremely widespread and can clearly play an important part in ecological strategy. Vegetative reproduction can provide an alternative to sexual reproduction through the seed, as also may agamospermy; and in a similar fashion the pollination mode may alternate from one favoring outbreeding to one establishing mainly inbreeding (Fryxell 1957).

These versatile reproductive systems are often subject to control by environmental factors, and where this is so, the genetic system itself may be said to exhibit phenotypic plasticity (Heslop-Harrison 1966). Pollination systems provide several examples. Many species of *Viola* have the capacity to produce self-pollinating, closed (cleistogamous) flowers as well as cross-pollinating, open (chasmogamous) ones. In some of the earliest work on photoperiodism, Allard and Garner (1940) showed that with *Viola fimbriatula,* short days promote chasmogamy and long days cleistogamy, and similar reactions have been demonstrated with other species of the genus. It seems that the responses are part of the normal, daylength-related time-keeping of the species; and the practical consequence is that the open pollinating flowers are formed in the short days of spring before the development of leaf canopies, and the cleistogamous flowers in the long days of summer when the plants tend to be shaded.

Table 1.3 Versatile Reproduction in the Andropogoneae

A. Alternation between cleistogamy, chasmogamy, with vegetative propagation as an adjunct (e.g., ***Rottboellia exaltata; Bothriochloa decipiens***, Heslop-Harrison 1959a, b, 1961, 1972)

Daylength	Reproductive mode	Propagation method
<12 hr	Male florets sterile; hermaphrodite florets cleistogamous. Inbreeding	Caryopses lacking inherent dormancy
<12->12 hr	Male florets fertile; hermaphrodite florets chasmogamous; outbreeding	Caryopses dormant; nitrate and light required for germination
>12 hr	Tillering; no flowering	Dispersal by fragmentation

B. Alternation between sexuality and aposporous apomixis (e.g., ***Dichanthium aristatum***, Knox and Heslop-Harrison 1963; ***Themeda australis***, Evans and Knox 1969)

Daylength	Reproductive mode	Propagation method
SD (<12 hr)	Up to 75% of the embryo sacs aposporous	Polyembryonic caryopses; most embryos apomictic
Minimal SD induction, then LD (>16 hr)	Up to 72% of the embryo sacs sexual; outbreeding	Occasionally polyembryonic; most embryos sexual

Some years ago we investigated pollination behavior in different daylengths in various grasses of the tribe Andropogoneae and found that several show a converse reaction (Table 1.3). In *Bothriochloa decipiens,* cleistogamous flowers form in short days, while chasmogamy is the rule in long days (Heslop-Harrison 1961). *Rottboellia exaltata* has a similar response, but here the versatility extends further: not only does the pollination mode change in the transition from short to long days, but also the amount of pollen produced, the dormancy behavior of the seeds, and the propensity for vegetative reproduction (Table 1.3A; Heslop-Harrison 1959a, b 1964).

Grasses of the Andropogoneae also provide examples of environmental control of the balance between sexual and apomictic modes of reproduction. *Dichanthium aristatum,* like many grasses of the Panicoideae and Andropogoneae (Brown and Emery 1958), reproduces apomictically by an aposporous method. The balance between apomixis and sexuality is subject to control by photoperiod, short days favoring the aposporous embryo sacs, and long days the sexual (Table 1.3B; Knox and Heslop-Harrison 1963). Similar behavior has been found in *Themeda australis* (Evans and Knox 1969), and the predictable shift in reproductive behavior has been shown to occur in field populations, locked to the natural cycle of day length.

The common features of the examples in Table 1.3 are the tendencies to adopt a reproductive mode restricting the release of variation in the early flowering season and for this to be replaced by an outbreeding mode with a free release of variation in the later season. The plants in question are short-lived perennials of an opportunistic, colonizing habit; and the strategies adopted seem well enough adapted to their life style. In *Rottboellia exaltata,* for instance, the early-season, inbred seed provides the means for speedy reoccupation of a habitat to which the population has already proved itself to be adapted. The late-season, outbred seed could evidently contribute the colonizing capacity, its innate dormancy allowing the survival of unfavorable conditions, and the greater genetic variability providing a wider range of genotypes for the colonization of new types of habitat, where once again partial stability would be attained with the subsequent cycle of autogamy.

SOME EVOLUTIONARY ASPECTS

I began with a quotation from Darlington; we might consider now another from the same text (Darlington 1939, p. 131): "Our understanding of the whole evolution of genetic systems depends in fact on the assumption that variations may survive merely because they favour posterity." Darlington was here offering a solution to the problem that Darwin (1877) had found baffling, namely, how changes in the reproductive system and its various appurtenances could be subject to natural selection when their beneficial effects were not expressed in the individual but in its more-or-less remote progeny. The essence of Darlington's solution lay in group selection (Wright 1931, 1956). Breeding populations, not individuals, are the units of evolutionary change; breeding populations may therefore be the targets of selection. If longer-term adaptive success is what matters, the premium will be on this, and not on the properties of the individuals existing at any one time that predicate that the success of the group in the future must be selected retrospectively. The model offered by Wright (1956) indicates the circumstances in which group selection might be expected to occur—namely, within a mosaic of populations in which local experimentation might produce adaptive innovations, the fate of which is to be tested in competition with other populations, not in a single generation, but over an indefinite passage of time. The reasonableness of the general proposition has been widely accepted (e.g., Mather 1943; Stebbins 1950; Dobzhansky 1951; Grant 1964; Lewontin 1970), but latterly it has become fashionable to challenge group selection as the basis for the evolution of genetic systems (e.g., Williams 1975; Solbrig 1976). Natural selection, so the argument goes, operates on individuals; therefore the components of genetic systems must be selected for or against in individuals. I do not wish to digress into a discussion of this view, beyond saying that it seems incapable of providing rational accounts of the evolution of genetic systems without an absurd amount of bolstering by secondary hypotheses. In illustration, we might consider just one example contained in another comment by Darlington (1939) concerning meiosis itself, of which he says, "No improvement . . . can benefit the individual in which it first arises." Indeed; the dividends to be derived from meiosis can only accrue in later generations, and the improvement must accordingly be selected for retrospectively. However else?

So the general thrust of the argument "that variations survive merely because they favour posterity" can be perceived. And yet, in contemplating the diversification in angiosperm reproductive systems, one is impressed by a puzzling element of redundancy. It has clearly not been a question of the establishment of a few pervasive processes, for the reproductive devices are strikingly heterogeneous both morphologically and physiologically, particularly in the herbaceous and short-lived, colonizing, woody groups. How, then, is it to be accounted for?

A picture of the possible situation in which the major trends were initially established can be derived from the proposition that the characteristics which in sum we regard as angiospermic were molded in a mosaic of partly isolated, probably sparse populations during some period in the Mesozoic era (Heslop-Harrison 1958). The absence of any well-defined sequences in what there is of an early fossil record may indicate that almost all aspects of the organization of the reproductive system in these proto-angiosperms were more or less fluid, apart from certain features that had already become well defined and broadly established, like the organization of the embryo sac. If so, we may envisage that various trends were built up in different populations according to their degree of isolation, much in the manner proposed in Wright's (1956) model, the products competing thereafter as, with increase of population size, they invaded contiguous territories. One difficulty in our accepting this starting point arises simply from the way that we are conditioned to the idea that reproductive structures in the flowering plants do *not* vary in a protean manner, either during the development of the individual, or within a breeding population, or within species or higher taxa. Indeed, our taxonomic system rests squarely on the conception that the major characteristics—mostly of the reproductive system—upon which the principal subdivisions into orders and families depend are invariable; that is after all why they were selected as primary differentiae. The circumstances we see today color our view of the past, and so we find it difficult to conceive of a situation so fluid that a single breeding population might have been heterogeneous in characters now regarded as of familial significance. Yet this may have been the starting point for the diversification, and in this fact may lie the solution to Darwin's "abominable mystery."

At another level, one might apply a similar line of argument in relation to the evolution of diversity in the more superficial elements of the reproductive systems of the flowering plants, not simply in contemplating the possible events of a remote past, but throughout the history of the group. Take as an instructive example the bizarre pollination method that has come to be adopted by certain orchid genera, namely pseudocopulation. Here the flowers have assumed the guise of female bees and lure male bees into transferring pollen as they vainly attempt to discharge their own parental responsibilities in the early-season absence of a fair ration of genuine female bees. Embarkation on this evolutionary by-way has clearly had a radical effect on the variation within the genus *Ophrys,* and probably a rather rapid one, for in the Mediterranean region the genus is now fragmented into numerous often sympatric species, essentially undifferentiated ecologically or in vegetative structure, yet held apart by the fidelity of the bees to the flower types that most closely stimulate the females of their own species. Why has it all happened? The answer seems to be, simply because it could, once certain preliminary conditions had been fulfilled—the main one being, presumably, some initial chance resemblance that began a process of assortative pollination and started a trend of diversification and breeding-group specialization.

And this is why I offer the example as an instructive one: it suggests how a trend of change can be initiated in the breeding system, and can therefore bowl along without necessarily conferring any immediate advantage—or for that matter any special long-term advantage either, if this be measured by increasing usurpation of the habitat. All that appears to have happened in the case of the ophryses is that the available territory has become divided out among the different "species," no single one of which has found a formula that gives supremacy in reproductive performance.

Perhaps we might view the broad sweep of diversity in angiosperm reproductive systems in the same way. To stand any chance at all of success, each variation must obviously preserve the basic elements of a functional system; yet beyond this, experimentation will be possible where the immediate selective pressures are not too intense. Over the longer term, certain trends may be selected and amplified and others rejected; but heterogeneity will be retained so long as there are several possible routes whereby much the same result can be achieved, and if the various processes balance out reasonably well, or are not tested out against each other in any very rigorous way. This proposition assumes a certain looseness about the evolutionary process, and suggests that the complexity we view today reflects little more than a chaotic mélange of compromises and partial solutions. But such ideas are familiar enough. As Medawar (1957) remarked in discussing the imperfections of man's own adaptations, "Evolution is very much a fallible, makeshift affair, and . . . loss of fitness in one regard is often the charge for some more than compensating gain."

LITERATURE CITED

Allard, H. A., and W. W. Garner. 1940. *Further observations on the response of various species of plants to length of day.* U.S. Dept. Agric. Tech. Bull. 727.

Brown, W. V., and W. H. P. Emery. 1958. *Apomixis in the Gramineae:Panicoideae.* Amer. J. Bot. 45:253–63.

Carroll, T. W., and D. E. Mayhew. 1976. *Occurrence of virions in developing ovules and embryo sacs of barley in relation to seed transmissibility of barley stripe mosaic virus.* Can. J. Bot. 54:2497–2512.

Darlington, C. D. 1937. *Recent Advances in Cytology.* 2nd ed. Churchill, London.

———. 1939. *The Evolution of Genetic Systems.* Cambridge Univ. Press, Cambridge.

Darwin, C. 1876. *The Effects of Cross- and Self-fertilization in the Vegetable Kingdom.* John Murray, London.

———. 1877. *The Different Forms of Flowers on Plants of the Same Species.* John Murray, London.
Dickinson, H. G., and J. Heslop-Harrison. 1970. *The behaviour of plastids during meiosis in the microsporocyte of Lilium longiflorum.* Cytobios 6:103–18.
———. 1977. *Ribosomes, membranes and organelles during meiosis in angiosperms.* Phil. Trans. Roy. Soc. Lond. B 277:327–42.
Dobzhansky, T. 1951. *Genetics and the Origin of Species.* 3rd ed. Columbia Univ. Press, New York.
Evans, L. T., and R. B. Knox. 1969. *Environmental control of reproduction in Themeda australis.* Austral. J. Bot. 17:375–89.
Fryxell, P. A. 1957. *Mode of reproduction in higher plants.* Bot. Rev. 135–233.
Grant, V. 1958. *"The regulation of recombination in plants."* Cold Spring Harbor Symp. Quant. Biol. 23:337–63.
———. 1964. *The architecture of the germplasm.* John Wiley, New York.
———. 1971. *Plant Speciation.* Columbia Univ. Press, New York.
Gustafsson, A. 1947. *Apomixis in higher plants.* Lunds Univ. Arrsskr. 42:1–67; 43:71–197, 183–370.
Haldane, J. B. S. 1954. *Biochemical Genetics.* Allen & Unwin, London.
Helsop-Harrison, J. 1958. *The unisexual flower.* Phytomorphology 8:177–88.
———. 1959a. *The influence of day length on the breeding system of grasses.* New Sci. 1959; 881–84.
———. 1959b. *Photoperiodic effects on sexuality, breeding system and seed germination in Rottboellia exaltata.* Proc. 11th Int. Bot. Cong. Montreal: 162–63.
———. 1961. *The function of the glume pit and the control of cleistogamy in Bothriochloa decipiens (Hack.) C. E. Hubbard.* Phytomorphology 11:378–83.
———. 1964. *Cell walls, cell membranes and protoplasmic connections during meiosis and pollen development.* Pages 39–47 in H. F. Linskens, ed. *Pollen Physiology and Fertilisation.* North Holland Publ. Co., Amsterdam.
———. 1966. *Reflections on the role of environmentally-governed reproductive versatility in the adaptation of plant populations.* Trans. Proc. Bot. Soc. Edinburgh 40:159–68.
———. 1971. *The cytoplasm and its organelles during meiosis.* Pages 16–31 in J. Heslop-Harrison, ed. *Pollen: Development and Physiology.* Butterworth, London.
———. 1972. *The sexuality of angiosperms.* Pages 133–289 in F. C. Steward, ed. *Plant Physiology: a Treatise.* Academic Press, New York and London.
———. 1979a. *The forgotten generation: some thoughts on the genetics and physiology of angiosperm gametophytes.* Proc. 4th John Innes Symp.:1–14.
———. 1979b. *Compartmentation in anther development and pollen-wall morphogenesis.* Pages 471–84 in L. Nover et al., eds. *Cell Compartmentation and Metabolic Channeling.* Gustav Fischer Verlag, Jena, and Elsevier/North Holland, Amsterdam.
Heslop-Harrison, J., and A. Mackenzie. 1967. *Autoradiography of soluble (2- ^{14}C)-thymidine derivatives during meiosis and microsporogenesis in Lilium anthers.* J. Cell Sci. 2:387–400.
Knox, R. B., and J. Heslop-Harrison. 1963. *Experimental control of apomixis in a grass of the Andropogoneae.* Bot. Notis. 116:127–41.
Lerner, I. M. 1954. *Genetic Homeostasis.* Oliver and Boyd, Edinburgh.
Lewontin, R. C. 1970. *The units of selection.* Ann. Rev. Ecol. Syst. 1:1–18.
Mackenzie, A., J. Heslop-Harrison, and H. G. Dickinson. *Elimination of ribosomes during meiotic prophase.* Nature 215:997–99.
Maheshwari, P. 1948. *The angiosperm embryo sac.* Bot Rev. 14:1–56.
Mather, K. 1943. *Polygenic inheritance and natural selection.* Biol. Rev. 18:32–64.
———. 1973. *Genetical Structure of Populations.* Chapman & Hall, London.
Medawar, P. B. 1957. *The Uniqueness of the Individual.* Methuen & Co., London.
Schwemmle, J. 1968. *Selective fertilisation in Oenothera.* Adv. Genet. 14:225–324.
Solbrig, O. T. 1976. *Plant population biology: an overview.* Syst. Bot. 1:202–8.

Stebbins, G. L. 1950. *Variation and Evolution in Plants*. Columbia Univ. Press, New York.
———. 1971. *Chromosomal Evolution in Higher Plants*. Addison-Wesley, Reading, Mass.
Williams, G. C. 1975. *Sex and Evolution*. Princeton Univ. Press. Princeton.
Wright, S. 1931. *Evolution in Mendelian populations*. Genetics 16:97–159.
———. 1956. *Modes of selection*. Amer. Nat. 90:5–24.
———. 1977. *Evolution and the Genetics of Populations*. Vol. 3. *Experimental Results and Evolutionary Deductions*. Chicago Univ. Press, Chicago.
Yampolsky, C., and H. Yampolsky. 1922. *Distribution of sex forms in the phanerogamic flora*. Bibl. Genet. 3:1–62.

two
TRANSFER OF INFORMATION AND GENETIC INTERPLAY

2] Interpretations of Sex in Higher Plants

by ROBERT ORNDUFF*

ABSTRACT

In recent years a paradox has been posed: sex is expensive, yet most higher plants and animals routinely indulge in it. David Lloyd has attempted to resolve this paradox by suggesting that, among its benefits, sex enables a parent to protect its descendants against biological opponents by allowing it to recruit temporarily appropriate alleles rapidly. He argues that the benefits of sex must be viewed over successive generations, rather than in the short-term parent-offspring period, and that the natural history of organisms must be given greater attention when evaluating the advantages and disadvantages of sex for a species. Most theorizers on this topic have been zoologists, but since higher plants differ from higher animals in a number of important respects it is possible that the advantages of sex for a plant may be very different than those for an animal. In addition, many of the models dealing with the cost and benefits of sex are overly simplistic or have been based on erroneous assumptions.

Most of what we know about sexual reproductive systems in flowering plants is based on studies of temperate herbs with dioecy, heterostyly, or autogamy, all systems that are relatively uncommon. Thus, this information is largely based on studies of atypical plants, since most angiosperm species have monomorphic, perfect flowers and are woody plants of the tropics. More attention should be devoted to reproductive factors such as pollinator behavior, spatial distribution of plants in a population, flower phenology, and seed dispersal patterns, since these also influence the relative costs and benefits of sexual reproduction. Ideally, studies of the reproductive system of a species should be made on more than a single population of that species and over a period of successive years, since such studies have revealed striking differences over time in such traits as population composition and size and in pollen flow patterns. Thus, single observations may lead to erroneous generalizations about the reproductive patterns of a plant species. Facultative apomixis, often

*Department of Botany, University of California, Berkeley, California 94720.

STRATEGIES OF PLANT REPRODUCTION (BARC Symposium number 6—Werner J. Meudt, ed.)
Allanheld, Osmun, Totowa

viewed as leading to an evolutionary dead end, has many appealing features since it allows a plant to combine the benefits of asexual and sexual reproduction. It has been suggested that sex is a disadvantageous historical artifact with which higher animals and plants have been saddled. This seems highly unlikely, but if it is true, then we must look for the compensations adopted by these sexual organisms that allow them to prosper in spite of this artifact. If it is not true, we must continue to seek the positive values conferred by sex on sexual organisms.

INTRODUCTION

In the past decade, there have been numerous theoretical and empirical studies concerned with various aspects of sex in organisms. During this period there have also been some important monographic studies on the subject as well. In 1974, Ghiselin published his book *The Economy of Nature and the Evolution of Sex,* which is impressive for the vast terrain it covers (there are 69 pages of references). Its thesis, when understood, was variously received; the book has been characterized as "provocative" (Nyberg 1975) and "infuriating" (Maynard Smith 1975). Ghiselin's book contains an excellent and wide-ranging review of theoretical as well as experimental studies of sex in plants and animals. It seems, however, to have made less of an impact than two more recent books with similar titles: George C. Williams's *Sex and Evolution* (1977) and J. Maynard Smith's *The Evolution of Sex* (1978). Williams's and Maynard Smith's books are shorter, less data ridden, narrower in scope, less idiosyncratic, and more easily read than Ghiselin's book. Both are concerned with the adaptive significance of sexual reproduction. Williams attempts to reconcile the advantages of sexual reproduction with disadvantages of the attendant "cost of meiosis." With a relentless and logical development of his ideas, Williams concludes that sexual reproduction puts most plants and animals that indulge in it at a disadvantage compared with those that do not. In a review of Williams's book, Nyberg (1976) wrote that the "weaknesses of this book are Williams' notion of the environment and the lack of integration of his models and ideas with classical population genetics," but he concluded that since the "conventional wisdom" concerning the advantages of sex is dubious, new ideas will supersede this supposed wisdom.

Explanations of sex. The primary function of sex is not that it results in the production of offspring. Many organisms produce offspring by asexual methods that are in fact less dependent on chance than are sexual methods. In various flowering plants that typically reproduce by asexual methods, chance events such as pollination, syngamy, seed production, seed dispersal, seed germination, and seedling establishment may be by-passed. Wilson (1975), in a review of Williams's book, contrasts what he calls the "long-term explanation" for sex with the "short-term explanation," explanations which he views as competing. The long-term explanation can be traced back to August Weismann and has been refined by more recent workers. It maintains that sexually reproducing populations evolve more rapidly and are at a competitive advantage over otherwise comparable asexual populations. This advantage results from the more rapid incorporation and distribution of mutations in sexual populations via meiosis and recombination, processes which do not occur in asexual populations. The short-term explanation, promulgated by Williams, views the advantages of sexual reproduction in quite different terms, although the consequences are overlapping: sexual reproduction evolves because it allows each successive generation to be different from its parents and thus to respond positively to environmental changes that occur over short periods of time. In the latter view, gene combinations are viewed as short-term ones that are constantly dismantled and reassembled in myriad different ways in the next generation. This represents a form of gambling in which the winning combinations are few but are important in survival success. In other words, sex allows for the efficient operation of natural selection. In a third commentary on Williams's book, Scudo (1976) analyzed in detail the arguments presented therein and referred to it as "intense and provocative," although he considered it in part to be nihilistic because of its apparent lack of a set of conclusive, unequivocal advantages to sex.

Maynard Smith's book likewise is concerned with the maintenance of sex, although it is disconcertingly inconclusive as to why sex is so commonly maintained. In a critique of this book Nyberg (1979) referred to the numerous models presented by Maynard Smith but concluded that for various reasons an empirical evaluation of them would seem to be almost impossible. As pointed out by Egbert Leigh (1978), sex is "useless unless either the genotype or the environment is changing." He mentions that Maynard Smith agrees with R. A. Fisher in concluding that the "only possible use of sex is to enable the simultaneous fixation of different mutations," a procedure which is infinitely more time-consuming and more costly for asexual organisms. In view of the scale of meiosis and recombination in organisms producing large numbers of offspring, the availability of mutations may be the limiting factor in the rate of evolution in sexual organisms. This, indeed, is the classical view of the flaw of asexuality—the sluggishness of genotypic response by asexual organisms may constitute a fatal flaw leading to their extinction when environmental change is too rapid. This is not to say that environmental change cannot "do in" sexual organisms as well, of course.

There is no doubt that the three books mentioned here are thoughtful and provocative, and those interested in the problems they cover must read them and judge their arguments for themselves.

Resolving a paradox. The major theorizers of the 1970s who were concerned with the adaptive advantages of sex are zoologists who have questioned the advantages of sex. This has left us with a paradox of grand proportions: there are no obvious advantages to sex, yet it prevails. We are fortunate that this paradox has not remained unnoticed and that a major botanical theorizer—David Lloyd in New Zealand (1980)—has recently reviewed the benefits and handicaps of sexual reproduction, with particular reference to plants. Lloyd's 1980 paper on the subject is valuable not only for its review of Williams's and

Maynard Smith's ideas, but for its concise account of contributions on the subject that have been published since the 1970s by these workers and by others. Most important, it presents new and positive perspectives on the subject of sex. Lloyd's most interesting suggestion is that "sex enables a parent to recruit temporarily appropriate alleles of major loci, which individually protect its descendants against biological opponents." The biological opponents referred to are not the unpredictable physical environment, but are "biological adversaries" such as parasites, predators, and competitors. The idea here is that a mechanism such as sex allows the rapid production of diverse genotypes, some of which are resistant or superior to these biological adversaries. Clearly, however, the resultant increase in fitness of these successive progenies must be sufficiently great to match or exceed the cost of sex. A second point is that "sex may be advantageous for a female when her male and female descendants have nonidentical opportunities for survival and reproduction." Here, Lloyd refers to evidence that in dioecious organisms, males and females within a population may occupy different ecological niches. This has also been suggested for some heterostylous plants (Levin 1975) in which the different morphs are believed to occupy different microhabitats in the site occupied by a population. A third point is that in hermaphroditic plants the individuals comprising progeny of a given seed parent are more likely to be dispersed in the vicinity of the seed parent than they are in the vicinity of the pollen parents, which may be several and scattered different distances from the seed parent. The offspring in a progeny thus offer more potential competition with the seed parent than with the more remotely situated pollen parents. This suggestion is open to empirical verification and would certainly constitute a most interesting and enlightening study. Whereas previous authors have emphasized mechanisms supplying the optimal supply of recombinants, Lloyd views the benefits of sex in a broader context, considering factors such as development, behavior, and ecological conditions. He realizes that the differences between higher plants and animals have an important bearing on the adaptive role of sex in these two groups of organisms, a consideration not given sufficient attention by previous workers.

Lloyd is quite right in pointing out that the models of individual selection that compare sexual and asexual reproduction must consider fitnesses of successive progenies over a series of generations, rather than of offspring versus parents. He also suggests that the mixed-sex progenies that are produced by sexual organisms but not by asexual ones might have an advantage over the single-sex progenies produced by the latter. In recognizing that sex and recombination are not synonymous and are thus subject to different selective forces, Lloyd makes a plea for greater consideration of the "natural history" of organisms and also suggests that we may achieve further enlightenment by studies of the "lower" groups of plants and animals, groups that have not been given sufficient attention to date.

Possibilities that I wish Lloyd would have explored further are the different consequences that result from being a plant rather than an animal. Higher plants differ from higher animals in that they are prevalently hermaphroditic organisms, they are sessile, most progenies of a single-seed parent contain individuals with different male parents, and the number of mating groups is generally very high within an individual population. There are other differences as well, and all of these differences should be taken into account when considering the advantages and disadvantages of sex in higher plants when comparing them with higher animals.

PITFALLS AND MODELS

Williams and Maynard Smith both relied heavily on models in presenting their arguments. One problem with even the most complicated models is that they still risk being too simplistic in accounting for the complexity of the natural situations for which they attempt to account. Models also risk making unwarranted assumptions, or even erroneous assumptions about these natural situations. Maynard Smith and Williams, for example, assert that hermaphroditic plants must be spending resources equally on pollen production and distribution and on ovule and seed production. I am not surprised that a recent, interesting review of experimental studies of reproductive energy allocation in plants suggests that this is probably not the case (Evenson, forthcoming). As a consequence, I am skeptical of any model incorporating this assumption, in view of the likelihood that it is incorrect. Nevertheless, statements of such assumptions—whether erroneous or not—make a positive contribution to evolutionary biology because they stimulate not only thought, but empiricism, and we must now agree that Williams's and Maynard Smith's books have both been provocative in this regard.

Models often have an absolutist, either-or quality. One can perceive of models dealing with various features of the reproductive systems of sexual organisms and of asexual organisms, yet the reproductive phase constitutes only a portion of the life history of an organism. Williams quite correctly realized that such extremes of complete sexuality or complete asexuality do not exist in many organisms. He considers a mixture of asexual and sexual reproduction in his strawberry-coral model, both of which organisms have the ability to reproduce sexually and asexually. Strawberries produce sexual flowers and eventually seeds, as well as runners (stolons) that bear asexual plantlets at their tips. Thus, strawberries have a means for producing sexual offspring as well as an energetically costly mechanism for producing exact duplicates of the parents. Some species of *Fragaria* are dioecious, a condition which has doubtless been derived from hermaphroditism. These dioecious strawberries have acquired an extreme sexual outbreeding system that is coupled with one of the most energy-expensive but effective methods of producing genetic duplicates in one's home turf by asexual methods.

Dioecism and heterostyly as minority conditions. I am impressed with the recent flurry of investigative activity and reporting of studies concerned with various aspects of the reproductive behavior of plants. At the same time, I am impressed with the intensity with which these studies are concerned with uncommon or minority conditions in plants. Within the past five years there has been considerable attention given to dioecious plants, yet such plants constitute only a minor portion of the world's flora. Studies of the reproductive systems and associated behavior of dioecious plants have ranged over a wide variety of taxa and topics. The appeal of dioecious plants is that the number of mating types within a population is only two, and that membership to a mating type is evidenced by floral morphology. What is commonplace in higher animals is rare in higher plants.

I can afford to offer this criticism of those preoccupied with rare breeding systems since I am guilty of this preoccupation myself. For the past two decades I have been interested in reproductive behavior of plants that are heterostylous, a phenomenon that occurs in only 10% of the families of flowering plants. Darwin (1877) postulated that the floral heteromorphism of heterostylous plants represents a mechanism that promotes cross-pollination between the two or three morphs of heterostylous species. This seems to be an eminently reasonable conclusion in view of the usual cross-compatibility of the morphs and the reciprocal positions of anthers and stigmas of the various morphs of heterostylous plants. Yet it took nine decades for someone to examine this hypothesis in natural populations. At about the same time, Levin (1968), Mulcahy and Caporello (1970), and I (1970) independently published the results of our studies on pollen flow in the heterostylous *Lithospermum caroliniense* in the Midwest, *Lythrum salicaria* in New England, and *Jepsonia parryi* in California, respectively. None of us was brash enough to suggest that Darwin was wrong, partly because of the unquantitative terms in which the Darwinian hypothesis was originally presented. But, I at least, believed that the results of these and later studies were sufficiently unexpected that Darwin might not have been right, either. The other workers I have mentioned had the wisdom to turn later to other problems, but I continued investigating an additional array of heterostylous plants. That the morphological-physiological syndrome of characters associated with heterostyly in genera of different families or orders is repeated so imitatively in these unrelated taxa would argue that there is a single explanation for the origin and significance of heterostyly in the life histories of these plants. Indeed, heterostyly represents a classical case of convergent evolution. An understanding of its significance, however, can come only by comparing the reproductive systems, particularly pollen flow patterns, in monomorphic (that is, nonheterostylous) species that are related to heterostylous ones. Since heteromorphism is doubtless derived from homomorphism, insights into the advantages of the former seem best achieved by such comparative studies. Once we have this information for a few homomorphic species, we may have the necessary comparative data that will allow for an explanation of what heterostyly is all about. Ganders (1979), in an excellent review of the occurrence of and biology of heterostyly, has attempted an alternative explanation to the Darwinian one, though one that is tentative and speculative. Complete understanding of the origins and advantages of any extant derivative breeding system may not necessarily provide us with any insights into how this system came about, since the best key to the past is not the present, but is the past, which is lost and which can be reconstructed only tentatively.

MATING PATTERNS IN FLOWERING PLANTS AND AREAS FOR FUTURE STUDY

Despite the emphasis on these reproductive systems, most plants are neither dioecious nor heterostylous. Most are monomorphic with hermaphroditic flowers. For those monomorphic species that lack self-incompatibility systems, perhaps the idea of mating types or mating groups is irrelevant, since a plant may mate with itself as well as with a variable number of other individuals in its population, theoretically up to 100% of them. For those plants that are self-incompatible, the number of mating types present in a population may be equal to the number of individuals present in that population since, though no individual can mate with itself, it can mate with all other individuals in its population. For sessile individuals such as flowering plants, other factors beside incompatibility or the lack of it may enter into mating patterns. Pollinator behavior is one factor, since pollinating agents are responsible for the distances pollen travels from an anther to a stigma, and they may also influence the choice of stigma independently of distance. Movement patterns of different pollinators differ. The movements of some groups—such as thrips—may be largely intrafloral, whereas tropical bees may "trap line" from plant to plant over long distances.

Mating patterns in plants are also determined by patterns of flower phenology. In certain populations, particularly those of annual plants in patchy habitats, some individuals in a population have finished flowering when others are just beginning to flower. Other annuals, particularly those of the desert, are opportunistic and may germinate, grow, and flower after a heavy local shower despite the season. It would be interesting to know whether certain genotypes in these areas respond positively to summer rains but not to winter rains, whereas other genotypes respond in the reverse manner. If this is so, two conspecific, sympatric populations will occur sympatrically that are genetically isolated by differences in temporal behavior. In the desert candle *Yucca whipplei,* of our southwestern deserts, not all mature individuals in a populations may flower during a given season. An individual will flower several times during its lifetime, but not in successive years. In the New Guinea shrub *Styphelia suaveolens,* matters may be even more complex, with the flowering of different individuals said to be out of phase with each other (Smith 1980). Are there, in these species, several subgroups of sympatric individuals whose flowering is synchronized, but only within the group, so that the population on a given site consists of subsets of individuals who are in genetic contact only with the several individuals within their particular flowering group? If so, the genetic structure of such populations merits study.

A factor that may affect the degree of relatedness among individuals in a population of sessile plants that is not directly related to breeding patterns is the nature of seed dispersal. Carlquist (1966a, b) in his admirable discussions of insular biotas, has mentioned two features of island plants that distinguish them as a phytogeographical group from mainland plants and from their presumed mainland ancestors in particular. Dioecism and other extreme derivative outbreeding mechanisms are believed to have been selected for in the descendants of insular immigrants as a means of maximizing and distributing genetic mutations in the small populations derived from very small numbers of initial immigrants. At the same time and for several reasons, the descendants of insular immigrants have adopted various mechanisms that reduce the dispersibility of their propagules, leading to a condition called precinctiveness in which the morphological characteristics of seeds leads them to be dispersed close to the parent that produced them. Because a genetic consequence of precinctiveness is inbreeding, the various derivative outbreeding mechanisms of insular plants might also have resulted, at least in part, from selection that counteracts this genetic consequence inherent in precinctiveness. Here, then, is a specialized dispersal mechanism that is seemingly unrelated to a genetic system, but which may be very closely related to this genetic system on closer inspection.

Above, I mentioned the large number of experimental as well as theoretical investigations of minority breeding systems in flowering plants. Dioecism and heterostyly were mentioned. The percentage of species in various floras that are dioecious varies, ranging from a high of about 27% in Hawaii to a low and probably approximate average for continental floras 2%–4% (Godley 1979). Heterostyly occurs in fewer than 10% of the families of flowering plants. In some families, such as the large, mostly tropical family Rubiaceae, it is common, but in many large families, such as Iridaceae and Saxifragaceae, it is known from single genera only, so I would estimate that heterostyly occurs in fewer species than does dioecism. A third minority breeding system, predominant self-pollination, may occur at higher levels and is particularly common in some families; Allard (1975) has suggested that approximately a third of flowering-plant species are predominantly self-pollinated. After scanning Fryxell's (1957) review of modes of reproduction in flowering plants, I believe that this figure is much too high, and I would cut his estimate by half and still arrive at a too-generous figure of 15%. Thus, these three intensively studied breeding systems at best characterize collectively no more than 25% of the species of flowering plants and probably considerably fewer. Most of the species studied are, in addition, herbs of temperate regions. Those pedestrian angiosperms with monomorphic, outcrossed hermaphroditic flowers remain relatively unstudied. Moreover, most plant species in the world do not grow in temperate areas nor are they herbaceous; they are woody plants—mostly trees—of tropical regions. Such plants, with a few notable exceptions, remain unstudied. So our view of reproductive systems in flowering plants is based on a biased sample. A global view based on such a bias seems premature.

FACTORS INFLUENCING LEVELS OF INBREEDING

Another point upon which I wish to comment concerns the use of the terms "inbreeding" and "selfing" as if they are interchangeable. This is reflected, at least in part, in Jain's very interesting and stimulating review of "inbreeding," in which he does not clearly discriminate between a description of reproductive events or mechanisms (such as apomixis, self-pollination) and terms that characterize genetic qualities of populations that are consequences of these events or mechanisms (such as inbreeding). When we speak of the evolution of inbreeding, what is generally being discussed is the evolution of selfing, a breeding system that ultimately results in various levels of inbreeding. Inbreeding can result from mechanisms other than selfing, however. Certain genetic systems in outcrossed plants, particularly those involving chromosomal behavior at meiosis, may lead to higher levels of apparent inbreeding than would be estimated on the basis of mating patterns alone. Perhaps the term "inbreeding" is not appropriately applied to such instances, but since the genetic consequences are similar I think it is a fair use of the term. Population size also has an effect on the randomness of mating. Many plant species restricted to localized edaphic situations (such as unusual soils, seeps, small islands, etc.) are characterized by small population sizes. Although the individuals in these populations have probably not been selected for inbreeding in the way that selfers have, it is possible that their genetic systems have been selected to tolerate levels of inbreeding that are higher than those in related populations, which are characteristically much larger. It would be most interesting to compare the onset and degree of inbreeding depression in selfed lines of outcrossers from populations that are typically very small with those of close relatives that typically occur in much larger populations. In the western United States, several species that occur in small populations on serpentine soils or in vernal pools are closely related to those that occur in much less restricted habitats. Species of the goldfield genus *Lasthenia* in the Compositae come immediately to mind, but there are others. In the East, several species that occur only on small granite outcrops are closely related to those that occur in large populations in other habitats. *Senecio tomentosus* and *S. smallii* come to mind as examples. These merit study.

Compatibility and incompatibility. An apparent paradox is posed by the fact that a large number of flowering plant species have flowers that are clearly adapted for outcrossing, but are at the same time self-compatible. Often, individuals of these species produce many flowers simultaneously. Under such conditions one would expect a great deal of geitonogamy, or pollination among flowers on the same plant, which is the genetic equivalent of selfing. Yet, plants with this syndrome of traits that have been investigated show high levels of outcrossing. Observations of pollinator behavior indicate that pollinators may not visit many open flowers on a plant during a visit, but instead move from plant to plant, sampling only a few of the available flowers on each plant, thus effecting apparent high levels of outcrossing. Still other pollinators do not behave this way and may visit most or all flowers on a plant. Under these circumstances one might expect high levels of selfing, but genetic tests may indicate that this does not occur. What is the explanation for this contradiction? I suspect that cryptic self-incompatibility may operate here. The standard method for testing whether a plant is self-compatible is by simply selfing a plant and measuring seed-set as compared with seed-set following cross-pollination in another individual. If the two seed-sets are similar the plant is judged to be self-compatible. In *Amsinckia grandiflora,* a heterostyled borage now confined to a single small population in central California, artificial selfing produces as much seed as does crossing plants, and the species has been considered to be self-compatible (Ornduff 1976). Since its complicated floral dimorphism is presumably an adaptation for outcrossing, one might assume, if there were any logic to nature, that these plants should be self-incompatible. The presence of both long- and short-styled plants in the natural population of this species during a few years of sampling suggests that the plant is naturally outcrossed, but examination of the pollen grains deposited on stigmas by pollinators indicates that a considerable amount of intramorph pollination occurs as well. Here is an apparently self-compatible plant which receives much self-pollen on stigmas, yet the evidence is that it is mostly outcrossed. Stephen Weller and I (Weller and Ornduff 1977) attempted to mimic natural pollinations of this species by applying pollen from both morphs on individual stigmas of cultivated plants, collecting the seed, and growing the progeny. The progenies indicated that all plants were outcrossed. Examination of the styles of plants pollinated in various ways indicated that though self-pollen will produce a full seed-set, self-pollen tubes grow more slowly than intermorph pollen tubes, so that when both types of pollen are present on a stigma, the intermorph pollen produces the progeny at the expense of the intramorph pollen. The result is effective outcrossing despite the self-compatibility. This phenomenon of gametophytic competition may be very common in nature, and its presence should be watched for in homomorphic plant species as a means of explaining the apparent biological contradictions between self-compatibility and observed outcrossing.

NECESSITY FOR LONG-TERM STUDIES

Studies of reproductive systems of plants in the wild ideally should be based on more than one year's fieldwork and preferably on more than one population of the species. In my own studies of the annual *Amsinckia grandiflora,* its population composition fluctuated from a 1:1 pin:thrum ratio to a 2:1 pin:thrum ratio and then back to a 1:1 pin:thrum ratio. The mechanics and explanation of this fluctuation are not clear, but the amount of pin pollen available for insects varies by a factor of 100% from generation to generation. Despite this, analysis of stigmatic pollen loads collected from both types of populations indicated very little difference between pollen flow patterns, indicating that the floral dimorphism of this species serves as a homeostatic mechanism that assures a specific pattern of pollen flow despite drastic and sudden alterations in the representation of the two floral types in the population. A similar observation was made on the dimorphic *Lythrum californicum* (Ornduff 1978), which had a 13.5:1 pin:thrum ratio in one population in 1976 and a 1.6:1 pin:thrum ratio in the same population in 1977. In this population, the stigmatic pollen loads maintained a fairly consistent composition despite the great differences in pollen availability of the two morphs during two successive years. Studies of the perennial herb *Hedyotis caerulea* in a small area of North Carolina

(Ornduff 1980) indicate that there are populations of this species with a 1:1 pin:thrum ratio and others with a 1.5:1 pin:thrum ratio. At any given season, even over fairly small areas, both types of populations are found. Furthermore, the composition of a single population is not necesssarily stable, since from year to year there are gradual changes from one morph ratio to another. The spatial distribution of pin and thrum flowers is random in some populations and nonrandom in others. In some populations the two morphs produce equal numbers of pollen grains per flower; in others they do not. This series of studies on a single apparently unremarkable species in a small portion of its total range over a period of several years has turned up some interesting demographic and reproductive features that still require explanation. Identifying these problems is a necessary first step in solving them. In this instance many of these problematical features would not have been evident with only a single season of study, or with study of only a single population.

BALANCES BETWEEN SEXUALITY AND ASEXUALITY

It has been pointed out that in the books mentioned above both Ghiselin and Williams were concerned with the question of why sexual reproduction is not replaced by asexual reproduction in those species where the cost of sex is high. Williams was aware of the fact that agamospermous apomixis—the production of seeds without a sexual process—seems to result largely as a matter of accident. That is, it is not a latent, obvious asexual method that can appear merely as a consequence of the presence of selective pressures favoring it. In many angiospermous lines there is evidence that the asexual forms have arisen from sexual ancestors and represent hybrids, often polyploid ones. The presumption is that had apomixis not arisen, these hybrids would be highly sterile and would ultimately disappear. This view lead C. D. Darlington (1939) to refer to agamospermy as an "escape from sterility," a phrase which refers best to those agamosperms that have no other method of reproducing. As Verne Grant (1971) and others have pointed out, however, some agamospermous plants retain the sexual apparatus as well, so that their reproductive system is one combining sexuality with asexuality. Williams has referred to an ideal balance between the two—one system producing genetically variable offspring, the other system producing duplicates on the parent. Although a balance between the two might seem to be an ideal, it is unlikely to be achieved because the environmental demands differ over time, on occasion favoring parental duplicates, on occasion favoring the genetic experiments that result from sexuality. In some plants the balance between sexual and apomictic reproduction at any one time is apparently under environmental control. It is not clear whether the conditions that induce the plant to produce high numbers of sexual offspring are those in which the fitness of these offspring is superior to that of asexually produced offspring. Since we have a tendency to think of apomixis as the consequence of a biological accident, we cannot take it seriously as a bona fide breeding system. Still, it occurs in a small number of flowering plants of diverse relationships and seems to "work" in many if not all of these. More than ten years ago I referred to apomixis (Ornduff 1969) as an "ideal mode of reproduction" since it provides the facultative apomict with the opportunity of producing a mix of sexually and asexually produced offspring, with the mix of the two determined by environmental conditions. We need to know a great deal more about the biology of apomixis before we can decide that it is an ideal breeding system despite its "accidental" origins, but it has many appealing features and hardly merits dismissal as an abnormality leading to extinction.

CONCLUSION

Doubtless we will look back on the 1970s and 1980s as decades in which important advances have been made in the way we view sexual reproduction and the values it has for higher organisms. We must thank Ghiselin, Williams, Maynard Smith, and Lloyd for stimulating so many of us to think about our "conventional wisdom" and to question it. I cannot accept Williams's view that sex is an historical artifact that is no longer adaptive, and that sexual organisms are stuck with this anachronism because they lack the "preadaptations for ridding themselves" of it. But if this is true, we must seek the compensations that have been adopted by organisms that are saddled with this anachronism—compensations that allow them to thrive in spite of this. Or we may take the attitude of Lloyd, who suggests that we must assume that sex is of positive value. We must then proceed to determine what these positive values are. They involve not only consideration of the genetic structure of populations and the cytogenetics of meiosis, but of pollination biology, demography, seed dispersal, and other features of organisms that affect their success. We must go even beyond these limits to consider the developmental biology of plants and its physiological and biochemical basis. Once these factors are understood and integrated, we will arrive at an understanding of the basis for the strategies of plant reproduction.

ACKNOWLEDGMENTS

I am indebted to Daniel Schoen for stimulating discussions on the subject of sex and to William Evenson for providing a copy of his unpublished review article.

LITERATURE CITED

Allard, R. W. 1975. *The mating system and microevolution*. Genetics 79:115–26.

Carlquist, S. 1966a. *The biota of long-distance dispersal. III. Loss of dispersibility in the Hawaiian flora*. Brittonia 18:310–35.

———. 1966b. *The biota of long-distance dispersal. IV. Genetic systems in the floras of oceanic islands*. Evolution 20:433–55.

Darlington, C. D. 1939. *The Evolution of Genetic Systems*. Cambridge Univ. Press, Cambridge.
Darwin, C. 1877. *The Different Forms of Flowers on Plants of the Same Species*. John Murray, London.
Evenson, W. E. (forthcoming) *Experimental studies of reproductive energy allocation in plants*. *In* C. E. Jones and R. J. Little, eds. *Handbook of Experimental Pollination Biology*. Van Nostrand Reinhold Co., New York.
Fryxell, P. 1957. *Mode of reproduction in higher plants*. Bot. Rev. 23:135–233.
Ganders, F. R. 1979. *The biology of heterostyly*. New Zealand J. Bot. 17:607–35.
Ghiselin, M. 1974. *The Economy of Nature and the Evolution of Sex*. Univ. Calif. Press, Berkeley.
Godley, E. J. 1979. *Flower biology in New Zealand*. New Zealand J. Bot. 17:441–66.
Grant, V. 1971. *Plant Speciation*. Columbia Univ. Press, New York.
Jain, S. K. 1976. *The evolution of inbreeding in plants*. Ann. Rev. Ecol. Syst. 7:469–95.
Leigh, E. G. 1978. *Accounting for sexual reproduction* (review). Science 202:1274–75.
Levin, D. A. 1968. *The breeding system of Lithospermum caroliniense: adaptation and counter adaptation*. Amer. Natural. 102:427–41.
———. 1975. *Spatial segregation in pins and thrums in populations of Hedyotis nigricans*. Evolution 28:648–55.
Lloyd, D. G. 1980. *Benefits and handicaps of sexual reproduction*. Evol. Biol. 13:69–111.
Maynard Smith, J. 1975. *Evolution of sex* (review). Nature 254:221.
———. 1978. *The Evolution of Sex*. Cambridge Univ. Press, Cambridge.
Mulcahy, D. L., and D. Caporello. 1970. *Pollen flow within a tristylous species: Lythrum salicaria*. Amer. J. Bot. 57:1027–30.
Nyberg, D. 1975. *Laissez-faire individualism* (review). Paleobiology 1:220–23.
———. 1976. *Sex and evolution* (review). Evolution 30:194–96.
———. 1979. *Some problems with the evolution of sex* (review). Evolution 33:776.
Ornduff, R. 1969. *The systematics of populations in plants*. Pages 104–28 in *Systematic Biology*. Nat. Acad. of Sci. Publ. 1692, Washington, D.C.
———. 1970. *Incompatibility and the pollen economy of Jepsonia parryi*. Amer. J. Bot. 57:1036–41.
———. 1976. *The reproductive system of Amsinckia grandiflora, a distylous species*. Syst. Bot. 1:57–66.
———. 1978. *Features of pollen flow in dimorphic species of Lythrum section Euhyssopifolia*. Amer. J. Bot. 65:1077–83.
———. 1980. *Heterostyly, population composition, and pollen flow in Hedyotis caerulea*. Amer. J. Bot. 67:95–103.
Scudo, F. M. 1976. *Sex and evolution* (review). Syst. Zool. 25:95–97.
Smith, J. M. B. 1980. *Ecology of the high mountains of New Guinea*. Pages 111–32 in P. van Royen, *The Alpine Flora of New Guinea*, vol 1. J. Cramer, Vaduz.
Weller, S. G., and R. Ornduff. 1977. *Cryptic self-incompatibility in Amsinckia grandiflora*. Evolution 31:47–51.
Williams, G. C. 1975. *Sex and Evolution*. Princeton Univ. Press, Princeton.
Wilson, E. O. 1975. *The origin of sex* (review). Science 188:139–40.

3] Pollination Processes: Understanding Fertilization and Limits to Hybridization

by H. F. LINSKENS*

ABSTRACT

Pollen grains of different plant species are everywhere in the biosphere. Pollination in higher plants can be considered as a selective transfer of genetic information. Abiotic vectors effect an at-random transfer without improving the chances of matching adequate partners. Biotic pollination is more selective and can be regarded as a first step in avoiding such chaotic hybridization and spillage of energy. Pollinators of various kinds enhance the chance of complementary matching by their specific relationship to a flower species. At the same time a first selection filter is erected to maintain the individuality of the species. The importance of attractants, pollinator specificity, stickiness of pollen grains and of stigmatic surface as components of this first filter mechanism is stressed.

The second filter mechanism is of a genetical or physiological nature and becomes a fertilization barrier during the progamic phase. Whereas incongruity is characterized by a lack of genetic interplay, incompatibility mechanisms stop outbreeding in many angiosperms. In sporophytic incompatibility, localized mostly on the surface of the stigma, pollen behaves according to the genetic information of its parents; in gametophytic incompatibility, localized on the contact surface between pollen tube and transmitting tissue, the control is according to the pollen's own genotype. The highly specific filter of the incompatibility barrier suggests participation of specific proteins, but the participation of lipidlike molecules in the cell-cell interaction cannot be excluded.

INTRODUCTION

In the biosphere we are surrounded by a curtain of male diaspores of spermatophytes. Pollen countings during the last decade in the service of hay

*Department of Botany, Section Molecular Developmental Biology, University Nijmegen, The Netherlands.

STRATEGIES OF PLANT REPRODUCTION (BARC Symposium number 6—Werner J. Meudt, ed.)
Allanheld, Osmun, Totowa

fever prognosis demonstrated the omnipresence of pollen grains (Lejoly-Gabriel 1980). During the flowering season up to 15,000 pollen grains per cubic meter are found on the average, with maxima dependent on the appearance of the main component species in the local flora. Contemporary airborne pollen amounts also vary with variations in climate and the local vegetation, so that average annual sedimentation can reach up to 4700 grains per square centimeter (Bonny 1980). This omnipresence of a complex pollen mixture could result in chaotic hybridization, with wastage of pollen that arrives in improper places, spoilage of receptive stigmata and even whole ovules, or the induction of outgrowth without fertilization, frequently followed by postpollination degeneration or abortion. If no selection and speciation mechanisms had been present to maintain the individuality of the species, evolution would have gone another way.

POLLINATION MECHANISMS

A first filter mechanism that functions to maintain the individuality is the specificity of pollination mechanisms. Different pollination mechanisms guarantee or improve the matching of the proper functional male sperm cells, locked in the motile pollen grain, with the appropriate sessile female cell, hidden in the complicated apparatus of the ovule.

Autogamy. Autogamy is the transfer of pollen from the anthers to the stigma in the same flower. In many cases morphological structures and the sequence of opening in the flower enhance autogamy without any external pollinating agent. Autogamy can happen by chance if the male flowers are arranged above the female flowers on the same (monoecious) plant, so that pollen encounters the stigma because of gravity. It also can take place when a corolla causes, as it drops, an anther presenting mature pollen to brush against a stigma (Hagerup 1954). In other types of flowers autogamy has to be induced, e.g., by a gust of wind or shaking by an animal pollinator. In this way autogamy can be the most efficient mechanism that guarantees pollination within the species. Nevertheless, autogamy can be considered as a last resort for flowers which have, for some reason or other, not been properly pollinated (Faegri and van der Pijl 1979). Autogamy can therefore also be considered as a retrogade development.

It is an interesting fact that during domestication many cultivated crops, such as wheat, barley, oats, and beans, have aquired complete autogamy through unplanned selection by early man. Other old cultivated plants, such as rye and corn, still have barriers that make autogamy less likely, although autogamy is not excluded, and natural crossing seems favored.

Genetically speaking, autogamy originated in isolated populations as a mechanism of increasing homozygosity. It is no wonder, then, that autogamy is rare in perennials but frequent in isolated populations and annuals.

Allogamy. A second way to promote matching of male and female sexual cells belonging to the same species is allogamy, when pollen of one unisexual flower is carried to the stigma of another (Faegri and van der Pijl 1979). If two flowers on an individual plant are participating in an allogamic fertilization it is called geitonogamy; if different plants are included it is called xenogamy. In terms of a mixture of genomes, geitonogamy is equivalent to autogamy. If plants of the same clone are involved, xenogamy likewise is genetically equivalent to autogamy.

In many cases of allogamy the pollen transferred to a stigma is a selection made from the omnipresent offer of pollen grains based decisively on the specific relationships the pollen vectors have to the flowers to be visited. This relationship at least within the steps of one pollination event should be between flowers of the same species in order to prevent mixed pollination.

In the case of allogamy, the selection process that insures intraspecies pollination is guaranteed by means of some kind of direct or indirect attraction (Faegri and van der Pijl 1979) for biotic pollinators. Allogamy is therefore linked with adaptive mechanisms of pollination.

Looking for pollinators of one species means at the same time looking for the special relationships that orginated between the vector and the two partners that should be brought together.

Pollination by physical forces is nonselective, for there is no question of mutual adaptation; there is only a one-sided, nonstable relationship between the sexual units to be transported and the pollinating agents (Faegri and van der Pijl 1979). Because the transfer is undirected, all types of abiotic pollination must be considered as lavish or sumptuous. The chance that the appropriate male pollen grain reaches the proper receptive organ of the same species is low. It is determined by the proximity in space, the degree of synchrony between maturity and receptivity, and the adaptation of morphological structures to the physical transport medium, since in abiotic pollination the dispersal units are scattered and subjected to the caprice of fortune.

ABIOTIC POLLINATION

Anemophily. It is supposed that more than 90% of all species of flowering plants in a certain area are pollinated by wind. According to Faegri and van der Pijl (1979) anemophily is derived from entomophily, the more primitive condition. Arguments supporting this conclusion are the occurrence of nectaries in flowers of many wind-pollinated species and the presence of specific scent in flowers that are normally not visited by biotic pollinators.

Wind pollination means wasting material. Pohl (1937) calculated the number of pollen grains produced in relation to the number of ovules of the same species: generally speaking, anemophilous plants produce far more pollen grains, in the range of half a million to 3 million pollen grains per ovule, in comparison with entomophilous species, which produce between 5 and 100 thousand grains per ovule. Wind-pollinated stigmata have developed two adaptations to improve their chance of catching pollen grains: the enlargement of the effective stigmatic surface, and, the stickiness of the receptive surface. Since the pollen grains of anemophilous species in general are dry (Hesse 1978,

1979, 1980) and powdery, the stigma surface is extended by featherlike or brushlike structures. At the same time reduction of the area of other flower organs, such as perianth or perigon, occurs so that an inert surface does not reduce the chance for attachment of flying pollen grains to the female organ. The aerodynamics of anemophilic flowers have not yet been investigated in detail. The extended stigma surface of anemophilous plants is in general wet and sticky, so that pollen grains once they are caught are held.

Hydrophily. Pollination by pollen floating on the surface of the water (ephydrophily) is similar to anemophily but takes place in a two-dimensional space, so that the pollen produced is used in a more efficient way than in the three-dimensional airspace.

Submersed pollination is the specialized method of pollination in plants living and producing flowers below the water surface, such as the sea grasses, which form an abundant and highly productive vegetation of the intertidal zones and deeper areas. Some detailed recent observations on this type of hyphydrophilic pollination can be found in de Cock (1980) and Pettitt, Ducker, and Knox (1981). The threadlike shape of the pollen of sea grasses increases the floating capacity. The sticky threads adher to each other and form a networklike structure. The pistils protrude their styles as soon as the stigmata become receptive, and the style remains in this raised position until pollination takes place. As soon as the reticle of pollen touches the stigmata, attachment occurs which is strengthened by the passive winding of the pollen threads round the stigma. Some authors report active entwining. On the other hand the specific structure of the style with its bifurcation enhances attachment of the pollen threads. After pollination, the styles bend back down under cover and the abscission of the stigmata takes place. Thus, the specific structure of the pollen grains and the structure and behavior of the stigmata plus the germination of the threads in several places enlarge the chance that at least part of an attached pollen thread arrives near the style channel. No wonder that pollination in the field is very efficient.

BIOTIC POLLINATION

Biotic pollination is characterized by a vector organism, which is indispensable for the interaction between pollen and style. The pollination process has then an increased efficiency because of the specific relationship that arose during evolution between flower and pollinator. The irony of this efficiency is that these mutually beneficial visits, which have become an integral part of the behavior of the pollinating animal, may have originated from an evolutionarily earlier parasitic relationship.

The pollinator can be defined (Covich 1974) as "a dispersal agent for highly specialized pollen grains which transmit genetic information and produce sexual recombination and heterozygosis." In biotic pollination the dispersal vector is an animal, either a vertebrate or an invertebrate. Very well known biotic pollen vectors among the invertebrates are beetles, flies Hymenopterans, Lepidopterans, Collembola, Hemiptera, and others. Vertebrate vectors are mostly found in the tropics and include birds and flying and nonflying mammals (Armstrong 1979).

We can distinguish between major and minor pollinators (Baker, Baker, and Opler 1971), the former being those which are most perfectly adapted and normally carry out the pollination. But plant-pollinator adaptations, however, are not necessarily the final product of a successful co-evolution. In the ever changing environment a continuous structural and functional modification of flower and pollinator(s) must occur in order to maintain the fruitful cooperation (Macior 1973). Most important is the constancy of the relationship between flowers of a certain species and the pollinator, that is, the fidelity of flower preference, which guarantees that the specific pollen reaches the appropriate stigmatic surface. This preference is established by means of attractants that trigger the chain reaction in the pollinator, resulting in the transfer of pollen to the receptive surface.

Attractants. Distinction can be made between six different primary and four different secondary attractants (Faegri and van der Pijl 1979): whereas secondary attractants act directly or indirectly on the sensory apparatus of the visitor, primary attractants satisfy the food demands of the flower visitor. Primary attractants include pollen, nectar, and oil as deliverers of energy. Apparently, pollen is a more selective attractant to the pollinator than nectar, probably because of its biochemically higher degree of usefulness as a source of high energy compounds (Stanley and Linskens 1977). Nectar in general is composed of different sugars and a selection of free amino acids, whereas pollen contains the broad spectrum of inorganic and organic compounds, including organic acids, lipids, sterols, nucleic acids, enzymes, pigments, hormones, and fragrances. Some species of bats derive essentially their entire food supply from nectar and pollen (Hevley 1979; Howell 1974). Protection and the opportunity for oviposition are also primary attractants, as well as is the apparent, fictive offer of intercourse, also called the chance of pseudocopulation. Besides sexually conditional pollination, pollination occurs after a rendezvous between pollinator and a flower caused by these secondary attractants: odor, visual (color, shape, texture), temperature and motion.

The receptive surface. The antagonist of the pollen within the pollination process is the stigmatic surface. A classification of the stigma in angiosperms can be based on the amount of wetness or dryness, and the anatomy and morphology of the stigmatic plane or lobes (Heslop-Harrison and Shivanna 1977). The processes by which pollen sticks is determined to a high degree by the wetness of the stigma. This wetness and the nature of the stigmatic exudate are not only of crucial importance for the pollen germination, but also for the continued adherence of the settled pollen grain. Woittiez and Willemse, however, recently (1979) made an attempt to introduce a physical approach to ascertain the main factor involved in the adherence of pollen to the stigma. They found surface tension to be the main factor in final sticking of the pollen grain to the stigmas. The degree of sticking does not depend on the composition of the fluid, whether the stigma covering is of lipid nature as in *Petunia* (Konar and Linskens 1966a,b), is a water solution of proteins or carbohydrates (Martin 1969), or is a liquidlike waxy coating. In addition to surface tension, electrostatical forces can act as short-time fastening factors. Because of the crucial role which stigmatic exudates on the stigmatic surface play in the sticking process, Woittiez and Willemse came to the conclusion that real, dry stigmata do not exist.

The stigma exudate also plays an important role for the primary steps of pollen-tube growth during the progamic phase (Martin and Brewbaker 1971; Baker, Baker, and Opler 1973; Dickinson and Lewis 1975; Shivanna, Heslop-Harrison, and Heslop-Harrison 1978).

Several other factors responsible for the final adhesion of the airborne pollen grains on the receptive surface are gravity, sedimentation velocity, and bio-electrical forces. A high percentage of different species of pollen carries a negative electrostatic charge, which may contribute to an electropotential gradient between pollen and receptive surface (see Stanley and Linskens 1974). Further investigation in this field is necessary.

Pollen energetics. There are some indications (Colin and Jones 1980) that pollen energetics is linked with pollination modes.

Smith and Evenson (1978) have found during an investigation of energy distribution in reproductive structures a striking difference in the caloric value between insect-pollinated and wind-pollinated monocot plants. Although there is a large range of variation in the caloric values of various pollen species, statistically significant differences have been found between tribes within one family (Asteraceae) which have been shifted from insect to wind pollination. Furthermore, Stebbins (1974) has found that wind-pollinated flowers differ from related insect-pollinated species in their tendency to be clustered in defense infloresences.

The total energy invested in pollen seems to be higher in wind pollen, but the per pollen investment deposited in the amount of reserve material is lower because of aerodynamic demands. A general hypothesis could be that because of a saving of energy by the reduction of the perianth as an attractant apparatus, a greater part of the total flower energy can be contributed to the pollen. Since wind pollination is less efficient than biotic pollination it makes sense that there is a higher investment into pollen. But the fraction of the energy invested in pollen is partly used for aerodynamic pollen structures in order to reduce the sedimentation velocity, and partly for increasing the number of pollen grains produced.

The close adaptation of the various pollen vectors guarantees the highest degree of interaction within a species. In economically important plant species, excluding asexual and apomictic reproduction, the majority of species have hermaphroditic flowers while unisexual plants are relatively scarce. This guarantees greater flexibility, efficiency, and economy in gene-flow than does unisexuality (Frankel and Galun 1977). In terms of reproductive strategies, the directed biotic pollination is the safest one for combining the sexes of the same species, and the most efficient one in terms of the use of energy input in the reproductive system.

FERTILIZATION BARRIERS

The various pollination mechanisms intend to bring together pollen and stigma, the sperm-cell–carrying gametophyte to the receptive surface of the female organ. The pollination syndrome makes the selection in the complex mixture of pollen present in the space between both partners of the fertilization process. On the other hand, the genotypical variation within a species is usually maintained and inbreeding is prevented. Nature built in a second group of filter mechanisms, which makes a positive selection for the fertilization partners possible, whereas the selective transfer of pollen implies a negative selection, the exclusion of the hybridizing partners. These second filter mechanisms we call fertilization barriers.

There are at least two different barriers in the progamic phase; A complex of barrier genes is active during the interaction between pollen and stigma. These genes can build up an inhibiting bar, which can be called an active or positive barrier, or can act as a positive or negative barrier. In this relationship the genes in the pollen can be considered as a complex of penetration genes that enable the pollen grains to react in an adequate manner on the barriers present in the pistil (Hoogenboom 1975). For a successful completion of the progamic phase, including a successful removal of the fertilization barriers, it is necessary that each gene or gene complex in the style has its corresponding gene or gene complex in the pollen so that the desired fertilization occurs. These complexes can be called matching genetic systems in pollen and pistil. If the pistil and the pollen, in addition to this complementary genetic system, also possess a special genetic system called S-genes, we have self-incompatibility. This is defined as an active fertilization barrier that comes forward in spite of strong biochemical congeniality or affinity of the partners of the mating process.

Incongruity is a fertilization barrier between partners that are less related, as in interspecific matching, when the partners belong to different species or genera; incompatibility is found only within intraspecific crosses, if the partner belongs to the identical species (Linskens 1980a, b).

If incongruity can be called interspecific-nonknowing, a third situation also seems to occur between closely related species and may be called interspecific-knowing, since here pollen-tube growth is more ruthlessly inhibited than after a self-incompatible pollination (Ascher and Drewlow 1975).

Incongruity. Incongruity prevents interspecific mixture by physiological mechanisms. If the selection of the pollen in the *mèr-à-boire* does not function and a pollen species reaches a stigma of a species that is not matching, an interspecific pistil-pollen relationship is present that is called incongruity (Hoogenboom 1972a, b). This can be defined as a bar to crossing that acts as an isolating mechanism between species or populations. Such barriers can take many forms (Mather 1975). This mechanism of non-functioning of pollen-pistil relationships in crosses between populations has to be considered as a separate and independent mechanism, different from incompatibility. Incongruity can be described as the "principle of non-functioning" as a consequence of incompleteness of the relationship between both sex partners. For example, if the pollen does not have all the necessary penetration genes, we have a fertilization barrier due to incongruity, the incomplete matching between the two partners in the crossing. On the other hand, if the pollen possesses all the necessary penetration genes we have complete matching or congruity.

The genetic and physiological background of incongruity is not well understood, although during the last ten years of interest in it by breeders (Grun and Aubertin 1966; Hermsen et al. 1974; Hoogenboom 1972a, b, 1975), it became clear that incongruity is not controlled by S-genes. Preliminary results (Jans 1981a, b) showed that neither RNA- nor protein-synthesis in the stigma are necessary for creating the incongruity barrier. It is a built-in system, with its expressivity depending on the degree of congeniality between the two partners at a taxonomic level of species or higher order.

Interspecific physiological barriers to fertilization are much like host-parasite relationships, and a host-parasite study is not essentially different from a study of interspecific barriers in a sexual relationship (Linskens 1976; Bushnell 1979). The inability of a specific disease organism to attack most plant species must be based on a non-recognition—the incompleteness of a relationship—and would thus be like incongruity. But the host-parasite relationship in which there is a positive pressure to cause non-growth, such as between wheat and rust, must be one based on recognition and would therefore be like interspecific-knowing between sexual partners. For research strategies that intend to obtain interspecific crosses in plants, "It is important that this new insight into the genetics of non-functioning may lead to a more deliberated research to overcome non-functioning" (Hoogenboom 1975). An important application of the incongruity barrier seems to be plant breeding. Beside male sterility, incongruity can be used by plant breeders in hybrid seed production, as proposed by Hoogenboom (1975).

There are two strategies to overcome both interspecific knowing and non-knowing. One is the lowering of the barrier capacity by inactivation of the interspecific-knowing barrier genes, the removing of gene products (e.g. by physical treatments), or the removal of the bar carrying stylar parts (Gardella 1950). The other strategy is the enhancement of the penetration capacity of pollen by X-ray treatment (Swaminathan and Murty 1959) or the application of the mentor pollen method as demonstrated by Stettler (1968). This was confirmed by substituting the mentor pollen with partially purified pollen-wall proteins from congruent pollen (Knox, Willing, and Ashford 1972). The treatment of pollen with solvents thus renders it congruent; the treatment of stigmas with solvents thus makes them receptive for incongruent pollen tubes (Willing and Pryor 1976; Stettler, Koster, and Steenackers 1980; and as suggested by Pandey 1979).

Incompatibility. While the biological function of interspecific barriers to fertilization can be seen in the maintenance of the species character, incompatibility functions within a species in the opposite way. In species crosses, incongruity is the rule, congruity the exception. But within a species it may be desirable that inbreeding should be prevented in order to maintain in a monoecious system a certain genetic flexibility. Incompatibility is a physiological mechanism steered by the S-gene system with multiple alleles. Charles Darwin in his book *The Effects of Cross- and Self-Fertilisation in the Vegetable Kingdom,* which came out more than a century ago, described the phenomenon: "Protected flowers, which had their own pollen placed on the stigmas, never yielded nearly a full complement of seeds; whilst those left uncovered produced fine capsules, showing that pollen from other plants must have been brought to them, probably by moths. Plants growing vigorously and flowering in pots in the greenhouse never yielded a single capsule" (Darwin 1878).

Investigation on both the homomorphic and heteromorphic incompatibility systems have been carried out during the last century. Many theories have been proposed to explain how incompatibility works, starting with the sexual-affinity theory by Strasburger in 1886, the line-stuff theory of Correns (1913, 1919), the immunity theory by Jost (1907) and East (1929, 1934), the ovarial-substance theory of Yasuda (1934), the consumption theory by Straub (1947), the bipartite theory of Lewis (1960) and the dimer hypothesis of the same author (Lewis 1965), linked with the antibody theory (Linskens 1955, 1965; Bhattacharjya and Linskens 1955) the velocity theory of Ascher (1966), the peroxydase-isoenzyme theory of Pandey (1967) now shown to be untenable (Bredemeijer and Blaas 1980), the application of the clonal-selection theory to incompatibility (Linskens 1962), the accelerator theory (Linskens 1968), the anti-repressor theory of Hoffmann (1971), the enzyme theory (Kroes 1973), the regulator theory of Ferrari and Wallace (1977), the tri-partite theory of de Nettancourt (1977), the master-gene theory of van der Donk (1975). They all try to explain the high degree of specificity of the incompatibility reaction, the S-gene control, the phenomenon of self-recognition, as well as the fact that one can distinguish between the recognition as an interation on the molecular level, and the rejection reaction, which is more or less aspecific.

The rejection reaction leads, in the case of sporophytic inhibition, to arrest of pollen germination and the impossibility of penetrance into the stigmatic tissue, and in the case of gametophytic inhibition, to a slowing down of the pollen-tube growth during the passage in the stylar canal or transmitting tissue, so that the tubes are arrested before they reach the base of the style and the ovule.

It became quite clear that the recognition reaction is the result of an interaction between proteins. The wall-born proteins of pollen grains are best investigated in the sporophytic system (Nasrallah, Barber, and Wallace 1970; Heslop-Harrison, Knox, and Heslop-Harrison 1974; Heslop-Harrison, et al. 1975; Heslop-Harrison, Heslop-Harrison, and Barber 1975; Ferrari and Wallace 1976). The recognition events involve proteins localized in the intine and exine (Knox and Heslop-Harrison 1971), which interact with the extracuticular proteinaceous receptor pellicle of the walls of the stigmatic papillae (Heslop-Harrison 1978b).

Also in the gametophytic system, proteins are linked with the recognition reaction (van der Donk 1975; Linskens 1953, 1955; Mäkinen and Lewis 1962; Nasrallah, Barber, and Wallace 1970; Herrero and Dickinson 1980; Nishio and

Hinata 1980; Bredemijer and Blaas 1981). The S-specific proteins occur in both the stigma and style of the pistil. Generally speaking, there are arguments for seeking homologies between sporophytic and gametophytic incompatibility systems (Heslop-Harrison 1978a).

There are some recent new observations on the gametophytic incompatibility system. These concern the influence of season, plant age, and temperature during the preprogamic phase of incompatible plants. Whereas temperature dependence of the incompatibility reaction was observed in several species (Lewis 1942; Ascher and Peloquin 1966; Dane and Melton 1973) it became evident that temperature pretreatment influences the expression of the S-gene too (van Herpen and Linskens 1981).

After self- and cross-pollination, a change in the degree of saturation of fatty acids in the lipid pool has been found, and the distribution of the fatty acids among the various glycosphingolipids in self- and cross-pollinated styles is different (Delbart et al. 1980[c]). It is a well-known fact that plants can adapt to changes in environmental temperature by altering the degree of saturation in their fatty acid side chains. So it seems possible that the environmental conditions before and during the progamic phase have a distinct influence on the pollen tube-style interaction via a change in structure and/or distribution of lipids in the membranes. This means that environmental factors are not only decisive during, but also before, the progamic phase. So one has to include the molecular events of the membranes in any biochemical explanation of incompatibility since the contact recognition surface of the incompatibility reaction must be at the membrane (Linskens 1980a). The characteristic changes which the glycosphingolipids undergo in pollen tubes during the incompatibility reaction (Delbart et al. 1980[a], [b], [c]) confirm the recent findings that show that the lipid-protein interaction has a genetic base (Sandhoff 1980).

Genetic defects in an activator protein determine the activity and the amount of lysosomal enzymes which catalyze the breakdown of the membrane-bound glycolipids. Depending on the activity in the lysosomes, the genetic defect can lead to an accumulation of the substrates of the defective enzymes, which then result in an increase in the amount of the lipid which cannot be degraded.

Another new aspect concerns the long-distance post-pollination effects of an incompatible or compatible pollination. The induction of corolla wilting is controlled by the style (Gilissen 1978), which means that there is a transport of information about the presence and the type of pollen. Another fact yet to be explained is that the ovaries receive information about both the presence and the compatible or incompatible nature of the pollen tubes long before the tube tips have reached the base of the style (Deurenberg 1977). Also, it seems likely that changes of the bioelectric potential have some relation to incompatibility: different types of response after cross- and after self-pollination suggest that the recognition and discrimination of both takes place (Spanjers 1981).

CONCLUSION

Pollination and fertilization barriers are both active parts of the reproductive strategy in flowering plants. Higher plants developed a delicate tuning system during evolution which, one guarantees or improves the combination of both sex partners within a species. And two, evolved mechanisms which enhance cross-pollination within a species in order to maintain the genetic variability that helps species adapt to a changing environment. It is a process involving two necessities: maintaining the species character and maintaining the flexibility for survival.

ACKNOWLEDGMENTS

I thank R. J. Campbell and L. van der Pijl for discussing and reviewing this chapter.

LITERATURE CITED

Armstrong, J. A. 1979. *Biotic pollination mechanisms in Australia, a review.* New Zealand J. Bot. 17:467–508.

Ascher, P. D. 1966. *A gene action model to explain gametophytic self-incompatibility.* Euphytica 15:179–83.

Ascher, P. D., and L. W. Drewlow. 1975. *The effect of prepollination injection with stigmatic exudate on interspecific pollen tube growth in Lilium longiflorum Thunb. styles.* Plant Sci. Lett. 4:401–5.

Ascher, P. D., and S. J. Peloquin. 1966. *Influence of temperature on incompatible and compatible pollen tube growth in Lilium longiflorum.* Can. J. Genet. Cytol. 8:661–64.

Baker, H. G., I. Baker, and P. A. Opler. 1973. *The stigmatic exudate and pollination.* Pages 47–60 in N. B. M. Brantjes, and H. F. Linskens, eds.: *Pollination and Dispersal.* Nijmegen.

Bhattacharjya, S. S., and H. F. Linskens. 1955. *Recent advances in the physiology of self-sterility in plants.* Sci. and Cult. 20:370–73.

Bonny, A. P. 1980. *Seasonal and annual variation over 5 years in contemporary airborne pollen trapped at a cumbrian lake.* J. Evol. 68:421–41.

Bredemeijer, G. M. M., and J. Blaas. 1980. *Do S-allele–specific peroxydase isoenzymes exist in self-incompatible Nicotiana alata?* Theor. Appl. Genet. 57:119–23.

———. 1981. *S-specific proteins in styles of self-incompatible Nicotiana alata.* Theor. Appl. Genet. 59:185–90.

Bushnell, W. R. 1979. *The nature of basic compatibility: Comparison between pistil-pollen and host-parasite interaction.* Pages 211-27 in J. M. Daly and I. Uritani, eds. *Recognition and Specificity in Plant Host-Parasite Interactions.* Japan. Sci. Soc. Press, Tokyo.

Colin, L. J., and C. E. Jones. 1980. *Pollen energetics and pollination modes.* Amer. J. Bot. 67:210–15.

Correns, C. 1913. *Individuum und Individualstoffe.* Biol. Cbl. 33:389–401.

———. 1919. *Selbststerilität und Individualstoffe.* Naturwissenschaften 4:183–98, 210–13.

Covich, A. 1974. *Ecological economics for foraging among coevolving animals and plants.* Ann. Missouri Bot. Garden 61:794–805.

Dane, F., and B. Melton. 1973. *Effect of temperature on self- and cross-incompatibility and in vitro pollen growth characteristics in Alfalfa.* Crop Sci. 13:587–91.

Darwin, C. 1878. *The Effects of Cross- and Self-fertilisation in the Plant Kingdom.* 2nd ed., 1900. John Murray, London.

De Cock, A. W. A. M. 1980. *Flowering biology of the seagrass Zostera marina L.* Ph.D. dissertation, University Nijmegen.

Delbart, C., H. F. Linskens, B. Bris, Y. Moschetto, and D. Coustaut. 1980. [a]. *Analysis of glycosphingolipids of Petunia hybrida, a self incompatible species. I. Composition in fatty acids and in long chained bases of pollen and unpollinated styles*. Proc. Kon. Nederl. Akad. Wet., Amsterdam, C 83:229–40.

Delbart, C., B. Bris, H. F. Linskens, R. Linder, and D. Coustaut. 1980. [b]. *Analysis of glycosphingolipids of Petunia hybrida, a self incompatible species. II. Evolution of the fatty acids composition after cross- and self-pollination*. Proc. Kon. Nederl. Akad. Wet., Amsterdam, C 83:241–54.

Delbart, C., B. Bris, H. F. Linskens, and D. Coustaut. 1980. [c]. *Analysis of glycosphingolipids of Petunia hybrida, a self incompatible species. III. Evolution of the long chained bases after self- and cross-pollination*. Proc. Kon. Nederl. Akad. Wet., Amsterdam, C 83:255–70.

De Nettancourt, D. 1977. *Incompatibility in Angiosperms. Monographs on Theoretical and Applied Genetics*, vol. 3. Springer, Berlin-Heidelberg-New York.

Deurenberg, J. J. M. 1977. *Aktivering van het vruchtbeginsel van Petunia*. Ph.D. dissertation, University Nijmegen.

Dickinson, H. G., and D. Lewis. 1975. *Interaction between the pollen grain coating and the stigmatic surface during compatible and incompatible interspecific pollinations in Raphanus*. Pages 166–74 in J. G. Ducket, and P. A. Racey, eds. *The Biology of the Male Gamete*. Biol. J. Linnean Soc. 7 (suppl.).

East, E. M. 1929. *Self-sterility*. Bibliogr. Genet. 5:331–68.

———. 1934. *The reaction of the stigmatic tissue against pollen tube growth in selfed self-sterile plants*. Proc. Nat. Acad. Sci. USA 20:364.

Faegri, K., and L. van der Pijl. 1979. *The Principles of Pollination Ecology*. 3rd ed. Pergamon Press, Oxford.

Ferrari, T. E., and D. H. Wallace. 1976. *Pollen protein synthesis and control of incompatibility in Brassica*. Theor. Appl. Genet. 48:243–49.

———. 1977. *A model for self-recognition and regulation of the incompatibility response of pollen*. Theor. Appl. Genet. 50:211–25.

Frankel, R., and E. Galun. 1977. *Pollination Mechanisms, Production and Plant Breeding. Monographs on Theoretical and Applied Genetics*, vol. 2. Springer, Berlin-Heidelberg-New York.

Gardella, C. 1950. *Overcoming barriers to crossability due to style length*. Amer. J. Bot. 37:219–24.

Gilissen, L. J. W. 1978. *Bevruchtingsbiologische aspecten van zelf-incompatibele planten van Petunia hybrida L*. Ph.D. dissertation, University Nijmegen.

Grun, P., and M. Aubertin. 1966. *The inheritance and expression unilateral incompatibility in Solanum*. Heredity 21:131–38.

Hagerup, O. 1954. *Autogamy in some drooping Bicornes flowers*. Bot. Tidskr. 51:103–6.

Hermsen, J. G. Th. J. Olsder, P. Jansen, and E. Hoving. 1974. *Acceptance of self-compatible pollen from Solanum verrucosum in dihaploids from S. tuberosum*. Pages 37–40, in H. F. Linskens, ed. *Fertilization in Higher Plants*. North-Holland Publ. Co., Amsterdam.

Herrero, M., and H. G. Dickinson. 1980. *Ultrastructural and physiological differences between buds and mature flowers of Petunia hybrida prior to and following pollination*. Planta 148:138–45.

Heslop-Harrison, J. 1978a. *Genetics and physiology of angiosperm incompatibility systems*. Proc. Roy. Soc. (Lond.) B 202:73–92.

Heslop-Harrison, J. 1978b. *Cellular recognition systems in plants. Studies in Biology*, vol. 100. Edward Arnold, London.

Heslop-Harrison, J., Y. Heslop-Harrison, and J. Barber, 1975. *The stigma surface in the incompatibility response*. Proc. Roy. Soc. (Lond.) B 188:287–97.

Heslop-Harrison, J., R. B. Knox, and Y. Heslop-Harrison. 1974. *Pollen-wall proteins:*

exine-held fractions associated with the incompatibility response in Cruciferae. Theor. Appl. Genet. 44:133–37.
Heslop-Harrison, J., R. B. Knox, Y. Heslop-Harrison, and O. Mattsson. 1975. *Pollen-wall proteins: emission and role in incompatibility response*. Biol. J. Linnean Soc. 7, suppl. 1:189–202.
Heslop-Harrison, J., and K. R. Shivanna. 1977. *The receptive surface of the angiosperm stigma*. Ann. Bot. 41:1233–58.
Hesse, M. 1978. *Vergleichende Untersuchungen zur Entwicklungsgeschichte und Ultrastruktur von Pollenklebstoffen bei verschiedenen Angiospermen*. Linzer biol. Beitr. 9:237–58.
———. 1979. *Entstehung und Auswirkungen der unterschiedlichen Pollenklebrigkeit von Sangiusorba officinalis und S. minor*. Pollen et Spores 21:399–413.
———. 1980. *Pollenkitt in relation to pollination ecology*. Calicut Univ. Res. J. 1:29–33.
Hevley, R. H. 1979. *Dietary habits of two nectar and pollen feeding bats in southern Arizona and northern Mexico*. J. Arizona-Nevada Acad. Sci. 14:13–18.
Hoffmann, M. 1971. *Induktion und Analyse von selbstkompatiblen Mutanten bei Lycopersicon peruvianum L. Mill. III. Analyse der selbstkompatiblen Mutanten*. Biol. Zbl. 90:33–41.
Hoogenboom, N. G. 1972a. *Breaking breeding barriers in Lycopersicon, 1–3*. Euphytica 21: 221–56.
———. 1972b. *Breaking breeding barriers in Lycopersicon, 4–5*. Euphytica 21:397–414.
———. 1975. *Incompatibility and incongruity: Two different mechanisms for the non-functioning of intimate partner relationship*. Proc. Roy. Soc. (Lond.) B 188:361–75.
Howell, D. J. 1974. *Bats and pollen: Physiological aspects of the syndrome of chiropterophily*. Comp. Biochem. Physiol. 48A:263–76.
Jans, P. J. H. TH. 1981a. *Effect of inhibition of gene expression in the style of Petunia on pollen germination and tube growth after intergeneric pollination*. Incomp. Newsl. 13 (in press).
———. 1981b. *De fysiologische achtergronden van incongruentie bij bloemplanten*. Vakbl. Biol. (in press).
Jost, L. 1907. *Uber die Selbststerilität einiger Blüten*. Bot. Ztg. 65:77–92.
Knox, R. B., and J. Heslop-Harrison. 1971. *Pollen-wall proteins: The fate of intine-held-antigens on the stigma in compatible and incompatible pollinations of Phalaris tuberosa*. J. Cell Sci. 9:239–51.
Knox, R. B., R. R. Willing, and A. E. Ashford. 1972. *Role of pollen-wall proteins as recognition substances in interspecific incompatibility in poplar*. Nature 237:381–83.
Konar, R. N., and H. F. Linskens. 1966a. *The morphology and anatomy of the stigma of Petunia hybrida*. Planta 71:356–71.
———. 1966b. *Physiology and biochemistry of the stigmatic fluid of Petunia hybrida*. Planta 71:372–87.
Kroes, H. W. 1973. *An enzyme theory of self-incompatibility*. Incomp. News. 2:5–14.
Lejoly-Gabriel, M. 1980. *Kwalitatieve en kwantitatieve pollenvorming in de Belgische atmosfeer in 1978 en 1979*. Université de Louvain.
Lewis, D. 1942. *The physiology of incompatibility in plants. I. The effect of temperature*. Proc. Roy. Soc. (Lond.) B 140:127–35.
———. 1960. *Genetic control of specificity and activity of the S antigen in plants*. Proc. Roy. Soc. (Lond.) B 161:468–77.
———. 1965. *A protein dimer hypothesis of incompatibility*. Genetics To-day 3: 657–63.
Linskens, H. F. 1953. *Physiologische und chemische Unterschiede zwischen selbst- und fremdbestäubten Petunien-Griffeln*. Naturwissenschaften 40:28–29.
———. 1955. *Physiologische Untersuchungen der Pollenschlauch-Hemmung selbststeriler Petunien*. Z. Bot. 43:1–44.
———. 1962. *Die Anwendung der "Clonal Selection Theory" auf Erscheinungen der*

Selbstinkompatibilität bei der Befruchtung der Blütenpflanzen. Portug. Acta Biol. Ser. A, 6:231–38.
———. 1965. *Biochemistry of incompatibility*. Genetics To-day 3:629–36.
———. 1968. *Egg-sperm interactions in higher plants*. Quad. Acc. Naz. die Lincei (Rome) 104:47–56.
———. 1976. *Specific interactions in higher plants*. Pages 311–26 in B. K. S. Wood, and A. Graniti, eds. *Specificity in Plant Diseases*. *NATO* ADV. STUDIES INST. Ser. A, 10.
———. 1980a. *Physiology of fertilization and fertilization barriers in higher plants*. Pages 113–26 in S. Subtelny, and N. K. Wessels eds. *The Cell Surface: Mediator in Developmental Processes*. 38th Symposium Soc. Devel. Biol. (Vancouver 1979).
———. 1980b. *Befruchtingsbarrieren bei höheren Pflanzen*. Naturwiss. Rdsch. 33:11–20.
Macior, L. W. 1973. *Pollination ecology, the study of cooperative interactions in evolution*. Pages 101–10 in N. B. M. Brantjes, and H. F. Linskens, eds. *Pollination and Dispersal*. Publ. Dept. Bot., University Nijmegen.
Mäkinen, Y. L. A., and D. Lewis, 1962. *Immunological analysis of incompatibility (S) proteins and of cross-reacting material in a self-compatible mutant of Oenothera organensis*. Genet. Res. 3:352–63.
Martin, F. W. 1969. *Compounds from the stigmas of ten species*. Amer. J. Bot. 56: 1023–27.
Martin, F. W., and J. L. Brewbaker. 1971. *The nature of the stigmatic exudate and its role in pollen germination*. Pages 262–66 in J. Heslop-Harrison, ed. *Pollen Development and Physiology*. Butterworth, London.
Mather, K. 1975. *Comment: Incompatibility and incongruity*. Proc. Roy. Soc. (Lond.) B 188:374–75.
Nasrallah, M. E., J. T. Barber, and D. H. Wallace. 1970. *Self-incompatibility proteins in plants: Detection, genetics, and possible mode of action*. Heredity 25:23–27.
Nishio, T., and K. Hinata. 1980. *Rapid detection of S-glycoproteins of self-incompatible crucifers using Con-A-Reaction*. Euphytica 29:217-21.
Pandey, K. K. 1967. *Origin of genetic variability: Combination of peroxydase isoenzymes determine multiple allelism of the S gene*. Nature 213:669–72.
———. 1979. *Overcoming incompatibility and promoting genetic recombination in flowering plants*. New Zealand J. Bot. 17:645–63.
Pettitt, J., S. Ducker, and B. Knox. *Submarine pollination*. Sci. Amer. 244:135–43.
Pohl, F. 1937. *Die Pollenerzeugung der Windblüter*. Beih. Bot. Zbl. 56:365–70.
Sandhoff, K. 1980. *Lipid-Protein-Wechselwirkungen. Mechanismen des enzymatischen Glykolipid-Abbaus und seine genetisch bedingten Entgleisungen*. Naturwissenschaften 67:431–41.
Shivanna, K. R., Y. Heslop-Harrison, and J. Heslop-Harrison. 1978. *The pollen stigma interactions: Bud pollination in Cruciferae*. Acta Bot. Neerl. 27:107–19.
Smith, C. A., and W. E. Evenson. 1978. *Energy distribution in reproductive structures of Amaryllis*. Amer. J. Bot. 65:714–15.
Spanjers, A. W. 1981. *Bioelectric potential changes in the style of Lilium longiflorum Thunb, after pollination*. Planta, (in press).
Stanley, R. G., and H. F. Linskens. 1974. *Pollen, Biology, Biochemistry Management*. Springer, Berlin-Heidelberg-New York.
Stebbins, G. L. 1974. *Flowering Plants' Evolution above the Species Level*. Harvard Univ. Press, Cambridge, Mass.
Stettler, R. F. 1968. *Irradiated mentor pollen: Its use in remote hybridization of black cottonwood*. Nature 219:746–47.
Stettler, R. F., R. Koster, and V. Steenackers. 1980. *Interspecific crossability studies in poplar*. Theor. Appl. Genet. 58:273–82.
Strasburger, E. 1886. *Uber fremdartige Bestäubung*. Jb. wiss. Bot. 17:50.

Straub, J. 1947. *Zur Entwicklungsphysiologie der Selbststerilität von Petunia II. Das Prinzip des Hemmungsmechanismus*. Z. Naturforsch. 2 b:433–44.

Swaminathan, M. S., and B. R. Murty. 1959. *Effect of x-radiation on pollen tube growth and seed setting in crosses between Nicotiana tabacum and N. rustica*. Z. Vererbungsl. 90:393–99.

Van der Donk, J. A. W. M. 1975. *Molecular biological aspects of the incompatibility reaction in Petunia*. Ph.D. dissertation, University Nijmegen.

Van Herpen, M. M. A., and H. F. Linskens. 1981. *Effect of season, plant age and temperature during plant growth on compatible and incompatible pollen tube growth in Petunia hybrida*. Acta Bot. Neerl. (in press).

Willing, R. R., and L. D. Pryor. 1976. *Interspecific hybridization in poplar*. Theor. Appl. Genet. 47:141–51.

Woittiez, R. D., and M.T.M. Willemse. 1979. *Sticking of pollen on stigmas: The factors and a model*. Phytomorphology 29: 57-63.

Yasuda, S. 1934. *Physiological research on self incompatibility in Petunia violacea*. Bull. Imp. Coll. Agricult. Forst. Morioka/Japan 20: 1-82.

4] Directing the Accelerated Evolution of Crop Plants

by JACK R. HARLAN*

ABSTRACT

The processes of evolution are generally held to be gene mutation, changes in chromosome structure and number, genetic recombination, natural selection, and reproductive isolation, together with the accesory processes of migration, hybridization, and chance. The accessory processes may be seen as special considerations of recombination. To these basic processes gene regulation may be added as something distinct from gene mutation, and it may turn out to be the most important of all in the acceleration of evolution. Each process is briefly discussed, but the process of gene regulation is featured in this chapter. All the processes can be manipulated artificially and some can be directed. It is proposed that in certain kinds of wide crosses two different processes operate in transfer of genetic information: (a) alien chromosomes may pair and crossover, exchanging chromosome segments in a conventional way and (b) mobile, repetitive noncoding DNA may move from genome to genome without chromosome pairing. Such DNAs may cause radical changes in gene action without exchange of structural genes. Rapid evolution is more easily explained on the basis of genetic regulation than by gene mutation and selection.

"Through the animal and vegetable kingdoms, nature has scattered the seeds of life abroad with the most profuse and liberal hand; but has been comparatively sparing in the room and nourishment necessary to rear them."

T. R. Malthus
An Essay on the Principle of Population, 1798

*Professor of Plant Genetics, Crop Evolution Laboratory, Agronomy Department, University of Illinois, Urbana, Illinois 61801.

STRATEGIES OF PLANT REPRODUCTION (BARC Symposium number 6—Werner J. Meudt, ed.)
Allanheld, Osmun, Totowa

INTRODUCTION

Malthus was credited by both Darwin and Wallace for providing a means to explain evolution. The "wastage" of reproductive effort resulted in strong natural selection pressures, and only the fittest survive. Indeed, the selection is so extreme that some biologists prefer to believe that only the lucky survive. An individual tree, for example, may produce millions of seeds in its lifetime, yet all that is required to maintain the population is a single, successful, replacement seedling, and this over a matter of several to many decades. The "surplus" reproductive effort is not wasted on the ecosystem as a whole, for many other organisms are supported by it. Seeds are consumed by herbivores of several phyla, while others decay and nourish microorganisms which play their role in recycling elements and organic compounds. Prominent among scavengers of excess reproductive effort is *Homo sapiens* L. The reproductive wastage, so conspicuous in natural systems, can be directed to our immediate advantage, and one of the goals of applied biology is to increase it for our benefit. Most of the recent advances in seed crop breeding have come from increases in harvest index. That is to say, we are breeding plants to put more and more energy into reproductive effort. We are enhancing a natural evolutionary trend in order to feed the population. Under good management, a maize plant needs only one good seed to replace itself; we can eat the rest. Nor does maize need anything like the amount of pollen produced in wild or primitive maize, and we have bred hybrids with much reduced tassels and smaller pollen production. Plant breeding, in short, consists in directing and accelerating evolution toward human objectives.

Since I am an ancient student of G. L. Stebbins, Jr., and have much admired his understanding of evolutionary processes, I shall follow his models for this discussion. Some ten years ago, Stebbins (1971) listed the processes of evolution as follows: (a) gene mutation, (b) changes in chromosome structure and number, (c) genetic recombination, (d) natural selection, and (e) reproductive isolation, together with three accessory processes, migration, hybridization and chance. The accessory processes can be recognized as special consideration of the recombination processes. We may now add a sixth major process, gene regulation. Very little was known about regulation systems ten years ago, but the decade of the 1970s was very active in genetic research and we now know more about them. The process is still not at all well understood, but it is slowly emerging as a major feature of evolution.

It is true that changes in gene regulation may involve changes in DNA sequences and, therefore, resemble gene mutation, but it is now becoming clear that the regulatory apparatus is something distinct from the structural gene, which encodes information for the production of a specific protein. We are sailing on uncharted seas when we deal with gene regulation, but I shall emphasize this process and its potential in this chapter and illustrate the reasons why I think it important with some of our current research on maize X *Tripsacum* interactions. But first, let us look briefly at the other, more familiar, processes to put the matter of accelerated evolution in perspective.

All the processes can be manipulated artifically so that we have some control over them. In some cases, our control is not very directional, but we can increase the frequency of mutation; we can reorganize chromosomes; we can direct and speed up genetic recombination; and we surely can substitute artificial selection for natural selection and provide reproductive isolation if it is desirable. In our hands, migration becomes plant introduction, hybridization is the heart and soul of plant breeding, and chance (drift) can be seduced in our favor by increasing the number of events and by improving the efficiency of selection screens.

GENE MUTATION

Gene mutation has been considered by many to be the ultimate and sole source of genetic variation. This is not altogether true since the effect of a gene depends on other genes in the neighborhood on the chromosome or in the genome, on regulatory systems, and so on. But gene mutation is an important process in evolution. In nature it is usually a slow process, but it can be speeded up and even directed to some extent by mutagenic agents. We have had successful results from X-rays, γ-rays, ultraviolet light, aging of seeds, rapid aging of seeds, mutator genes, and treatment by a wide assortment of mutagenic chemical agents.

At the DNA level, there are two basic kinds of events that can take place: (a) base substitution and (b) frame shift. A base substitution can result in changing the sense of the coding triplet so that a different amino acid is inserted into the protein product. This can be lethal, cause genetic defects, or have no effect at all depending on the nature and function of the protein. A frame shift can make nonsense of the code from the point of the shift to the end of the message (Baer 1977). Mutagens can be somewhat selective in action. Base analogues like 5-bromouracil found in caffeine can substitute for adenine and deaminating compounds such as nitrous acid, nitrogen oxides, and sulfur oxides which may change GC pairs to AT pairs. Frame shifts are readily induced by acridine orange and similar compounds (Baer 1977). The chlorinated hydrocarbons, such as DDT, dieldrin, Mirex,® polychlorinated biphenyls (PCBs), and vinyl chlorides, carbon tetrachloride, etc., tend to induce both substitutions and frame shifts (Fishbein 1978b). Hycanthone, used to treat schistosomiasis, induces both frame shifts and chromosome breakage (Baer 1977). Other agents such as ozone, SO_2, and the dichlorvos specialize in chromosome breakage (Fishbein 1978a,b).

While we are properly concerned about the insertion of artificial mutagens into our environment, many of them occur naturally and must play some role in evolution. The mycotoxins such as aflotoxin and sterigmatocsin and toxins of *Penicillium* and *Fusarium* are abundant under some conditions. The oxides of nitrogen and sulfur as well as ozone occur in nature (Fishbein 1978a).

Diversity may be easily generated by mutagens. Neuffer and Coe (1977) developed a technique for treating maize pollen with chemical mutagens. In one study using EMS, some 2,457 recessive mutants were obtained from 3,172 selfed ears; i.e., when pollen was treated and applied to normal miaze, 78% of the seed carried mutations that were revealed by self-fertilization. An estimated 535 loci were involved (Neuffer and Sheridan, 1980). In fact, mutation rates can be accelerated without mutagens. Hristov (1979) reported an increase in mutation rate in maize by simply storing seed at high temperature (50° C) in a high-moisture environment.

Yet diversity is not necessarily useful if it is deleterious in nature, and most mutations are damaging. Is mutation breeding a useful approach to the directed acceleration of plant evolution? We have had three decades of intensive and extensive experience with mutation breeding and should be able to answer the question. Volumes have been published on the subject and vast numbers of papers reporting good, or at least promising, results. Sigurbjörnsson and Micke (1974) summarized much of the information as of 1973. They listed the following reported occurrences: higher yield, 47; early maturity, 36; better quality, 27; lodging resistance, 26; disease resistance, 24; shorter stem, 16; improved plant type, 9; and a scattering of other miscellaneous improvements. It may be noted that the most frequently reported improvements by mutation breeding deal with a complex of features including early, short-stemmed, lodging resistant, higher yielding types. These are often correlated, especially in cereals and pulses. Konishi (1977) pointed out that dwarf or semi-dwarf mutants are easy to produce in barley, for example, but almost all are lower yielding than the parental type. It is a rare mutant that can be used. Similar results were obtained in rice (Okuno 1977). Gotoh (1977), in a review of the extensive use of Norin 10 in wheat breeding, pointed out that some induced mutants appear to be about as good as the naturally occuring semi-dwarf Norin 10 source, yet they have been seldom used, while Norin 10 has been incorporated into wheats all over the world.

I think that we can fairly state that mutation breeding has not been as fruitful as many early enthusiasts had hoped, but that it has become a useful methodology for plant improvement and can be added to the tools of the trade. Much depends on the goals of the breeding program. Perhaps the objective is the development of slow-growing dwarf plants, e.g., lawn grasses, dwarf fruit trees or ornamental shrubs, or semi-dwarf cereals and pulses as mentioned above. Ikeda (1977), for example, reported on induction of dwarf growth habit in apples. Improved "quality" or change in flower or grain color are often due to mutations that block a metabolic pathway, and these should readily be produced by mutagenic agents.

Enhancement of metabolic activity has also been reported. Rabie and Pasztor (1979) report heterosis in maize from an induced "corn grass" mutant. Chowdhury (1979) reported increased physiological efficiency induced in wheat by EMS and γ-rays. On the other hand, experienced and successful plant breeders like Frey (Jalani and Frey 1979) and Burton (Burton, Hanna, and Powell 1980), who have compared methodologies, have generally found conventional breeding approaches more efficient in getting desired results. Yet both would agree that mutation breeding has its place in the arsenal of tools for plant breeding.

CHANGES IN CHROMOSOME STRUCTURE AND NUMBER

The art of chromosome engineering has been advancing steadily; old techniques have been improved and new ones added to our arsenal. The use of naturally occurring genetic systems has increased noticeably in recent years. McClintock (1978) recently reviewed the potential use of the breakage-fusion-bridge system to reorganize rapidly the maize genome, Green (1978) has shown that the male recombination (MR) chromosome in *Drosophila* causes not only gene mutations, but translocations and inversions as well. MR chromosomes are widespread in the US, Europe, Asia, and Australia. Berg, Engels, and Kreber (1980) described rapid rearrangement of the X-chromosome of *Drosophila* in combinations that yield hybrid dysgenesis. Gerstel and Burns (1976) reviewed the occurrence of megachromosomes in plants and show figures of enlarged eukaryotic megachromosomes in the hybrid between *Nicotiana tabacum* and *N. plumbaginifolia*.

Structural reorganization of chromosomes can be enhanced by a variety of artificial techniques. Wienhues (1979) irradiated an additional line of winter wheat that carried a chromosome of *Agropyron intermedium* and induced a translocation, transferring resistance to stripe rust into the wheat genome. This is an old technique pioneered by Sears (1956). Mukade (1978) induced a similar transfer of rust resistance from rye to wheat by X-ray treatment. He introduced an additional element of rapid generation turnover to speed up elimination of undesirable genes carried with the translocated piece of chromosome. Ethylenediaminetetraacetic acid and dimethyl sulfide enhance genetic recombination (Ihrke and Kronsted 1975), and other complexing agents induce breakage and chromosome rearrangement in maize (Sukhapinda and Peterson 1980). The extent to which karyotypes can be restructured is illustrated by Kunzel and Nicoloff (1979). They maintained that the standard barley karyotype was not very useful because only three of the seven chromosomes can be recognized on sight. They reconstructed the karyotype so that all seven could be identified. But chromosome 1 had arms of equal length, so they changed that as well so that all arms can be recognized in their synthetic genome.

Radical changes in karyotype can be produced by alien chromosome substitution. One of the most spectacular examples is the substitution of chromosome 4H of barley for 4A of wheat (Islam, Shephered, and Sparrow 1978; Islam 1980). The combination is both viable and fertile. This might be taken to suggest that the differences in wheat and barley chromosomes lie more in non-coding DNA than in the coding sequences. This will be addressed later.

With respect to chromosome number, the production of hyploids as a means of developing homozygous lines has become quite popular in recent years. The Chinese have, perhaps, led the field in exploiting anther culture for the production of haploids (Harlan 1980). They were the first to do it with wheat and have conducted haploid-breeding programs on a large scale. Chang and Hong-Yuan (1981) report the induction of haploids of rice by ovary culture. For Europe, pollination by alien species is perhaps more popular. Adamski

(1979) and Simpson, Snape, and Finch (1980) describe the production of barley haploids through pollination by *Hordeum bulbosum*. This technique has long been used in potato breeding (Mok and Peloquin 1975). Dumas de Vaulx (1979) obtained haploid melons by pollinating *Cucumis melo* with *C. ficifolius*. The *ig* gene in maize that induces paternal haploids is a remarkable genetic system for producing haploids (Kermicle 1969).

Colchicine is still the most commonly used agent for increasing chromosome number, although other methods are also used. The natural way is primarily through non-reduction (Harlan and de Wet 1975), and this is used on an increasing scale, especially in wide crosses. In maize, the elongate gene *(el)* increases the frequency of non-reduction very sharply (Rhoades and Dempsey 1966) and has been used to develop tetraploid populations (Alexander 1958; Alexander and Beckett 1963).

Rapid changes in chromosome number and in genome reorganization are obtained from wide crosses. The production of new alloploid combinations has become routine in a number of plant groups. Among the many cases of genomic reconstitution that might be cited, I would like to call attention to the elegant study by Wagenaar (1969) of the cross *Triticum crassum* ($n=21$) × *T. turgidum* ($n=14$). The F_1 had $2n=35$ as expected. Some of the resulting seeds developed into $2n=70$ plants. In successive generations chromosomes were eliminated and, by F_8, all plants were $2n=56$) and fertile. The amphiploid behaves like a new species. Selective chromosome elimination is reported in *Nicotiana* species hybrids (Szilagyi and Nagy 1978) in *Zea* × *Sorghum* (CIMMYT 1978), *Saccharum* × *Sorghum* (Gupta, Harlan, and deWet 1978), *Hordeum* species hybrids (Schooler and Anderson 1979), and others.

GENETIC RECOMBINATION

Traditionally, plant and animal breeding has consisted almost entirely of the production, evaluation, and selection of new gene combinations. One may exploit genes that have accumulated over millennia by the assembly and introduction of exotic germplasm (migration in evolutionary terminology) or one may induce mutation as discussed above. New combinations are generated by hybridization and the operations of chance (genetic drift). Recombination can be artifically enhanced by the methods mentioned in the last section—irradiation, chemicals that cause chromosome breakage or enhance crossing over, genetic systems, shorter generation turnover, more generations per unit of time. Drift or genetic fixation may be speeded up by sib-mating, self-fertilization, apomixis, vegetative reproduction, haploidly followed by chromosome doubling, and so on.

One of the primary current challenges in plant breeding is to make more effective use of wild relatives of our cultivated plants. Examples of the utility of wild races and species have been reviewed recently (Harlan 1976, n.d.; Stalker 1980). The problems are formidable, in some cases, but the pay-offs have also been very great. It appears that crops such as sugarcane, tomato, and tobacco probably could not maintain commercial status without the genetic support of wild relatives, primarily as sources of disease resistance. Other crops have been protected to a lesser degree by transfers of genetic resistance from related forms. Other uses of wild relatives include improvement in yield, quality, and adaptation, and changes in cytoplasms, modes of reproduction, and methods for producing commercial hybrids.

While genetic recombination has been at the very heart of plant and animal breeding, it is one of the more conservative of the evolutionary processes. In a stimulating paper, Reanney observed: "One of the most striking things about generalized recombination is its conservative character. No information is gained or lost and gene order is not disturbed" (Reanney 1979, p.600). In his view, legitimate recombination acts as a brake on change, and that continuing evolution is nourished by more primitive and more radical processes such as RNA:RNA splicing and illegitimate recombination. This general view fits well with current thoughs about rapid evolution to be discussed shortly.

SELECTION

In the acceleration of evolution, both natural and artificial selection come into play. They can be very powerful forces. Natural selection sets the limits of genetic reach by determining which species can be hybridized and the limits of fertility in wide crosses. The lethal and defective combinations tend to be eliminated. Artificial selection is the most directed of all the evolutionary processes and the most under our control. It pays us to make maximum use of it.

The power of artificial selection is usually limited by the efficiency of the evaluation screen. If evaluation is easy and fast (as, for example, in seed size, oil content measured by NMR or near infrared light reflectance, certain disease reactions, gross morphology, and so on), large numbers of individuals can be screened and selection pressures can be very strong. If the evaluation is slow and expensive, as in measuring livestock gain on a given genotype of forage, or longevity, persistence, and the like, only a small number of tests can be conducted and most of the selection is usually based on some other criterion. Full use of the process of selection, then, depends a good deal on ingenuity and originality in developing evaluation techniques. As a matter of fact, the historic drift in genetics from peas to flies to *Neurospora,* yeast, bacteria, and on to viruses and plasmids was brought about as much by powerful selection screens as by shorter generation turnovers. The rare mutant or recombinant that could survive on a specific medium or was resistant to some chemical in the environment could be identified. An event that occurs once in a million times is not necessarily rare, if there are 100 million events and the selection screen is sufficiently sensitive to identify it.

Herein lie the hopes for tissue culture techniques (Sala et al. 1980). I am confident that a great deal of ingenuity and originality will be generated by those who try to exploit microbiological techniques for the improvement of higher organisms. Pollen selection may be another point at which very strong selection pressures could be applied. Pollen characters have already been useful for mapping of genetic fine structures (Nelson 1968), and other techniques will no doubt be developed.

REPRODUCTIVE ISOLATION

While reproductive isolation is important in speciation, the evolution of domestic plants and animals is almost always carried out at the infraspecific level. This does not mean that it is not frequently good strategy to separate populations and individuals to prevent interbreeding. Indeed, we do this routinely in controlled matings. This bull is mated to that cow, this boar to that sow, this inbred to that inbred, and so on. Physical, spacial, seasonal isolation, cloth bags, paper bags, detasseling, male sterility, and a variety of other techniques is used. In maize it is possible to insert gametophytic genes into stocks that effectively isolate them from certain other stocks. These are routinely inserted into popcorn and sweetcorn cultivars so they can be grown next to field corn and not be contaminated. But it is a rare and very special case where human manipulation has resulted in reproductive isolation as in speciation.

GENE REGULATION

At this point I will summarize, briefly, some of our work with maize X *Tripsacum* derivatives to illustrate the power of regulatory systems. We have made many hybrids between maize and *T.dactyloides,* but the material to be discussed here traces ultimately to a cross made by R. J. Lambert more than twenty years ago. The cytological interaction was remarkably complex, and a number of pathways were followed and published (Fig. 4.1) (Harlan and deWet 1977). In successive backcross generations the chromosome numbers rose and fell, but generally took a downward trend toward 2n = 20 as *Tripsacum* chromosomes were eliminated. By the eighth to tenth backcross, all *Tripsacum* chromosomes had been lost and the populations had the normal maize number. We had recovered maize, but it is very strange maize. Some of the tripsacoid characteristics are illustrated by Stalker, Harlan, and deWet (1977a).

The term "tripsacoid" was coined by Anderson and Erickson (1941), who referred to morphological traits that might be introduced into maize from either teosinte (*Zea mays* ssp *mexicana*) or *Tripsacum*. In their view, plants with short, stiff tassel branches and short pedicels subtending male flowers were tripsacoid. This seemed logical, considering the nature of the *Tripsacum* inflorescence and the fact that in *T. dactyloides* both members of the male spikelet pair are essentially sessile. When tripsacoidy is produced experimentally, however, the results are exactly reversed. Pedicel lengths are increased and tassel branches are longer, more lax, and less stiff. The overriding feature of tripsacoidy is a decondensation of the plant, and tripsacoid plants look much more like wild maize (teosinte) than like our familiar corn.

There are many tripsacoid traits, such as reduced cob length, increased

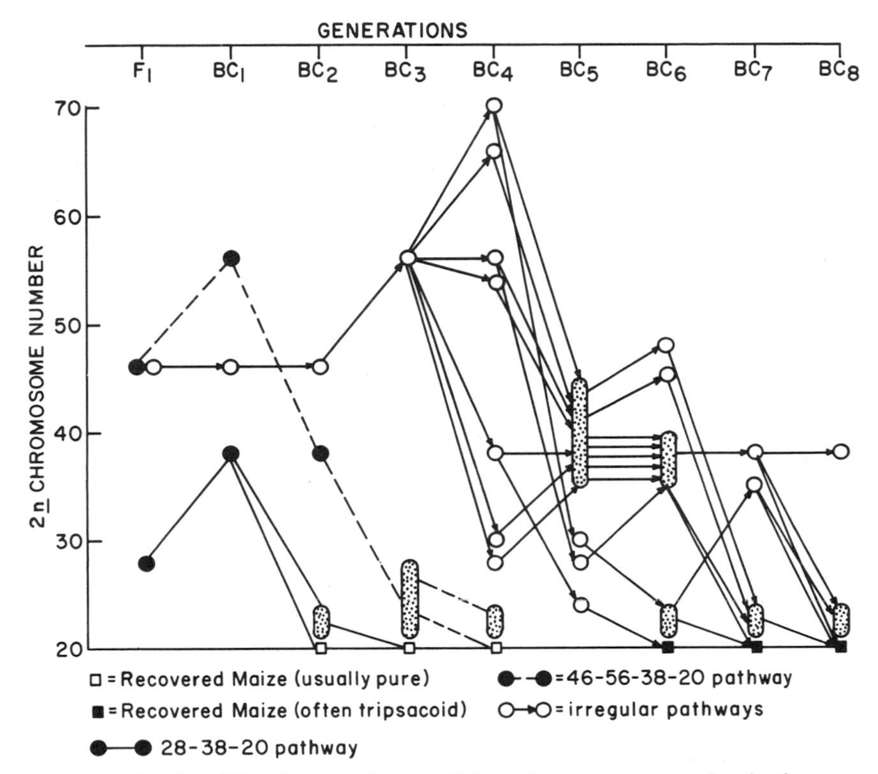

Figure 4.1. Simplified diagram of some of the pathways to recovered maize from *Zea* x *Tripsacum* hybrids. The stippled ovals represent populations with the indicated range of chromosome numbers varying from plant to plant within families.

number of nodes bearing ear shoots, increased depth and size of cupules, decreased pubescence of the cupule, increased frequency of ears with male tips, reduced diameter of pith, wider spacing between pairs of kernel rows, and wider spacing of kernels within a row (Stalker, Harlan, and deWet 1977b), but the length of the longest ear branch and number of ears on it seem to be the most convenient characteristics for measuring tripsacoidy. We have measured plants with ear branches more than two meters in length, and some have as many as 16 ears per branch. Of course, many of the ears do not produce grain, but the potential is there. Highly tripsacoid plants are dramatically different, morphologically, from corn belt corn.

What is the nature of the genetic transfer? We do not yet know. It must be *genetic* because tripsacoid characteristics are passed on from generation to generation, but it does not seem to be *genic*. The characters do not mendelize well. We have been trying to stabilize populations by selection so that we would have something repeatable and with an identity for experimental purposes. After four cycles of selection, the populations are more or less repeatable but are still extremely variable.

One would suppose that if alien chromosomes had paired and crossing over took place, the event would be visible cytologically. There should be loops or unpaired regions and especially unpaired chromosome ends. Maguire (1960) found a piece of *Tripsacum* had been translocated to the end of chromosome 2 in similar material. The translocation was clearly visible. We have looked reasonably carefully at pachytene cells of tripsacoid maize and have not seen anything obvious.

One would suppose that if there had been pairing and crossing over, pieces of *Tripsacum* chromosomes would be incorporated into the maize chromosomes and this would alter crossover map distances between linked genetic markers. Stalker (Stalker, Harlan, and deWet 1978) tested six pairs of linked markers distributed on what we consider to be the most likely chromosomes to pair with *Tripsacum,* i.e., chromosomes 2, 4, 7, and 9. He used 20 different lines of recovered maize. Some showed small but statistically significant shifts in crossover map distances between one to four of the marker pairs. In some cases the distances increased, and in some cases they decreased. We interpreted this to mean that segments of chromatin had been either inserted or removed, as the case may be (Stalker, Harlan, and deWet 1978). There are alternative interpretations, however.

We then conducted an isozyme analysis (unpublished). Without going into detail, it was found that several esterase and malate dehydrogenase bands were present in the parental clone of *Tripsacum dactyloides* and absent in our maize genetic stock controls. A few of the esterase bands and two of the malate dehydrogenase bands appeared in our tripsacoid maize. The work is incomplete and the maize controls may not have been broad enough, so we do not wish to make too much of this, but the evidence so far suggests that either genes were transferred, suppressed maize genes were activated, or maize genes were altered.

Bergquist (1979, 1981) extracted high levels of resistance to six maize diseases in BC_8 populations of recovered maize. The resistances have held up well and are especially striking in winter nurseries in Florida and Hawaii, where the leaves of most materials turn brown from leaf diseases and the tripsacoid derivatives remain dark green. Only one of the resistances (common rust) seems to be due to a gene (Bergquist 1981); all the others seem to be "polygenic" in nature. Simone and Hooker (1976) also report transfer of a gene for disease resistance from *Tripsacum* to maize; in this case, the disease was northern corn leaf blight and the exotic parent *T. floridanum*.

When tripsacoid plants are crossed to corn belt inbreds, the F_1s show a striking heterosis with respect to size of plant. On backcrossing to the same inbreds, the BC_1 plants are not as tall as the F_1s, but the longest ear branches are much reduced in length, the ears are longer, and there is usually only one ear per branch. The tripsacoid morphologies are readily lost on further backcrossing. There are some homologies between maize and *Tripsacum,* as Rao and Galinat (1974, 1976) have shown. The alien chromosomes do pair to some extent, and it seems that genes can be transferred from *Tripsacum* to maize, but such interaction is probably not the major cause of tripsacoidy.

I tentatively propose that there are at least two kinds of genetic transfer possible where alien genomes occur in the same nuclei: (a) pairing and crossing over can take place and transfer takes place in the classic cytogenetic sense, and (b) nomadic, repetitive DNA sequences can migrate from genome to genome without pairing of the chromosomes. The repetitive DNA inserted sequences can alter the regulatory apparatus with respect to gene function. It is my thesis here that gene regulation can bring about more rapid and dramatic morphological changes than can recombination, even of alien chromosomes. Regulation may become the most powerful tool of all for accelerating evolution. Whether it can be directed and stabilized remains to be seen.

Having lifted anchor on uncharted seas, it is necessary to look around for support. It comes from an unexpected source. In the summer of 1979, Mme. Guangyu Zhou of the Shanghai Institute of Biochemistry, Chinese Academy of Sciences, visited our Crop Evolution Laboratory at Urbana. She had read of our work and wanted to talk to us about some material she had been studying derived from a hybrid between rice (♀) and sorghum (♂). The usual reaction to mention of such a cross is, "You can't be serious!" It is easy to dismiss the idea on one or more of the following grounds: (a) communist propaganda, (b) national pride, (c) peasant naiveté, or (d) biological impossibility. But none of the above are operative. Mme. Zhou is a first-class sicentist and very professional. Politics has nothing to do with it. She has some strange material to study. She readily admits that the original "F_1" was not a hybrid in the conventional sense. It was a rice plant with respect to cytology, but it was highly sterile. The few shriveled seeds if produced developed into a strange array of plants and "segregation takes place crazily since F_2" (Zu et al. 1979, p. 420).

The photographs and descriptions do, indeed, indicate some kind of genetic transfer, but almost certainly not by the classical chromosome pairing and crossover system. Zhou, Gong, and Wang (1979) report an isozyme band present in some of the modified plants that is found in the sorghum parent and not in the rice parent. A subsequent report states that sorghum DNA shows some homology to rice DNA at the molecular level (Zhou, Zeng, and Yang 1980). More detailed studies may reveal the class of DNA involved. The Shaghai group is also of the opinion that they are dealing mostly with modification of regulatory systems (Zhou, personal communication).

The idea of mobile genetic elements is not new. Early evidence for it in higher plants was produced by McClintock (1950, 1951) some 30 years ago when she described controlling elements. Work on the Ds-Ac (dissociation-activator) systems has continued, and they have been used for a variety of genetic studies; e.g., Nelson (1968) used a transposed controlling element to map the waxy locus and more recently used the Ds-Ac system to regulate action of the bronze mutant locus (Dooner and Nelson 1979). The idea that rapid evolution is more likely to be based on gene regulation than gene mutation and that the regulation is likely to involve repetitive DNA sequences is relatively new but gaining in support. Flavell and Smith (1975) noted that rye *(Secale cereale)* has 50% more DNA than diploid wheat. When the DNA is extracted and digested with S_1 nuclease, about 50% of the resistant DNA of rye consists of very long duplexes and very little of wheat DNA is of this kind. The main difference between wheat and rye DNA is in these long repeat segments, much of which are located in heterochromatic regions. Among their conclusions from extensive studies of wheat DNA, they state: "Speciation is likely to have its genetic origin in control sequences rather than enzyme protein sequences" (Flavell and Smith 1975, p. 65).

On the wider evolutionary scene, some people may have been a bit dismayed when King and Wilson (1975) reported in a well-researched paper that the proteins of man and chimpanzee are about 99% identical by amino acid sequencing and that the nonrepeated DNAs differed by about 1.1%. Compared to other groups studied, *Pan* and *Homo* are sibling species. But the morphological and other differences are considerable. They attribute these largely to gene regulation. Elsewhere, Wilson (1975) has shown that amphibians evolve much more slowly than mammals and this slowness can be attributed to slow regulatory evolution. For example, there are two species of *Xenopus* so much alike morphologically that they have been put in the same species, yet DNA anealing shows more difference in unique DNA than is shown between man and New World monkeys. Wilson suggested: "It seems likely that evolution at the organismal level depends predominantly on regulatory mutations. Structural gene mutations may have a secondary role in organismal evolution" (Wilson 1975, p. 118).

Mobile, nomadic, dispersed, repeated DNA sequences are turning up everywhere these days now that we have tools to find them and know what to look for (Marx 1981). A group at the Institute of Molecular Genetics in Moscow has been working on what they call "mobile dispersed genetic elements" (MDGI) in *Drosophila* (Ilyin et al. 1980). Green (1977a, b, 1978) has been studying a class of mutations in *Drosophila* that revert to wild type at high frequencies, and he has reason to believe DNA insertion is involved. A whole class of retroviral genes and oncogenes has come into view and some of these are highly mobile. Todaro et al. (1980) point out that RNA tumor virus genes are contained in the chromosomal DNA of most vertebrates, and they may be transmitted from organism to organism as infectious particles. Restriction enzyme analyses of cellular type C viral DNA of mouse has shown great diversity in sequence among various strains (Canaani and Aaronson 1979). The authors attribute the instability of viral genes to DNA sequences that dissociate and reinsert at various sites on the genome.

In recent years we have become aware of "multigene families." Hood, Campbell, and Elgin (1975) define a multigene family: "as a group of nucleotide sequences or genes that exhibits four properties—multiplicity, close linkage, sequence homology, and related or overlapping phenotypic functions. In addition to these features, multigene families share a novel evolutionary characteristic: natural selection appears to operate on these gene families as a whole, and not upon the individual gene members" (p. 305–6). Examples are antibody genes, robisomal RNA genes, histone genes, tRNA genes, and satellite DNA. All are subject to rapid change with coincidental rapid evolution.

These discoveries may shed some light on a phenomenon I have called "genetic recall" (Harlan, deWet, and Price, 1973; Harlan 1981). Here, I refer to instant reversals of long-time evolutionary trends. For example, the genus *Hordeum* is characterized by an inflorescence that is a spike with three spikelets at each node, the central one sessile and female fertile and the lateral ones pedicellate, male or neuter. This is the nature of the whole genus, and the arrangement is presumed to have evolved over a long period of time. It can be undone by a single recessive mutation that returns fertility to the lateral spikelets and produces 6-rowed barley.

In the tribe Andropogoneae, the standard arrangement is for spikelets to occur in pairs, one sessile and one pedicellate. In the subtribe Saccharininae, both members of the spikelet pair are female fertile, and this is considered to be the primitive condition. In most of the other subtribes, the sessile spikelet is female fertile and the pedicellate one is male, neuter, reduced, or even wanting. Reversal to the primitive state and recovery of lost fertility is known in maize, sorghum, *Bothriochloa,* and *Dichanthium.* Where the inheritance is known, a single gene is responsible. Similarly, reduced and rudimentary florets can be rendered fertile in *Sorghum, Panicum,* and *Zea.* One gene is responsible in sorghum and two in maize.

Perhaps the most striking cases of genetic recall are in wheat and pearl millet. In the case of wheat, certain genotypes produce grain in the lower glumes, which are normally sterile. Sterile lower glumes is a characteristic of the whole family Gramineae, yet a combination of two genes can render them fertile (Harlan 1981). In pearl millet, sterile rudimentary branches of the inflorescence may develop flowers and produce grain. This is not a case of development of rudimentary structures but a *de novo* production of spikelets where none were produced in the ancestral forms.

Characters that distinguish genera, subtribes, or even a whole family seem to be reversible by the action of one or two genes. It seems reasonable to suggest that the genetic information required for development of rudimentary structures has been present all along, but in a suppressed state. The release of suppression is a regulatory function.

Recent studies on apes (Zimmer et al. 1980), monkeys (Singer 1979), *Drosophila* (McDonald et al. 1977), and bacteria (Ornston 1979), among many others, have stressed the importance of regulatory non-coding sequences in rapid evolution. Calos and Miller, in a review of transposable elements, state: "Studies of transposable elements have revealed that the arrangement of DNA in organisms is more fluid than had previously been recognized, and that considerable scope for adaptation can be gained from these rearrangements" (Calos and Miller 1980, p. 592). The partitioning and classification of plant DNAs give some insight into the evolutionary significance of repetitive versus single copy or few copy DNAs (Rimpau, Smith, and Flavell 1978, 1980). The literature on the field in general is massive, and I cannot begin to do it justice in a short chapter. I would like to add one more example, however, for consideration. Templeton (1979) and his group have been studying a remarkable system extracted from Hawaiian garbage dump fruit flies, *Drosophila mercatorum.* Females of these flies can be induced to reproduce parthenogenetically in the laboratory. The probability of parthenogenesis is only about 10^{-5}, but this is enough for them to establish several lines. The selective bottleneck to produce a line with a coadapted genome is very strong indeed. The life cycle passes through meiosis so that a line is established from *a single haploid genome,* which then doubles to produce completely homozygous flies. This provides maximum genetic drift and enormous selection pressure.

The female line K28-OIm differs from its ancestral population in appearance; the flies are larger and darker; there is a bend in the wing; they reproduce faster, and have a much higher egg-laying capacity; they forage more widely and are reproductively isolated from males of the population from which they were extracted. These differences meet all of the criteria for speciation. To produce a new species in one generation is about as fast as we can expect evolution to go. Templeton (1979) refers to the event as a "genetic revolution," the term coined by Mayr (1954) more than a quarter-century ago. He concludes that genetic revolutions are real phenomena, but they are obviously not based on gene mutation. He finds that the isozymes of K28-OIm are identical to those of the parent population. In his words: "Genes that underlie the 'genetic revolution' appear to be at loci having fundamental regulatory roles in the organism."

The current emphasis on gene regulation as a process in rapid evolution seems to lend support to at least one group of evolutionary theoreticians. The punctuated equilibria of Eldridge and Gould (1972), Gould (1977), and Gould and Eldridge (1977) indicate rapid bursts of evolution interrupted by long periods of stasis. Bush et al. (1977), Wilson (1975), Raup et al. (1973), and many others are in agreement with the general idea of evolutionary rates differing greatly with time and with taxonomic group. These studies were conducted mostly on animals, but Niklas (1978) and others find similar trends in plants. While quantum evolution has not gained universal acceptance, it has gained many strong and vocal supporters in the last decade, during which time there have been some fascinating attempts to integrate evidence from fossils, plate tectonics, chromosomes, morphology, proteins, and DNA. Ho and Saunders (1979), among others, have called for a new view of evolutionary processes. Natural selection on random mutations is not enough; they suggest an epigenetic approach.

These are very exciting times in evolutionary studies, and I feel more confident than I did that we will find ways to put some of our sophisticated methodologies to practical use. The protoplast fusion techniques, for example, have produced some extremely exotic combinations of chromosomes, but so far they have not led to much in the way of end products. Now, if migratory non-transcribing DNA can move freely from genome to genome and be inserted, as seems to be the case, we may find ways of modifying regulatory systems without producing true hybrids in the conventional sense. The recent discovery of the Ti plasmid (Belgian Crown-Gall Research Group 1979; Merlo 1979) leads us to hope that other plant plasmids may be found and that modification of DNAs can become a practical means of directing and accelerating plant evolution.

I believe that our maize X *Tripsacum* derivatives will provide excellent experimental material for the exploration of regulatory DNAs. It seems that in every case where rapid evolution can be demonstrated, regulatory systems are implicated. It is likely that the next breakthrough in plant or animal breeding will come from an understanding of such systems. One can certainly argue, in the case of our tripsacoid maize, that it does not help much to speed up evolution if we are going in the wrong direction, i.e., if we are going back toward primitive maize rather than onward and upward toward improvement of corn belt types. But the first step in solving a problem is to define it. If genetic regulation is where the action is, we need to know it, and if genes can be regulated in one direction, they probably can be regulated in another.

LITERATURE CITED

Adamski, T. 1979. *The obtaining of autodiploid barley lines using haploids from the cross Hordeum vulgare L. x Hordeum bulbosum L.* Genet. Pol. 20:31–42.

Alexander, D. E. 1958. *Genetic induction of autotetraploidy in maize.* Proc. 10th Int. Congr. Genet. Montreal: 3 (abst.).

Alexander, D. E., and J. B. Beckett. 1963. *Spontaneous triploidy and tetraploidy in maize.* J. Hered. 54:103–6.

Anderson, E., and R. O. Erickson. 1941. *Antithetical dominance in North American maize.* Proc. Nat. Acad. Sci. USA 27:436–40.

Baer, Adela S. 1977. *The Genetic Perspective.* W. B. Saunders Co., Philadelphia.

Belgian Crown-Gall Research Group. 1979. *Transfer of genes into plants via the Ti-plasmid of A. tumefaciens.* Pages 521–35 in S. Rosenthal, H. Bielka, Ch. Coutelle, and Ch. Zimmer, eds. *Gene Function.* Federation of Experimental Biological Sciences, vol. 51. Pergamon Press, New York.

Berg, Raissa, W. R. Engels, and R. A. Kreber. 1980. *Site-specific X-chromosome rearrangements from hybrid dysgenesis in Drosophila melanogaster.* Science 210:427–29.

Bergquist, R. R. 1979. *Selection for disease resistance in a maize breeding program.* Eucarpia 10: Meeting of the Maize and Sorghum section, Varna, Bulgaria.

———. 1981. *A major locus, Rp_tTd in Tripsacum dactyloides conditioning resistance to Puccinia sorghi.* Phytopathology (in press).

Burton, G. W., W. W. Hanna, and J. B. Powell. 1980. *Hybrid vigor in forage yields of crosses between pearl millet inbreds and their mutants.* Crop Sci. 20:744–47.

Bush, G. L., S. M. Case, A. C. Wilson, and J. L. Patton. 1977. *Rapid speciation and chromosomal evolution in mammals.* Proc. Natl. Acad. Sci. USA 74:3942–46.

Calos, M. P., and J. H. Miller. 1980. *Transposable elements.* Cell 20, pp. 579–95.

Canaani, E., and S. A. Aaronson. 1979. *Restriction enzyme analysis of mouse cellular type C viral DNA: Emergence of new viral sequences in spontaneous AKR/J hyphomas.* Proc. Nat. Acad. Sci. USA 76:1677–81.

Chang, Z., and Y. Hong-Yuan. 1981. *Induction of haploid rice plantlets by ovary culture.* Plant Sci. Let. 20:231–37.

Chowdhury, S. 1979. *Induction of variation for improved physiological efficiency.* Page 524–32 in S. Ramanujam, ed. *Proc. Fifth International Wheat Genetics Symposium,* vol. I. Indian Agricultural Research Institute, New Delhi.

CIMMYT. 1978. *Maize.* CIMMYT Review, p. 138.

Dooner, H. K., and O. E. Nelson, Jr. 1979. *Heterogeneous flavonoid glucosyltranferases in purple derivatives from a controlling element-suppressed bronze mutant in maize.* Proc. Nat. Acad. Sci. USA 76:2369–71.

Dumas de Vaulx, R. 1979. *Obtention de plantes haploides chez le melon (Cucumis melo L.) apres pollinisation par Cucumis ficifolius A. Rich.* C. R. Hebd. Seances Acad. Sci., Paris, Ser. D. 289:875–78.

Elridge, N., and S. J. Gould. 1972. *Punctuated equilibria: An alternative to phyletic gradualism*. Pages 82–115 in T. J. M. Schopf, ed. *Models in Paleobiology*. Freeman-Cooper, New York.

Fishbein, Lawrence. 1978a. *Environmental sources of chemical mutagens. I. Naturally occurring mutagens*. Pages 175–256 in W. G. Flamm and M. A. Mehlman, eds. *Mutagenesis, Advances in Modern Toxicology No. 5*. John Wiley, New York.

———. 1978b. *Environmental sources of chemical mutagens. II. Synthetic mutagens*. Pages 257–348 in W. G. Flamm and M. A. Mehlman, eds. *Mutagenesis, Advances in Modern Toxicology No. 5*. John Wiley, New York.

Flavell, R. B., and D. B. Smith. 1975. *Genome Organization in Higher Plants*. Stadler Symp. vol. 7, pp. 47–69.

Gerstel, D. U., and J. A. Burns. 1976. *Enlarged euchromatic chromosomes ("mega-chromosomes") in hybrids between Nicotiana tabacum and N. plumbaginifolia*. Genetica 46: 139–53.

Gotoh, Torao. 1977. *Semidwarf Norin 10 wheat and its contribution to the progress of wheat breeding*. Gamma-Field Symposia 16: 85–103.

Gould, S. J. 1977. *Ontogeny and Phylogeny*. Belknap Press, Cambridge, Mass.

Gould, S. J., and N. Eldridge. 1977. *Punctuated equilibria: The tempo and mode of evolution reconsidered*. Paleobiology 3: 115–51.

Green, M. M. 1977a. *The case for DNA insertion mutants in Drosophilia*. pages 437–45 in A. T. Bukhari, J. A. Shapiro, and S. L. Adhya, eds. *DNA Insertion Elements, Plasmids and Episomes*. Cold Spring Harbor Laboratory, Cold Spring Harbor, N.Y.

———. 1977b. *Genetic instability in Drosophila melanogaster: De novo induction of putative insertion mutants*. Proc. Nat. Acad. Sci. USA 74: 3490–93.

———. 1978. *The genetic control of mutation in Drosophila*. Stadler Symp. vol. 10, pp. 95–104.

Gupta, S. C., J. R. Harlan, and J.M.J. deWet. 1978. *Cytology and morphology of a tetraploid sorghum population recovered from a Saccharum x Sorghum hybrid*. Crop Sci. 18: 879–83.

Harlan, J. R. 1976. *Genetic resources in wild relatives of crops*. Crop Sci. 16: 329–33.

———. 1980. *Plant breeding and genetics*. pages 295–312 in Leo A. Orleans, ed. *Science in Contemporary China*. Stanford Univ. Press, Stanford, CA.

———. 1981. *Evaluation of Wild Relatives of Crop Plants*. FAO, Rome.

———. 1981. *Human Interference with grass systematics. In AIBS Symposium on Grass Systematics*. Oklahoma Univ. Press, Norman (forthcoming).

Harlan, J. R., and J.M.J. deWet. 1975. *On Ö. Winge and a prayer: The origins of polyploidy*. Bot. Rev. 41: 361–90.

———. 1977. *Pathways of genetic transfer from Tripsacum to Zea mays*. Proc. Nat. Acad. Sci. USA 74: 3494–97.

Harlan, J. R., J.M.J. deWet, and E. G. Price. 1973. *Comparative evolution of cereals*. Evolution 27: 311–25.

Ho, M. W., and P. T. Saunders. 1979. *Beyond neo-Darwinism—An epigenetic approach to evolution*. J. Theor. Biol. 78:573–91.

Hood, L., J. H. Campbell, and S.C.R. Elgin. 1975. *The organization, expression, and evolution of antibody genes and other multigene families*. Ann. Rev. Genet. 9: 305–53.

Hristov, K. 1979. *Accelerated aging of the seeds—a method for producing mutations in maize*. Genet. Sel. 12: 26–35. Bulgarian, English summary.

Ihrke, C. A., and W. E. Kronsted. 1975. *Genetic recombination in maize as affected by ethylenediaminetetraacetic acid and dimethyl sulfoxide*. Crop Sci. 15: 429–31.

Ikeda, Fukio. 1977. *Induced mutation in dwarf growth habits of apple trees by gamma rays and its evaluation in practical uses*. Gamma-Field Symposia 16: 63–83.

Ilyin, Y. V., V. G. Chmeliauskaite, E. V. Ananiev, N. V. Lyubomirskaya, V. V. Kulguskin, A. A. Bayev, Jr., and G. P. Georgiev. 1980. *Mobile dispersed genetic element MDGI of Drosophila melanogaster: Structural organization*. Nucleic Acids Res. 8: 5333–46.

Islam, A.K.M.R. 1980. *Identification of wheat-barley addition lines with N-branding of chromosomes*. Chromosoma 76: 365–73.

Islam, A.K.M.R., K. W. Shepherd, and D.H.B. Sparrow. 1978. *Production and characterization of wheat-barley addition lines*. Proc. 5th Int. Wheat Genet. Symp., New Delhi, pp. 365–71.

Jalani, B. S., and K. J. Frey. 1979. *Contribution of growth rate and harvest index to grain yield of oats (Avena sativa L.) following selfing and outcrossing of M_1 plants*. Euphytica 28: 219–25.

Kermicle, J. L. 1969. *Androgenesis conditioned by a mutation in maize*. Science 166: 1422–24.

King, M. C., and A. C. Wilson. 1975. *Evolution at two levels in humans and chimpanzees*. Science 188: 107–16.

Konishi, Takeo. 1977. *Effects of induced dwarf genes on agronomic characters in barley*. Gamma-Field Symposia 16: 21–38.

Kunzel, G., and H. Nicoloff. 1979. *Further results on karyotype reconstruction in barley*. Biol. Zentralbl. 98: 587–92.

McClintock, B. 1950. *The origin and behavior of mutable loci in maize*. Proc. Nat. Acad. Sci. USA 36: 344–55.

———. 1951. *Chromosome organization and genic expression*. Cold Spring Harbor Symp. Quant. Biol. 16: 13–47.

McClintock, Barbara. 1978. *Mechanisms that rapidly reorganize the genome*. Stadler Symp. vol. 10, pp. 25–47.

McDonald, J. F., G. K. Chambers, J. David, and F. J. Ayala. 1977. *Adaptive response due to changes in gene regulation: A study with Drosophila*. Proc. Nat. Acad. Sci. USA 74: 4562–66.

Maguire, M. P. 1960. *A study of homology between a terminal portion of Zea chromosome 2 and a segment derived from Tripsacum*. Genetics 45: 195–209.

Malthus, T. R. 1798. *An Essay on the Principle of Population, as It Affects the Future Improvement of Society*. J. Johnson, London, p. 396.

Marx, Jean L. 1981. *A movable feast in the eukaryotic genome*. Science 211: 153–55.

Mayr. E. 1954. *Changes of genetic environment and evolution*. pages 157–80 in J. Huxley, ed. *Evolution as a Process*. Allen & Unwin, London.

Merlo, D. J. 1979. *Ti plasmids of Agrobacterium: potentials of genetic engineering*. Stadler Symp. vol II: 69-90.

Mok, D.W.S., and S. J. Peloquin. 1975. *Breeding value of 2n pollen (diplandroids) in tetraploid x diploid crosses in potatoes*. Theor. Appl. Genet. 46: 307–14.

Mukade, K. 1978. *Chromosome engineering and acceleration of generation advancement in breeding rust resistant wheat*. Trop. Agric. Res. Ser., Japan, 11: 135–44.

Nakayama, Aogu. 1973. *Induction of the somatic mutations in tea plants by gamma irradiation*. Gamma-Field Symposia 12: 37–47.

Nelson, O. E. 1968, *The waxy locus in maize: II. The location of the controlling element alleles*. Genetics 60:507–24.

Neuffer, M. G., and E. H. Coe, Jr. 1977. *Paraffin oil technique for treating mature corn pollen with chemical mutagens*. Maydica 23:21–28.

Neuffer, M. G., and W. F. Sheridan. 1980. *Defective kernel mutants of maize: I. Genetic and lethality studies*. Genetics 95:929–44.

Niklas, K. J. 1978. *Morphometric relationships and rates of evolution among Paleozoic vascular plants*. Evol. Biol. 11:509–43.

Okuno, Kazutoshi, 1977. *Induction of short-culm mutations and inheritance of induced short-culm mutants in rice*. Gamma-field Symposia 16:39–62.

Ornston, L. N. 1979. *Origins of metabolic diversity: Evolutionary divergence by sequence repetition*. Proc. Nat. Acad. Sci. USA 76:3996–4000.

Rabie, H., and K. Pasztor. 1979. *Experimental results in the crossing of 'corn grass' type maize macromutants*. Novenytermeles 28:301–8. Hungarian, English summary.

Rao, B. G. S., and W. C. Galinat. 1974. *The evolution of the American Maydeae I. The characteristics of two Tripsacum chromosomes (Tr 7 and Tr 13) that are partial homeologs to maize chromosome 4*. J. Hered. 65:335–40.

———. 1976. *The evolution of the American Maydeae II. The characteristics of a Tripsacum chromosome (Tr 9) homoeologous to maize chromosome 2*. J. Hered. 67:235–40.

Raup, D. M., S. J. Gould, T. J. M. Schopf, and D. Simberloff. 1973. *Stochastic models of phylogeny and the evolution of diversity*. J. of Geol. 81:525–42.

Reanney, D. 1979. *RNA splicing and polynucleotide evolution*. Nature 277:598–600.

Rhoades, M. M., and Ellen Depsey. 1966. *Induction of chromosome doubling at meiosis by the elongate gene in maize*. Genetics 54:505–22.

Rimpau, J., D. B. Smith, and R. B. Flavell. 1978. *Sequence organization analysis of the wheat and rye genomes by interspecies DNA/DNA hybridization*. J. Molec. Biol. 123:327–359.

———. 1980. *Sequence organizatin in barley and oats chromosomes revealed by interspecies DNA/DNA hybridization*. Heredity 44:131–49.

Sala, F., B. Parisi, R. Cella, and O. Ciferri, eds. 1980. *Plant cell cultures: Results and Perspectives*. Elsevier North-Holland Biomedical Press, Amsterdam.

Schooler, A. B., and M. K. Anderson. 1979. *Interspecific hybrids between Hordeum brachyantherum L. × H. bogdanii (Wilensky) × H. vulgare L*. J. Hered. 70:70–72.

Sears, E. R. 1956. *The transfer of leaf-rust resistance from Aegilops umbellulata to wheat*. Brookhaven Symposia in Biology no. 9, pp. 1–22.

Sigurbjörnsson, R., and A. Micke. 1974. *Philosophy and accomplishments of mutation breeding. Pages 303–43 in Polyploidy and Induced Mutations in Plant Breeding*. IAEA, Vienna.

Simone, G. W., and A. L. Hooker. 1976. *Monogenic resistance in corn to Helminthosporium turcicum derived from Tripsacum floridanum*. Proc. Amer. Phytopath. Soc. 3:207.

Simpson, E., J. W. Snape, and R. A. Finch. 1980. *Variation between Hordeum bulbosum genotypes in their ability to produce haploids of barley, Hordeum vulgare*. Z.f. Pflanzenzüchtung 85:205–11.

Singer, D. S. 1979. *Arrangement of a highly repeated DNA sequence in the genome and chromatin of the African green monkey*. J. Biol. Chem. 254:5506–14.

Stalker, H. T. 1980. *Utilization of wild species for crop improvement*. Adv. in Agron. 33:111–47.

Stalker, H. T., J. R. Harlan, and J. M. J. deWet. 1977a. *Observations on introgression of Tripsacum into maize*. Amer. J. Bot. 64:1162–69.

———. 1977b. *Cytology and morphology of maize-Tripsacum introgression*. Crop Sci. 17:745–48.

———. 1978. *Genetics of maize -Tripsacum introgression*. Caryologia 31:271–82.

Stebbins, G. L. 1971. *Processes of Organic Evolution*. 2nd ed. Prentice-Hall, Englewood Cliffs, N.J. 193 pages.

Sukhapinda, K., and P. A. Peterson. 1980. *Enhancement of genetic exchange in maize: Itragenic recombination*. Can. J. Genet. & Cytol. 22:213–22.

Szilagyi, L., and A. H. Nagy. 1978. *Chromosome elimination in alloploid tobacco hybrids*. Acta. Bot. Acad. Sci. Hungary 24:343–50 (Engl.).

Templeton, A. R. 1979. *The unit of selection in Drosophila mercatorum II. Genetic revolution and the origin of coadapted genomes in parthenogenetic strains*. Genetics 92:1265–82.

Todaro, G., R. Callahan, U. Rapp, and J. DeLarco. 1980. *Genetic transmission of retroviral genes and cellular oncogenes*. Proc. Roy. Soc. (Lon.) Ser. B. 210P367–85.

Wagenaar, E. G. 1969. *Meiotic restitution and the origin of polyploidy III. The cytology and fertility of eight generations of the off-spring of a spontaneously produced amphipolyploid of Triticum crassum x T. turgidum*. Can. J. Genet. Cytol. 11: 729–38.

Wienhues, A. 1979. *Resistenz genen Gelbrost (Puccinia striiformis) aus Agropyrum intermedium ubertragen in den Winterweizen*. Z. Pflanzenzücht. 82: 201–11.

Wilson, A. C. 1975. *Evolutionary importance of gene regulation.* Stadler Symp. 7: 117–33.

Zhou, Guangyu, Zhenzhen Gong, and Zifen Wang. 1979. *The molecular basis of remote hybridization—an evidence of the hypothesis that DNA segments of distantly related plants may be hybridized.* Acta Genet. Sinica 6: 412–13.

Zhou, G. Y. Zeng, and W. Yang. 1980. *Molecular basis of remote hybridization. True recombination of sorghum DNA sequences with rice genomes during remote hybridization.* Acta Genet. Sinica 7: 119–22.

Zimmer, E. A., S. L. Martin, S. M. Beverley, Y. W. Kan, and A. C. Wilson. 1980. *Rapid duplication and loss of genes coding for α chains of hemoglobin.* Proc. Nat. Acad. Sci. USA 77: 2158–62.

Zu, Deming, Lanfang Dai, Shanbao Chen, Xianbin Song, and Xiaolan Duan. 1979. *Diversity and specific performance of progenies from distant hybridization between rice and sorghum.* Acta Genet. Sinica 6: 414–20. Chinese with English summary.

three
PHOTOCONTROL SYSTEMS

5] Photoperiodic Control of Plant Reproduction

by DAPHNE VINCE-PRUE*

ABSTRACT

Responsiveness to the natural photoperiods of any environment may determine whether or not a plant can complete its life cycle successfully in the prevailing conditions. Examples are given where a daylength response affecting the duration of the life cycle is important for the avoidance of environmental hazards, such as water stress or low temperature.

The advantages and disadvantages of responsiveness to photoperiod for crop plants are considered for field crops grown in natural environments and for glasshouse flower crops, where flowering time is manipulated with artificial photoperiods.

The possible mechanism for photoperiodic induction is discussed with reference to the transition from vegetative to reproductive growth in *Pharbitis nil*. In the cultivar, Violet, photoperiodic induction requires only a single cycle and is effected with a variety of experimental protocols. From these, it is concluded that induction consists of a photophase followed by a sufficiently long dark period; induction is prevented by a light pulse given at a particular time. The photophase is satisfied when continuous light is substituted for by a skeleton day beginning and ending with a brief pulse of light which establishes Pfr phytochrome; however, the pattern of response to a night-break differs after a skeleton photoperiod. The nature of the photoreactions during the photophase are not fully characterized. From studies of the inductive dark period, Pfr is required for floral induction but does not affect the time-measuring process; it is concluded that natural spectral shifts during twilight may not be important environmental cues for the beginning of the biological "night." The precise nature of this cue has not yet been identified.

The photophase is satisfied by a single pulse of red light when plants also receive benzyladenine, and this makes it possible to make detailed

*Glasshouse Crops Research Institute, Littlehampton, Sussex, England.

STRATEGIES OF PLANT REPRODUCTION (BARC Symposium number 6—Werner J. Meudt, ed.)
Allanheld, Osmun, Totowa

measurements of phytochrome and simultaneously to study physiological responses during an inductive dark period. At certain times, Pfr appears both to promote and to inhibit flowering, and the existence of more than one pool of phytochrome controlling flowering is inferred.

The possible diversity of photoperiodic mechanisms is illustrated by reference to the control of flowering in two commercial flower crops, carnation and chrysanthemum. Photoperiodic induction in the long-day plant, carnation, is dependent on prolonged exposures to light that contains both red and far-red wavelengths. In the short-day plant, chrysanthemum, the initiation of the inflorescence bud and its further development into open flowers show a number of differences which suggest that, although both require short days, the control of these two processes may involve different mechanisms.

A detailed understanding of the photoperiodic mechanism is most likely to be derived from simple model systems like *Pharbitis,* where phytochrome properties and physiological responses can be studied simultaneously. The diversity of photoperiodic behavior, however, emhasizes the need for continued studies with individual crop plant species under a variety of environmental conditions, if their photoperiodic responses are to be fully exploited.

INTRODUCTION

The photoperiodic control of reproduction has two major consequences for the organism: to locate the response in time and to limit the latitudinal distribution of the plant. Because daylength is the one environmental factor that invariably gives totally reliable information about the passage of the seasons, many organisms utilize this signal to synchronize their developmental patterns with particular times of the year. Survival of the individual may be ensured by daylength-dependent physiological and morphological changes that increase resistance to unfavorable conditions such as below-freezing temperatures or water stress. Survival of the species, however, depends on reproduction of the individual, and it is essential that this is effected under optimal conditions for the formation of progeny and for their survival. Some degree of responsiveness to daylength appears to be widespread in the plant kingdom; in the species studied so far (mostly monocarpic herbaceous plants of the temperate zone) about 80% have been reported to show some response to daylength in their flowering behavior (Krekule 1979).

Under natural conditions, responsiveness to photoperiod can confer a substantial advantage to the organism, enabling reproduction to be synchronized with a favorable environment or to avoid an unfavorable one. For example, a long-day (LD) response in high latitudes can synchronize flowering with the high light integral of summer and thus support the energetic demand of seed production; alternatively, a short-day (SD) response may enable a woodland species to complete its reproductive cycle before the canopy closes. Photoperiodic adaptation may also be important in relation to water availability, as in the kangaroo grass, *Themeda australis* (Evans 1975a). Northern tropical populations of this grass are SD plants which flower at the end of the wet monsoon, while those from southern Australia are mostly LD plants which flower toward the end of the winter-spring period of rainfall. The selective advantage of responsiveness to photoperiod is well illustrated by reference to populations of *Xanthium strumarium*. Plants of American origin, with a wide array of photoperiodic types, have been successful introductions in many parts of the world, showing pre-adaptation to a range of habitats; on the other hand, those of Eurasian origin are mostly day-neutral or nearly so and have proved to be at a selective disadvantage in comparison with American introductions, even in parts of the Old World (McMillan 1974 a,b).

Daylength is one of the major factors limiting the latitudinal spread of a species, and those with a wide distribution frequently show marked ecotypic variation with respect to the effects of daylength on their flowering (e.g., *Chenopodium rubrum* [Cumming 1969] and *Themeda australis* [Evans 1975a]). All ecotypes of *Chenopodium rubrum* are SD plants, but the most northern ecotypes flower in a wide range of daily photoperiods and tend toward a quantitative SD response type, while the more southern ecotypes flower only in a narrow range of daily photoperiods, tending toward an absolute SD requirement. The Australian populations of *Themeda,* on the other hand, include both SD and LD response types.

When crop plants are considered, daylength responsiveness is a major factor to be taken into account in breeding programs for improved productivity or wider geographical dispersal. It does not always follow, however, that the best approach is to breed for day neutrality, and too much emphasis may have been given to daylength indifference as a specific objective in breeding programs aimed at a wider dissemination of varieties. Adaptation to photoperiod can be an important component of high crop yield, and cultivars with developmental patterns tailored to make the best use of particular seasonal conditions offer an alternative strategy to the breeding of cultivars without daylength (and temperature) sensitivity and, therefore, are widely adapted throughout the world. Furthermore, there are some horticultural crop plants where a strong response to photoperiod enables their flowering time to be artificially manipulated and precisely controlled.

PHOTOCONTROL OF SEXUAL REPRODUCTION

The regulation of flowering by daylength has been studied in great detail, and many different types of response are known (see Evans 1969a; Vince-Prue 1975). The main groups of response generally considered are short-day plants (SDP), which only flower or flower more quickly with decrease in length of the daily photoperiod: long-day plants (LDP), which only flower or flower more quickly with increase in length of the daily photoperiod, and day-neutral plants with no response to daylength. Other response groups such as intermediate-day plants, which flower only within a restricted range of daily photoperiods, and those with dual daylength requirements may be combinations of the major ones (Vince-Prue 1975). Plants with absolute photoperiod requirements often show a sharp change from zero flowering to maximum response as the daylength changes through a "critical" value, but in others the response to a change in daylength is more quantitative and less dramatic. Thus the photoperiodic sensitivity (i.e., the change in magnitude of response per unit change in daylength) can vary widely. The absolute value of the critical daylength above (SDP), or below (LDP) which flowering is suppressed also varies widely; for example, in the SDP *Chenopodium rubrum,* ecotypes 60° 52' N and 34° 20' N have critical daylengths of 18 hr and 12 hr respectively (Cumming 1969). Where the response to daylength is a quantitative one (as in flower initiation in chrysanthemum [Cockshull 1976] and carnation [Harris and Ashford 1966]) the response is, of course, not completely suppressed at any daylength.

The most studied aspect of the photocontrol of sexual reproduction in plants is the switch from vegetative to reproductive growth, with the first recognizable morphological event of flowering being used as the criterion of this switch. In the course of forming a flower there is a serial initiation of primordia destined to enter a particular developmental pathway to become petals, anthers, etc. The fate of these primordia may depend on whether plants continue to be exposed to appropriate photoperiods. Thus daylength is often an important environmental factor for the continued growth and development of the flower, as well as for its inception. The initiation of flowers and their development into open blooms may be affected in different ways by daylength (Vince-Prue 1975). The autumn-flowering chrysanthemum, for example, will initiate flower buds readily if the dark period exceeds ca 9.5 hr and all cultivars appear to be capable of floral initiation, even in continuous light (Cockshull 1975). The further development of the inflorescence bud, however, not only requires longer nights (Post 1948; Cathey 1957), but also appears to have an absolute SD requirement (unless manipulated experimentally by removing lateral buds below the inflorescence, Schwabe 1951). This difference is illustrated by recent results obtained at Glasshouse Crops Research Institute (GCRI) with the cultivar Polaris (Fig. 5.1). The number of leaves formed below the terminal inflorescence bud decreased sharply from ~ 26 to ~ 14 as the nightlength was increased above 7 hr. With 10.5 hr. of darkness, the acceleration of initiation was almost saturated, but there was essentially no further development of the inflorescence bud until the duration of the dark period was increased further. Moreover, as we shall see later, these two stages of reproductive development may be controlled by different photoperiodic mechanisms. This cultivar is clearly well adapted to flowering under the conditions of shortening autumn days. The perpetual-flowering carnation, on the other hand, initiates flowers most readily in continuous light (Harris and Ashford 1966), but their further development is quite unaffected by daylength. Examples are also known where the daylength requirement for rapid floral initiation is completely opposite to that for flower development. The strawberry, *Fragaria* x *ananassa,* requires SD for the initiation of floral primordia, but their development into open flowers may be most rapid in LD (Guttridge 1969; Tafazoli and Vince-Prue 1978), while in the china aster, *Callistephus chinensis,* the opposite is true (Hughes and Cockshull 1965); these two strategies of response result in spring and autumn flowering, respectively.

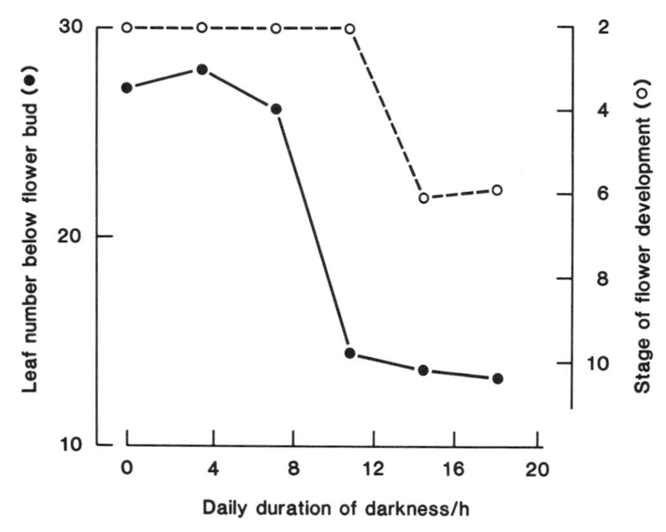

Figure 5.1. Effect of daily duration of darkness on initiation and development of the inflorescence in chrysanthemum cv Polaris. Plants received an 8-hr main light period extended with low-irradiance light from incandescent lamps. (K. E. Cockshull, unpublished data.)

The emphasis of interest in the switch from vegetative to reproductive growth, often using "model" systems which require only a single photoinductive cycle to effect the switch, has tended to obscure the fact that the primordia formed after one or a few inductive cycles frequently do not complete their development into fertile flowers. When induction is minimal (with only a few inductive cycles or daylengths near the critical), flowers often abort at an early stage or they may develop various degrees of leafiness of the flower (Vince-Prue 1975). When the induction is stronger, flowers usually continue their development but sporogenesis may be inhibited, microsporogenesis usually being more sensitive. The formation of both anthers and carpels, however, can be influenced by the daylength regime to which the plant is exposed, and either male or female fertility, or both, may be adversely affected.

Photoperiodic effects on flower initiation and development are known to be important determinants of yield in many of the major crop plants (Evans 1975b; Evans and King 1975; Summerfield and Wien 1979; Vince-Prue and Cockshull 1981), and a few examples are sufficient to illustrate the type of problem that may be encountered in the field.

All annual species of pasture plants, and also many perennials, rely on regeneration from seed. Their time of flowering, often under the control of daylength, is an important feature in their adaptation to climate and has been widely studied in many grasses (Evans 1964). A more recent study of the pasture legume, *Stylosanthes,* has considered its daylength response in relation to possible adaptation to a range of environments in Australia (Cameron and Mannetje 1977). The agronomic value of a pasture legume depends on its vegetative growth, most of which occurs before the plant is fully committed to reproduction; on the other hand, regeneration from seed is essential for both annual and perennial forms of *Stylosanthes*. Plants of different geographical origin and with different photoperiodic responses offer a choice of reproductive strategies for the varying climatic conditions of northern Australian pastures. Because of the lower temperatures south of the tropic, successions with LD response are thought likely to be most successful, since flowering will be induced in summer by the lengthening photoperiod. SD types are characterized by determinate flowering and could be used to select plants for a late-flowering response in regions with a long growing season. Day-neutral or quantitative SD response types are likely to be successful in areas where there is substantial within- and between-season variability as early flowering allows some seed production even in a very short growing season, whereas the capacity for continued vegetative growth results in heavy yields during long growing seasons.

Many of the leguminous crops are grown for grain rather than as pasture. Photoperiod plays a major role in controlling the earliness and rate of floral initiation and development in grain legumes, which are dependent for heavy yields on abundant flowering, pod set, and seed development. Daylength responses retain a strong correlation with taxonomic grouping and, where they retain photoperiodic sensitivity, plants in the Phaseoleae are SDP, while those in the Fabeae and Genisteae are LDP (Evans and King 1975). Among the Phaseoleae, soya bean *(Glycine max)* has been the subject of a great deal of research that has contributed substantially to our knowledge of the photoperiodic mechanism, in particular the involvement of circadian rhythmicity in time measurement (Hamner 1969). The effect of daylength on flowering in North American cultivars of soya bean is so marked that most are restricted to within about 4° of latitude (480 km) of their adapted area; outside this range plants fail to mature before the frosts in the North, or mature too early in the South (Summerfield and Wien 1979). A feature of the grain legumes is that, almost without exception, the photoperiodic requirements for flower development are more stringent than those for initiation (Shibles, Anderson, and Gibson 1975; Evans and King 1975). The period of anthesis and seed set is a critical stage in their development, and a substantial loss of buds, flowers, and immature pods can occur in unfavorable daylength environments (Hamner 1969; Shibles, Anderson, and Gibson 1975; Evans and King 1975; Zehni, Saad, and Morgan 1970).

Daylength is also an important regulatory factor in the initiation and development of inflorescences in the cereal crops, which include both LDP (temperate cereal grains, such as wheat and barley) and SDP (maize, sorghum, and rice). A response to photoperiod affecting the duration of the life cycle may be important for the avoidance of environmental hazards, such as low temperature or water stress (Aspinall 1966; Evans 1975a), and many cultivars are strongly adapted to the prevailing daylengths in their region of origin. Daylength is also known to influence individual components of yield, such as the number of spikelets in wheat (Wall and Cartwright 1974) and grain weight and spikelet fertility in spring barley (Kirby and Appleyard 1980). Overall grain yield, however, is not necessarily correlated in any simple way with the developmental response of a particular cultivar to photoperiod (Kirby and Appleyard 1980), and other environmental factors such as temperature often interact strongly with daylength in the final determination of yield.

PHOTOCONTROL OF VEGETATIVE REPRODUCTION

Many vegetative storage organs are important food crops, and their formation is, in many cases, under the control of daylength. With the important exception of the formation of bulbs in the genus *Allium,* most are favored by exposure to SD (Vince-Prue 1975). As with many other responses to daylength, the formation of storage organs may be quantitatively or qualitatively controlled by photoperiod. Tuberization in potato appears always to be hastened by SD, although cultivars differ considerably in their sensitivity. In SD conditions, early tuber formation is associated with early senescence of the haulm and relatively low yields (Pohjakallio 1953). Similar considerations apply to onion, where bulb formation is dependent on exposure to LD. In longer daylengths the onset of bulbing is accelerated, but the yield is low because the emergence of new leaves ceases rapidly and the first formed leaves collapse and senesce (Austin 1972). Cultivars are, therefore, best adapted when grown near their critical daylength because, under these conditions, leaf growth and bulb formation are prolonged and heavy yields arise from the large leaf area duration in the bulbing phase (Abe, Katsumata, and Nagayoshi 1955).

It is not uncommon to find horticultural crop plants propagated by means of either rooted stem cuttings or runners. The production of runners is sometimes directly under the control of photoperiod, as in strawberry where runners form more readily in LD (Guttridge 1960) and in *Chlorophytum* where they form more readily in SD (Hamner 1976). The regeneration of roots or buds on cuttings is also often influenced by daylength, and this effect may be exercised either through the stock plants or the cuttings themselves (Vince-Prue 1975; Whalley 1977).

EXPERIMENTAL APPROACHES: THE CONTROL OF FLOWERING IN *PHARBITIS NIL*

Although the diversity of responses to daylength and their ecological and agricultural importance are now well established, the mechanisms through which these responses are effected remain poorly understood. Attempts to elucidate the underlying mechanism in plants have largely been concentrated on a single phenomenon, namely the transition from vegetative to reproductive growth where this is effected by one photoperiodic cycle of appropriate duration. The SDP *Pharbitis nil* cv Violet is such a plant (Takimoto 1969) and offers many advantages as a model system for studying photoperiodism. In this

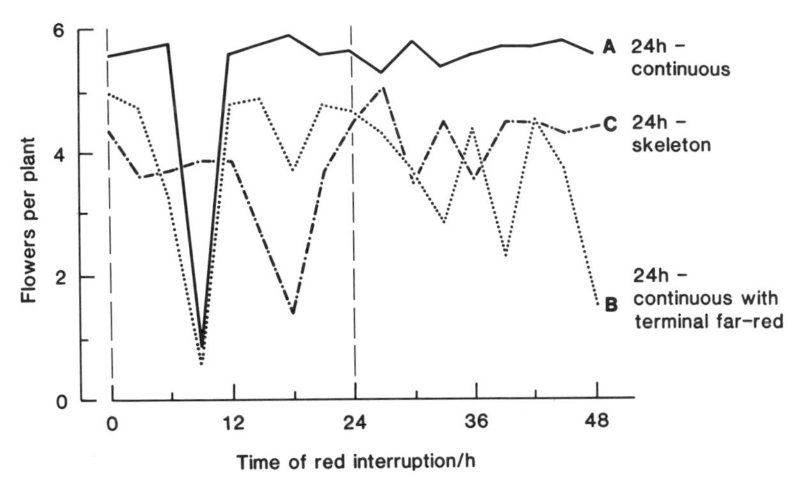

Figure 5.2. Effect of light pretreatment on the flowering response of Pharbitis nil to a 5-min night-break with red light given at different times during a 48-hr photoinductive night. The pretreatments were (A) 24-hr white fluorescent light, (B) 24-hr white fluorescent light terminated by 10-min far-red, (C) a skeleton 24-hr day beginning and ending with a 5-min red-light pulse. (B. Thomas, P. Lumsden, and D. Vince-Prue, unpublished data.)

section, *Pharbitis* will be used to illustrate some generally accepted concepts about the photoperiodic mechanism, and also to demonstrate some of the problems that remain unresolved.

In photoperiodism, plants respond to the duration and timing of light and dark periods in the daily cycle. Any mechanism must, therefore, involve at least two basic components: a photoreceptor that discriminates between light and darkness, and a system that measures time. In natural 24-hr cycles, short days are always associated with long nights, and vice-versa. A basic question for photoperiodism is, therefore, whether time is measured in light, or in darkness, or in both. *Pharbitis* seedlings raised in continuous light are already photoperiodically competent three to four days after sowing, and floral initiation in such seedlings can be effected by interrupting continuous light at any time with a single dark period of sufficient duration. Under these conditions, the onset of darkness appears to set in motion a process in the cotyledons that results in photoperiodic induction; this leads to the initiation of floral primordia even when plants are subsequently maintained in continuous light. Since floral induction depends only on the duration of a single dark period that interrupts continuous light, it is clear that photoperiodic time measurement here proceeds only in darkness and that induction occurs in the cotyledons when a critical duration is exceeded. The overriding importance of dark time measurement in the determination of flowering is emphasized by night-break experiments. As first demonstrated in the SDP *Xanthium strumarium* (Hamner and Bonner 1938), the effect of a photoinductive dark period is prevented by a short exposure to light given at a particular time; in *Pharbitis,* the time of maximum effectiveness of a night-break occurs about 9 hr after transfer from continuous light to darkness (Fig. 5.2, A), and is relatively unaffected by temperature (Takimoto and Hamner 1964).

The observation that a brief night-break prevented floral induction in SDP, and promoted it in LDP, allowed the first critical studies of the photoreceptor to be made at Beltsville (Borthwick, Hendricks and Parker 1952). In several SD and LD plants, the night-break was shown to act by the formation in red light of the active form of phytochrome, Pfr, since a subsequent exposure to far-red light at 730 nm (which would remove Pfr by photoconverting it back to the inactive, Pr form) reversed the effect of red light (Downs 1956). Initially it proved difficult to establish red/far-red reversibility of the night-break response in *Pharbitis* (Fredericq 1964). This was shown to be partly due to the fact that Pfr acts very rapidly and so the response quickly escapes from reversibility; however, provided that the exposures are kept short and far-red is given immediately, partial reversibility of a red night-break can be demonstrated under some conditions. The speed with which Pfr acts as a switch to prevent photoperiodic induction in *Pharbitis* suggests that the biochemical consequences of Pfr formation at this time may involve rapid responses such as changes in membrane permeability (Marmé 1977) rather than the direct regulation by Pfr of the synthesis of new compounds. Other plants that show a similarly rapid loss of reversibility (within 1 and 5 min respectively) are *Chenopodium album* (Borthwick 1964) and *Kalanchoe* (Fredericq 1965). In some plants such as chrysanthemum, however, Pfr appears to act much more slowly, and reversibility by far-red is retained for up to about 1.5 hr (Cathey and Borthwick 1971).

Although the photoperiodic induction of flowering in *Pharbitis,* as in other SDP, requires exposure to a sufficient duration of darkness, inductive dark periods are without effect unless preceded by an exposure to light. The nature of this requirement for light was first studied in *Pharbitis* by exposing young seedlings to various light treatments at the end of the photoperiod, immediately before transferring them to a photoinductive night (Nakayama, Borthwick, and Hendricks 1960). Flowering was shown to be depressed if a short exposure to far-red light was given, and this inhibition was prevented by a subsequent brief exposure to red. The action spectra for inhibition and reinduction were found to be the same as those for other phytochrome-mediated responses. In 24-hr cycles, the inhibitory effect of far-red was seen only after a very short photoperiod, and there was little or no effect with photoperiods of 5 hr or longer (Fredericq 1964), presumably because the flower promoting action of Pfr was completed during the photoperiod itself. These early physiological experiments demonstrated first that Pfr is present in the cotyledons at the end of the photoperiod, and second that, at least under some conditions, this Pfr has a flower-promoting function, since its removal early in the night depressed the flowering response (Fredericq 1964; Takimoto and Hamner 1965; Evans and King 1969; Fig 5.2, B). Nevertheless, a night-break that establishes Pfr acts to prevent flowering when given ca 9 hr later, and Pfr is presumed to have fallen below some threshold value during the intervening period in darkness.

From these and similar experiments with other plants, many investigators have concluded that the photoperiodic induction of flowering in SDP consists of a photophase during which Pfr is required, followed by a dark- or skotophase during which Pfr must be absent. This general hypothesis also includes the following assumptions about the behavior of phytochrome: Pfr is present in the leaf during the photoperiod but, on transfer to darkness, Pfr falls to a level that allows the dark process of induction to occur. The photoperiodic control of flowering in LDP is often assumed to operate via the same mechanism, except that the response is in the opposite direction.

This very simple model raises several questions about the photoperiodic mechanism which can be partly, but not completely, answered by reference to more recent experiments with *Pharbitis*.

The first question concerns the role of phytochrome in time measurement and the nature of the time-measuring system. When grown under conditions of continuous light and induced to flower by a single inductive dark period, time measurement in *Pharbitis* behaves as if it were an hour-glass, since the magnitude of the flowering response increases with the increase in duration of the interrupting dark period (Takimoto 1969). Moreover, when a long dark period that spans two or three circadian cycles (i.e., 48 or 72 hr) is interrupted by a single red pulse at different times, flowering is inhibited only when the night-break is given after about 9 hr in darkness (Fig 5.2, A), and there is little evidence for a rhythm in responsivity,such as would be expected if time measurement were effected by a circadian rhythm of sensitivity to light (Bünning 1967). An early hypothesis for photoperiodic timing developed at Beltsville by Sterling Hendricks (1960) suggested that an hour-glass type of timing might be a consequence of the properties of the phytochrome system itself. He proposed that, following the transfer of plants from light to darkness, the amount of Pfr in the leaf would decrease (through reversion to Pr, or by destruction) to a threshold level that no longer inhibited the process of floral induction and that the time taken to reach this threshold could be the factor which determined the critical nightlength.

Although the results from many physiological experiments have argued against this interpretation (see Vince-Prue 1975), a major problem has been to make direct measurements, since photoperiodic induction in response to a long dark period does not occur without prior exposure to light; this causes greening and prevents the usual spectrophotometric assay for phytochrome. In *Pharbitis,* however, phytochrome changes in the cotyledons can be followed in dark-grown plants that have been rendered competent to respond to a photoinductive night by light treatments that do not allow chlorophyll accumulation (King and Vince-Prue 1978). In one series of experiments, photoperiodic competence was achieved by exposing three-day-old, dark-grown seedlings to 24-hr, white fluorescent light at 17°–18°C (which photo-bleaches them), or to 24-hr, far-red light with a short terminal exposure to red light: these treatments resulted in substantially different amounts of both total phytochrome and Pfr at the beginning of the night, and the patterns of Pfr disappearance in darkness were very different (Fig. 5.3, A). Despite the marked difference in the time at which any particular threshold level of Pfr would have been reached in

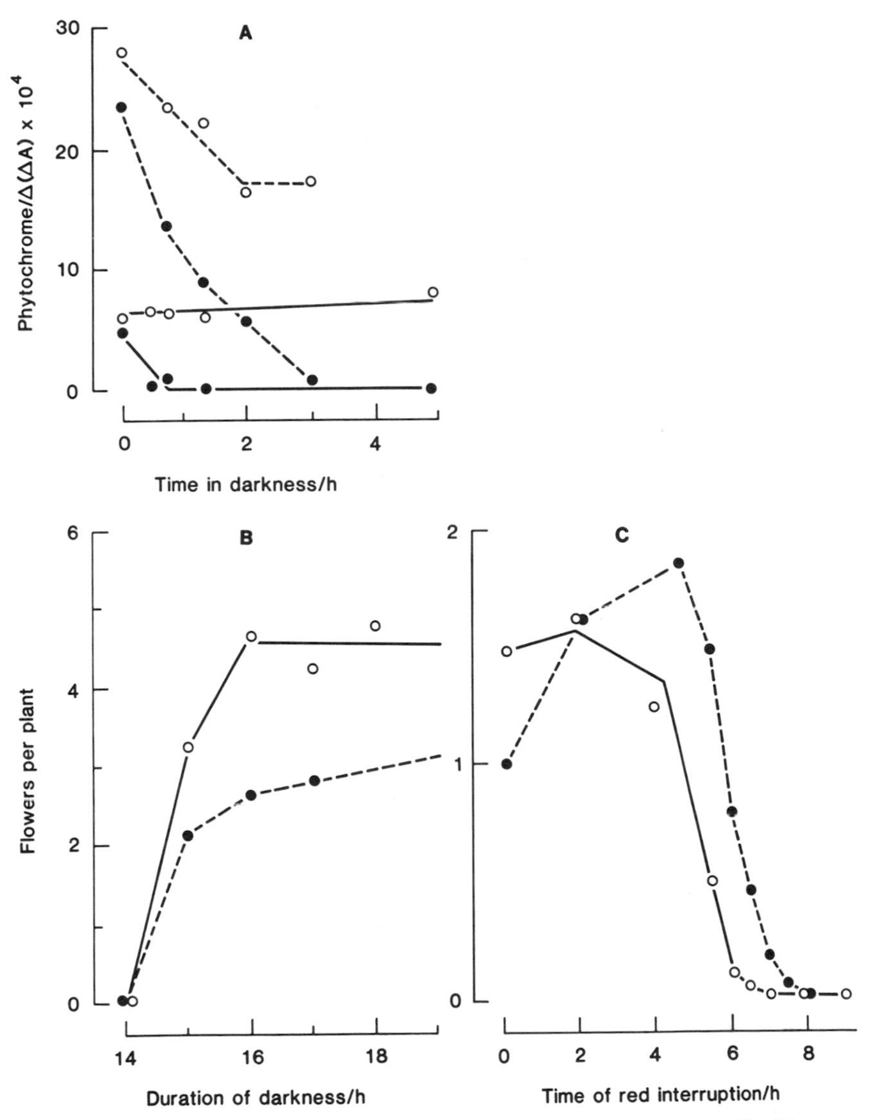

Figure 5.3. Phytochrome transformations and physiological responses in Pharbitis nil following different light pre-treatments given before transfer to a photoinductive dark period. The pretreatments were 24-hr far-red terminated with 5-min red (-----), or 24-hr white fluorescent light at 18°C (———). (A) phytochrome transformations; (B) critical nightlength; (C) response to a 5-min night-break. (From Vince-Prue, King, and Quail 1978; King et al 1978.)

darkness, the critical nightlength (Fig. 5.3, B) and the time of maximum sensitivity to a night-break (Fig. 5.3, C) were essentially the same in the two treatments. Estimations of the ratio of Pfr to total phytochrome were also made using a physiological "null" method based on the flowering response (King, Vince-Prue, and Quail 1978). The results from these experiments showed a reasonably close relationship between the spectrophotometrically measured Pfr/P ratio and the physiological estimate during the first few hours of darkness, demonstrating that, at least under these conditions, the kinetic properties of the "measured" phytochrome (in terms of dark loss) were similar to those of the "physiological" phytochrome. Especially with the far-red pretreatment, however, the spectrophotometric measurements indicated a more rapid fall in Pfr/P than did the physiological estimates.

There seemed to be no simple relationship between the disappearance of Pfr and time measurement in darkness in the experiments with nonchlorophyll-containing seedlings, and in green seedlings other physiological experiments indicated that Pfr may persist for many hours in darkness and has a flower-promoting role (Takimoto and Hamner 1965, Evans and King 1969). These apparent inconsistencies in the rate and pattern of Pfr loss following transfer of plants to darkness have led us to look again at the nature of the "light-off" signal that starts dark time measurement.

Under natural conditions,the irradiance decreases gradually at the end of the photoperiod, especially at high latitudes. There are also changes in the ratio of red:far-red light, which falls from about 1.1 in full sunlight to about 0.7 at the end of twilight in the evening (Holmes and McCartney 1976). Such changes in light quality would be expected to result in a photochemical lowering of the amount of Pfr present in the tissue during twilight, and this might be important in generating the "light-off" signal that couples to the beginning of dark time measurement under natural conditions. The nature of the environmental cue that begins the biological "night" has yet to be resolved satisfactorily, but the answer is important for use with daylength-screening protocols in breeding programs. These are almost always done with instantaneous light-to-dark transitions rather than with gradual ones, and there is usually no twilight period of changed light quality. At least one example can be given where results from controlled environment experiments were not in agreement with those obtained under natural conditions for, in sugar cane, the effective photoperiod in the field appeared to be some 26 min longer (Clements 1968); this period was equivalent to the time when the natural twilight illuminance exceeded 40 lx. In some early experiments carried out under natural conditions in Japan (Takimoto and Ikeda 1961), it was found that photoperiodic darkness began when a particular threshold irradiance was reached during twilight and that this irradiance differed in different plant species. None of these experiments, however, clearly distinguishes between quantity and quality of light as the natural environmental signal for dusk since, under natural conditions, these two components of light are changing more or less in parallel.

Under experimental conditions where the red:far-red ratio of the light was maintained constant, transfer of *Pharbitis* seedlings to 10 lx of white light ($\sim$ 40 W m^{-2}) allowed dark time measurement to proceed normally, even if plants remained in the light (Takimoto 1967), indicating that with respect to photoperiodic time measurement, light below a threshold level of irradiance is perceived as darkness. We recently examined the effect of lowering Pfr with an end-of-day exposure to far-red light on the time of maximum sensititity to a red night-break (Fig. 5.2). Somewhat to our surprise, we found that *Pharbitis* seedlings showed essentially the same night-break timing whether plants were transferred to darkness with little Pfr (i.e., after end-of-day far-red, Fig.5.2, B) or with a high level of Pfr (i.e., from white fluorescent light, (Fig. 5.2, A). Thus it appears that, at most, only a small acceleration of dark time measurement occurs in *Pharbitis* when the level of Pfr is lowered photochemically at the end of the day. It follows, therefore, that spectral shifts such as those occurring during twilight may not be important environmental cues for the beginning of the biological night. From recent studies with another SDP, *Xanthium strumarium,* Salisbury (1981) has also concluded that the environmental cue for the beginning of dark time measurement during twilight is most probably the attainment of a threshold irradiance. Further critical experimentation with other species is necessary before any general statement can be made, however. In view of the variation in the irradiance that was sensed as darkness in different plants when they were tested under natural conditions (Takimoto and Ikeda 1961; Salisbury 1981), further experimentation is also needed in order to determine the extent to which the dusk signal may vary with species and environment.

The experiment with end-of-day exposure to far-red confirmed the effect of Pfr on the magnitude of flowering in *Pharbitis* (Fig 5.2). It is clear from these results that, when seedlings are induced to flower with a single 24-hr light/48-hr dark cycle, photochemical lowering of Pfr at the end of the photoperiod reduced the flowering substantially if plants remained in darkness throughout the night (Fig 5.2, curve B, 48-hr point). Flowering was completely restored, however, by a red pulse given up to about 24 hr after the far-red exposure with the exception of the time when a red night-break also completely prevented flowering at the ninth hour of darkness. A red night-break at this time always resulted in low flowering irrespective of whether or not Pfr had been removed by an end-of-day exposure to far-red light (Fig. 5.2, curves A and B). Thus, Pfr phytochrome seems to be required for floral induction but does not need to be present for the time-measuring process itself.

These results with end-of-day far-red have led us to reexamine the question of the light requirement for the photophase of photoperiodic induction in *Pharbitis*. It has been demonstrated that photoperiodic competence can be induced by two 5-min exposures to red light when these are given 24 hours apart. Thus, in *Pharbitis* (as in *Lemna* [Hillman 1964; Oota and Nakashima 1978]) a complete photoperiod can be substituted for by a skeleton day beginning and ending with a brief pulse of light, which is a characteristic feature of some overt circadian rhythms (Pittendrigh 1966). In *Pharbitis,* both the "dawn" and "dusk" signals of the skeleton have been shown to operate through the formation of Pfr, since the effect of either of the two red pulses was reversible by far-red light (Friend 1975). In contrast, only the initial pulse was clearly identifiable with Pfr in the LDP *Lemna gibba* G3, while the final pulse was sensitive only to far-red or blue light (Oota and Nakashima 1978). Even though a skeleton 24-hr photoperiod beginning and ending with a red pulse can induce a high level of flowering in *Pharbitis* when followed by a 24- or 48-hr dark period, preliminary experiments indicate that the pattern of response to a night-break (Fig 5.2, curve C) differs from that obtained after a complete photoperiod (Fig 5.2, curve A). Thus, although the photophase of photoperiodic induction has clearly been idenfitied with phytochrome in *Pharbitis,* differences between continuous and skeleton photoperiods indicate that the nature of the photoreactions during the photophase have not yet been fully characterized.

More recently, it has been observed that the light requirement can be reduced to a single saturating exposure to red, providing that, at the same time, the seedlings are treated with the cytokinin, benzyladenine (Ogawa and King 1979); floral induction can then be achieved when the red pulse is followed by a 48- or 72-hr inductive dark period before transfer to continuous white fluorescent light for realization of the flowering response. Using this system, it has been possible for the first time to make detailed measurements of the amount and form of phytochrome present during an inductive dark period, and simultaneously to study the physiological responses of the system to various light treatments.

Some of the results obtained so far with this model system for *Pharbitis* are summarized in Figure 5.4. Spectrophotometric measurements of phytochrome in the cotyledons (Fig 5.4A) show that both Pfr and total phytochrome fell to a low value within about 4–6 hr in darkness but subsequently decreased only very slowly; some Pfr could still be measured after many hours in darkness. When tested physiologically, we found that both red and far-red light strongly inhibited flowering 14 hours after the initial red pulse, and there was no evidence for an optimum Pfr level at this time (Table 5.1). All wavelengths between 660 and 750 nm were inhibitory, including those which would establish a higher or lower Pfr/P ratio than was actually measured in the cotyledons at that time in darkness (ca 0.07). The physiological experiments (Fig. 5.4B) also indicated that phytochrome is present in the Pfr form during most or all of the night, since its removal by a far-red pulse at any time before the end of the critical dark period of about 46 hr depressed flowering. The effect of a second red pulse, on the other hand, varied with time after the first, and showed a circadian rhythm of promotion and inhibition of flowering. We consider that the most likely explanation for this apparent paradox, in which Pfr appears to be required for floral induction on the one hand and yet simultaneously interacts with a circadian timing system to prevent induction on the other, is that phytochrome is associated or compartmented in more than one pool in the cell. In one of these, Pfr appears to be relatively stable in darkness and, as indicated by the response to a far-red pulse, is required for the induction of flowering. Depending on conditions, Pfr in this pool may be required throughout most of the night (where the photoperiod is a single saturating red pulse, Fig 5.4B) or only during the photoperiod (in 24-hr cycles with daily photoperiods of more than 5 hr, Fredericq 1964). A second pool of phytochrome, in which Pfr appears to be less stable in darkness, may be more immediately

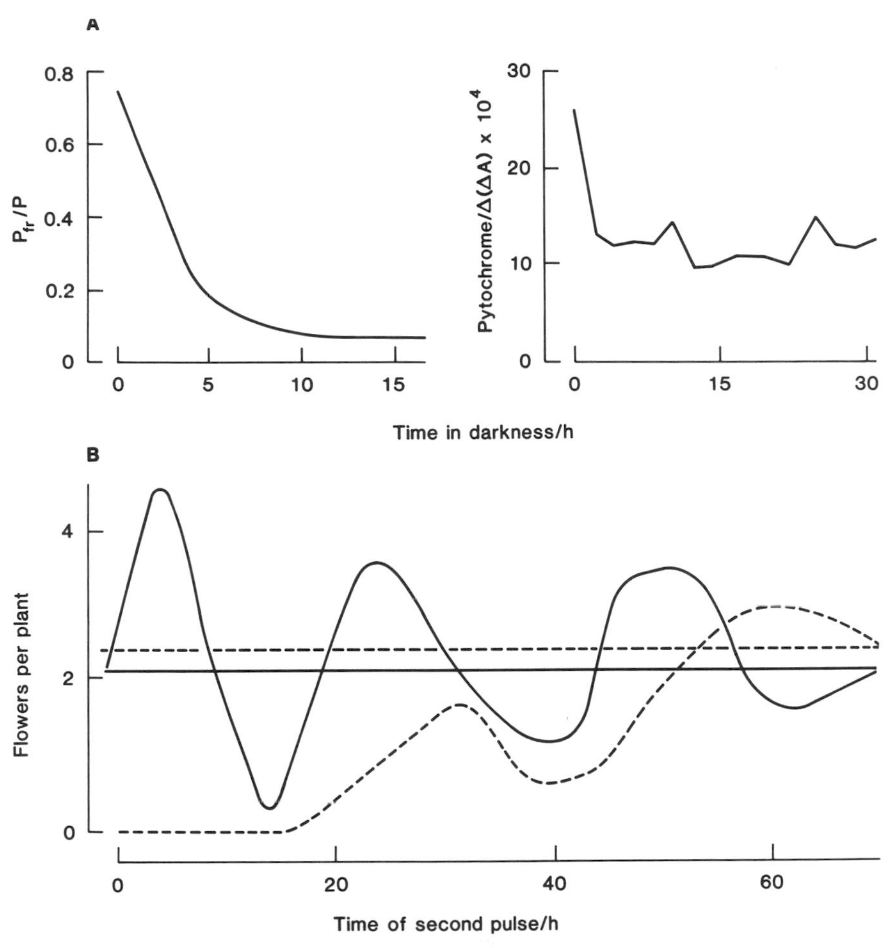

Figure 5.4. Phytochrome transformations and physiological responses in Pharbitis nil following a single saturating red light pulse. (A) changes in Pfr and total phytochrome in cotyledons, measured spectrophotometrically (B. Thomas unpublished data). (B) flowering response to a second pulse with red (———) or far-red (-----) light. (Schematic diagram based on unpublished data of R. W. King.)

involved in the perception of photoperiodic light signals and may be more closely associated with the timing aspects of photoperiodic induction. Similar conclusions have also been reached by Takimoto (see Vince-Prue, 1980, and Rombach, 1981).

The use of *Pharbitis* as a model system demonstrates the complexity of the photoperiodic response in even a single plant. Depending on conditions, time measurement has circadian or hour-glass properties (figs 5.2, 5.4B). Pfr may or may not be required during part or all of the inductive dark period, and at certain times Pfr appears both to promote and inhibit floral induction (Fig. 5.4),

Table 5.1 The Effect on Flowering in Pharbitis nil of Phytochrome Photoequilibrium (ϕ) Fourteen Hours after a First Saturating Red Light Pulse

Wavelength/nm	ϕ	Number of flower buds
660a	0.75	0
680	0.58	<0.2
688	0.47	0
699	0.24	0
708	0.13	0
720	0.04	0
750	0.03	0
750a	<0.01	0.1 ± 0.1
control	—	2.5 ± 0.4
controla	—	1.0 ± 0.3

Note: Seedlings were given a saturation exposure to light of various wavelengths at the 14th hour of a 48-hr dark period before transfer to continuous white fluorescent light.

Source: B. Thomas and D. Vince-Prue, unpublished data.

indicating the possibility that more than one pool of phytochrome may participate in the control of flowering. Nevertheless, it is from such model systems that fundamental information about the mechanism of photoperiodic induction is most likely to be gained.

THE CONTROL OF FLOWERING IN CHRYSANTHEMUM AND CARNATION

As shown in the previous section, one approach to the problem of understanding the photoperiodic mechanism is to study in detail a "simple" system such as *Pharbitis,* where the experimental treatments can be reduced to a single, short photophase coupled with a dark period that can be probed with light pulses to study both the timing and induction components of the mechanism. These experimental conditions are very different from those to which plants are exposed in the natural environment or in the glass-house, however. In addition, plants show a considerable diversity in their response to factors such as the duration, irradiance, and quality of light; interaction with temperature; number of inductive cycles; age; and genotype (cf. Evans 1969a). It is essential, therefore, to study the photoperiodic behavior of a wide range of plants under a variety of conditions. The diversity of photoperiod control can be exploited both in breeding programs and commercial-cropping schedules, providing that the individual plant response is known. This is well illustrated by reference to the use of artificial daylength regimes to control the time of flowering in commercial flower crops such as chrysanthemum and carnation. Although little is known about the mechanisms involved, the light requirements for the photoperiod control of flowering in these two plants differ in many respects from those seen in *Pharbitis* and may indicate important differences in the way in which photoperiodic cycles act to control flowering in different plants.

Carnation. The perpetual-flowering carnation, *Dianthus caryophyllus*, is a quantitative LDP, for floral initiation and artificial, long photoperiods are frequently used in commerce to accelerate the time of flowering and so achieve higher market prices (Koon 1974). Photoperiodic induction in LDP has often been assumed to be the mirror image of that in SDP, with the control of flowering being dependent on whether plants are exposed to an uninterrupted dark period longer than a critical duration. Nevertheless, the photoperiodic control of flowering in many LDP appears to operate through a different mechanism, requiring long daily exposures to light and being relatively insensitive to night-break treatments which completely prevent flowering in SDP. Such LDP show maximum flowering only when exposed to night-breaks of several hours duration, or to long photoperiods during which light is given more or less continuously (Vince-Prue 1975, 1976). The differences between these two types of daylength response has been recognized for many years. In one case, control seems largely to be dependent on exposure to a sufficiently long night, with light acting at a particular time to prevent the effect of darkness. In the other, prolonged exposures to light are necessary. The latter is well illustrated by reference to the perpetual-flowering carnation and may also operate in the control of floral initiation in chrysanthemum.

The carnation shows many of the response characteristics found in other LDP, and flowering is relatively little affected by a night-break treatment (Harris 1968). Although the same quantity of light given as a night-break is always more effective than when given as a day-extension, the acceleration of flowering increased with increase in duration of the daily light exposure to a maximum in continuous light. This requirement for light exposures far in excess of those needed to photoconvert phytochrome to Pfr raises the question whether phytochrome is the photoreceptor and, if so, why such long exposures are necessary (Vince-Prue 1975, 1976, 1980, 1981). The possibility that Pfr is rapidly lost in darkness and must be maintained by repeated exposures to light seems to be ruled out in carnation, where the flowering response is related only to the total light exposure, irrespective of whether this is given intermittently or continuously through a 16-hr day extension (Table 5.2).

The control of flowering in carnation illustrates another feature commonly, but not always (e.g., *Fuchsia* [Vince-Prue 1976] and *Calamintha* [Tcha, Jacques, and Jacques 1976]), found in LDP, namely that they are more responsive to light that contains a mixture of both red and far-red wavelengths than they are to red light alone (Vince, Blake, and Spencer 1964; Table 5.3). Also characteristic is a change in sensitivity to these two wavebands during the

Table 5.2 Flowering Response of Carnation cv White Sim to Incandescent Light Given Continuously or Intermittently Throughout a Sixteen-Hour Night

Lighting	Exposure/ lux. hr^{-1}	Leaf pairs below	Days to visible bud
None	—	24	130
80 lux intermittent	512	23	120
80 lux continuous	1280	18	84
200 lux intermittent	1280	18	86
200 lux continuous	3200	14	57

Note: 6 min light : 9 min dark.

Source: G. P. Harris (1972).

course of a long photoperiod (Vince-Prue 1976, 1980; see also Chapter 6, this volume). In carnation, however, the changes in responsivity to red and far-red light are relatively small (Harris 1968), and long-day induction is most effective when natural days are extended with incandescent lamps (which emit light throughout the red and far-red wavebands) and light is given during the whole of the natural night. Unfortunately, such dusk-to-dawn lighting with incandescent lamps not only accelerates flower initiation at the shoot apex, but also suppresses development of the basal laterals that are necessary for continued cropping (Heins, Wilkins, and Healy 1979; Powell 1979). There is also a tendency toward the production of weaker and more elongated stems due to the morphogenetic effect of the long daily exposures to low-intensity incandescent light (Vince-Prue 1977, and see 1975). A problem here is to devise an experimental protocol that will induce only the desired photoperiodic effect on flowering without the undesirable side-effects; this may be possible with better understanding of the mechanisms through which phytochrome controls these different physiological processes.

Photoperiodic induction in LDP is often strongly dependent on irradiance and may involve photosynthesis (cf. Bodson et al 1977); this is also well demonstrated in the carnation. Results from GCRI (Bunt, Powell, and Chanter 1981) have shown that the number of photo-inductive cycles necessary to initiate flowers in a given proportion of shoots decreased with increase in the daily light integral during the main light period, even with the optimum photo-inductive treatment of dusk-to-dawn lighting with incandescent lamps. Similar interactions have been observed in other LDP (Ballard 1969).

Chrysanthemum. The chrysanthemum *(Chrysanthemum morifolium)* is perhaps the most important crop in which flowering time is regulated by artificial daylength schedules, and there is a large industry for the production of all-

Table 5.3 Flowering Response of Carnation cv White Sim to Light Quality during a Four-Hour Night-break Given in the Middle of a Sixteen-Hour Dark Period

Night-break treatment	Leaf Pairs below bud	Days to visible bud
red	22.0	81
far-red	21.0	69
red + far-red	19.7	59
red + far-red (x2)[a]	18.4	49
None	22.5	82

[a] Irradiance twice that given in other treatments.

Source: G. P. Harris (1968).

year-round flowers and pot plants. It is necessary to simulate both long days for vegetative growth and short days to induce flowering in normal cropping programs (Vince-Prue and Cockshull 1981). There is considerable genetic variation in terms of the flowering response of chrysanthemums to daylength (Langton 1977) and, although types which will flower during summer are known, cultivars for controlled year-round cropping have been selected from among the autumn-flowering types, which have a pronounced response to short days for floral induction. Chrysanthemums are quantitative SDP, and all will initiate flower buds ultimately in long days (Fig. 5.1); cultivars with strong photoperiod response, however, initiate flower buds only after forming a large number of leaves and are consequently less prone to form buds prematurely during the LD phase of the cropping schedule (Cockshull 1975). Selection under long photoperiods for a high leaf number before the inflorescence is initiated is, therefore, a useful practice in breeding or selection programs (Langton and Cockshull 1979).

In chrysanthemum, both the initiation of flowers and their further development are under daylength control, and both require short days (Fig. 5.1). In recent experiments at GCRI, a comparison of the effectiveness of various lighting treatments on these two SD-dependent processes has, however, revealed differences which suggest that their control by daylength may not operate in the same way.

When the response to a night-break treatment was examined, the delay of flower initiation, as measured by the number of leaves formed below the inflorescence primordium, required a considerably higher irradiance from incandescent lamps than was needed to prevent the further development of the flower (Fig. 5.5). Flower development beyond stage 2 was already completely suppressed at 0.1 W m^{-2}. (400–700 nm), but the number of leaves below the bud continued to increase with increasing irradiance, reaching saturation at 0.5 W m^{-2}. Incandescent lamps are the usual light source for the LD phase in

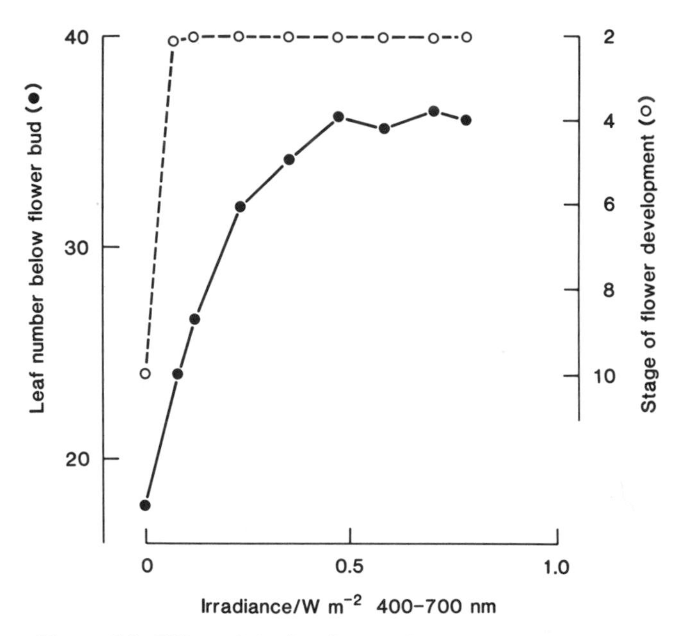

Figure 5.5. Effect of the irradiance of a 5-hr night-break with incandescent lamps on initiation and development of the inflorescence in chrysanthemum cv Polaris. (K. E. Cockshull, unpublished data.)

commercial production and are not effective in delaying initiation unless given over a relatively long period of four to five hours each night. Thus, at least with incandescent lamps, control of initiation requires a relatively long night-break and is irradiance-dependent, both features commonly found in LDP. The saturation irradiance of 0.5 W m^{-2} is much lower than is found in many other LDP for day-extension and night-break treatments (for example up to 30 W m^{-2} in *Brassica campestris,* Friend 1969) but is comparable with *Lolium temulentum,* where the saturation irradiance is about 1.0 W m^{-2} (Evans 1969b). Unlike carnation, flowering in chrysanthemum is not dependent on the total light exposure; the same quantity of light given intermittently is more effective than when given continously and may indicate that the maintenance of Pfr above a threshold value is an important factor. When light is used for a shorter fraction of the intermittent light/dark cycle, however, a higher irradiance appears to be necessary (Cathey 1961). The greater effectiveness of intermittent light is exploited commercially in various "cyclic" lighting schedules.

As already described for carnation, a characteristic feature of the photoperiodic response of LDP is a change in the optimum ratio of red to far-red light during the course of the long photoperiod; red promotes flowering during the first half of a 16-hr day, but in the second half red light is inhibitory (Vince-Prue 1976, 1980). The possibility that a similar pattern of response might occur in chrysanthemum has been examined both in completed controlled environ-

Table 5.4 The Effect on Flowering in Chrysanthemum cv Polaris of a Low Irradiance Eight-Hour Day Extension Given before or after an Eight-Hour Main Light Period

Treatment sequence			Number of leaves below flower bud	Floral stage
08–16hr	16–14hr	24–08hr		
A. sunlight	dark	R	33.6	2
sunlight	R	dark	17.0	2
B. sunlight	dark	R + FR	33.8	2
sunlight	R + FR	dark	32.2	2

Note: The day-extension treatments were with grolux fluorescent lamps (R) or incandescent lamps (R + FR); the main light period was mixed warm white fluorescent and incandescent lamps.

Source: K. E. Cockshull and D. Vince-Prue (1980).

ments and using low-irradiance day extensions in conjunction with 8-hr photoperiods. The responses have been similar under both conditions, and results from a representative experiment in controlled environments with the cultivar Polaris are shown in Table 5.4.

In all treatments, plants were given an 8-hr main light period in mixed fluorescent and incandescent light (i.e., with both red and far-red wavelengths present). This was immediately preceded or followed by an 8-hr day-extension with low irradiance red light (grolux lamps). The results (Table 5.4,A) show that, although both sets of plants were exposed to a 16-hr photoperiod combined with an 8-hr night, only when the 8-hr extension with red light was given *before* the main light period, (i.e., plants received red plus far-red in the second half of the day) was floral initiation delayed, with 33.6 leaves formed below the terminal inflorescence bud. Red given *after* the main light period (i.e., plants received only red light in the second half of the day) was relatively ineffective, and only 17.0 leaves were formed before initiation. Day-extensions with incandescent lamps (red plus far-red) were equally effective at both times (Table 5.4,B).

From these results, it is concluded that initiation in chrysanthemum cannot be dependent only on exposure to a sufficiently long dark period, since it is not prevented simply by shortening the night to eight hours (e.g., Fig. 5.1). Initiation rather seems to be inhibited by prolonged daily exposures to light such as those required for the induction of flowering in LDP. Characteristically, such treatments are relatively ineffective as "long" days when they are deficient in far-red light, especially during the latter part of the photoperiod. Flower development, on the other hand, appears to be promoted by exposure to long dark periods since all LD treatments prevented development beyond stage 2, irrespective of the wavelength sequence during the long photoperiod.

CONCLUSIONS

Daylength has a number of effects on all aspects of plant reproduction; it can modify all stages of development of the flower and is often an important factor controlling the onset and development of vegetative reproduction. The photoperiodic mechanism is not well understood and involves a complex interaction of a photoreceptor, phytochrome, with a timing system, which is most probably a circadian rhythm. Differences between plants, however, indicate that more than one type of control mechanism may operate in the control of photoperiodic responses.

There is need for a greater knowledge of the properties and mode of action of both the photoreceptor and the timing system, as well as the nature of the interactions between the two. A better understanding of the mechanism(s) through which photoperiodism functions to control reproduction may allow improved exploitation in breeding programs and in the artificial manipulation of desired responses. It may also lead to novel methods of controlling these responses.

ACKNOWLEDGMENTS

I wish to thank my colleagues K. E. Cockshull, B. Thomas, and P. Lumsden of the Glasshouse Crops Research Institute, and R. W. King of CSIRO, Canberra, for permission to reproduce some of their unpublished data.

LITERATURE CITED

Abe, S., H. Katsumata, and H. Nagayoshi. 1955. *Studies on the photoperiodic requirements for bulb formation in Japanese varieties of onions with special reference to their ecological differentiation*. J. Hort. Assoc. Japan 24:6–16.

Aspinall, O. 1966. *Effects of daylength and light intensity on growth of barley. IV. Genetically controlled variation in response to photoperiod*. Austral. J. Biol. Sci. 19:517–34.

Austin, R. B. 1972. *Bulb formation in onions as affected by photoperiod and spectral quality of light*. J. Hort. Sci. 47:493–504.

Ballard, L.A.T. 1969. *Anagallis arvensis L*. Pages 376–92 in *The Induction of Flowering*. Macmillan of Australia, Melbourne.

Bodson, M., R. W. King, L. T. Evans, and G. Bernier 1977. *Photosynthesis and flowering in the long day plant Sinapis alba*. Austral. J. of Plant Physio. 4:467–78.

Borthwick, H. A. 1964. *Phytochrome action and its time displays*. Amer. Natural. 95:347–55.

Borthwick, H. A., S. B. Hendricks, and M. W. Parker 1952. *The reaction controlling floral initiation*. Proc. Nat. Acad. Sci. 38:662–66.

Bünning, E. 1967. *The Physiological Clock*. Academic Press, New York.

Bunt, A. C., M. C. Powell, and D. O. Chanter. 1981. *Effects of shoot size, number of continuous light cycles and solar radiation on flower initiation in the carnation (Dianthus caryophyllus L.)*. Scientia Horticulturae (in press).

Cameron, D. F., and L't. Mannetje. 1977. *Effects of photoperiod and temperature on flowering of twelve Stylosanthes species.* Austral. J. Exper. Agric. and Anim. Husbandry 17:417–24.
Cathey, H. M. 1957. *Chrysanthemum temperature study. F. The effect of temperature upon the critical photoperiod necessary for the initiation and development of flowers of Chrysanthemum morifolium.* Proc. Amer. Soc. Hort. Sci. 69:485–91.
———. 1961. *Cyclic lighting for controlling flowering of chrysanthemums.* Proc. Amer. Soc. Hort. Sci. 78:545–52.
Cathey, H. M., and H. A. Borthwick. 1971. *Phytochrome control of flowering of Chrysanthemum morifolium on very short photoperiods.* J. Amer. Soc. Hort. Sci. 96:544–46.
Clements, H. F. 1968. *Lengthening vs shortening dark periods and blossoming in sugar cane as affected by temperature.* Plant Physiol. 43:57–60.
Cockshull, K. E. 1975. *Premature budding in year-round chrysanthemums.* Ann. Rep. Glasshouse Crops Research Institute 1974: 128–36.
Cockshull, K. E. 1976. *Flower and leaf initiation by Chrysanthemum morifolium Ramat. in long days.* J. Hort. Sci. 51:441–50.
Cockshull, K. E., and D. Vince-Prue. 1980. *Effects of the timing and light quality of daylength extensions on flower initiation.* Ann. Rep. Glasshouse Crops Research Institute 1979:64–67.
Cumming, B. G. 1969. *Chenopodium rubrum L. and related species.* Pages 328–49 in *The Induction of Flowering.* Macmillan of Australia, Melbourne.
Downs, R. J. 1956. *Photoreversibility of flower initiation.* Plant Physiol. 31:279–84.
Evans, L. T. 1964. *Reproduction.* Pages 126-53 in *Grasses and Grasslands.* Macmillan, New York.
———. 1969a. *The Induction of Flowering.* Macmillan of Australia, Melbourne.
———. 1969b. *Lolium temulentum L.* Pages 328–49 in *The Induction of Flowering.* Macmillan of Australia, Melbourne.
———. 1975a. *Daylength and the Flowering of Plants.* W. A. Benjamin, Menlo Park, California.
———. 1975b. *The physiological basis of crop yield.* Pages 327-55 in *Crop Physiology.* Cambridge Univ. Press, London.
Evans, L. T., and R. W. King. 1969. *Role of phytochrome in photoperiodic induction of Pharbitis nil.* Zeitschrift f. Pflanzenphysiol. 60:277–88.
———. 1975. *Factors affecting flowering and reproduction in the grain legumes.* In Report of the TAC Working Group on the Biology of Yield of Grain Legumes. Publi. No. DDDR:IAR/75/2. FAO, Rome.
Fredericq, H. 1964. *Conditions determining effects of far-red and red light on flowering response in Pharbitis nil.* Plant Physiol. 39:812–16.
———. 1965. *Action of red and far-red light at the end of short-day and in the middle of the night on flower induction in Kalanchoe blossfeldiana.* Bio. Jaarboek 33:66–91.
Friend, D. J. C. 1969. *Brassica campestris L.* Pages 364–75 in *The Induction of Flowering.* Macmillan of Australia, Melbourne.
———. 1975. *Light requirements for photoperiodic sensitivity in cotyledons of dark-grown Pharbitis nil.* Physiol. Plant. 35:286–96.
Guttridge, C. G. 1960. *The physiology of flower formation and vegetative growth in the strawberry.* Bulletin de L'Institut Agronomique et des Stations Recherches de Gembloux, Hors serie 2:941–48.
———. 1969. *Fragaria.* Pages 247–67 in *The Induction of Flowering.* Macmillan of Australia, Melbourne.
Hamner, K. C. 1969. *Glycine max (L.) Merrill.* Pages 62–89 in *The Induction of Flowering.* Macmillan of Australia, Melbourne.
Hamner, K. C., and J. Bonner. 1938. *Photoperiodism in relation to hormones as factors in floral initiation and development.* Bot. Gaz. 100:388–431.

Hamner, P. A. 1976. *Stolon formation in Chlorophytum*. Hort. Sci. 11:570–72.

Harris, G. P. 1968. *Photoperiodism in the glasshouse carnation: the effectiveness of different light sources in promoting flower initiation*. Ann. Bot. 32:187–97.

———. 1972. *Intermittent illumination and the photoperiodic control of flowering in carnation*. Ann. Bot. 36:345–52.

Harris, G. P., and M. Ashford. 1966. *Promotion of flower initiation in the glasshouse carnation by continuous light*. J. Hort. Sci. 41:397–406.

Heins, R. D., H. F. Wilkins, and W. E. Healy. 1979. *The effect of photoperiod on lateral shoot development in Dianthus caryophyllus L. cv Improved White Sim*. J. Amer. Soc. Hort. Sci. 104:314–19.

Hendricks, S. B. 1960. *Rate of change of phytochrome as an essential factor determining photoperiodism in plants*. Cold Spring Harbor Symposia in Quantitative Biology 25:245–48.

Hillman, W. S. 1964. *Endogenous circadian rhythms and the response of Lemna perpusilla to skeleton photoperiods*. Amer. Natural. 98:323–28.

Holmes, M. G., and H. A. McCartney. 1976. *Spectral energy distribution in the natural environment and its implications for phytochrome function*. Pages 467-76 in *Light and Plant Development*. Butterworth, London.

Hughes, A. P., and K. E. Cockshull. 1965. *Interrelations of flowering and vegetative growth in Callistephus chinensis (var. Queen of the Market)*. Ann. Bot. 29:131–51.

King, R. W., and D. Vince-Prue. 1978. *Light requirement, phytochrome and photoperiodic induction of flowering of Pharbitis nil Chois. I. No correlation between photomorphogenetic and photoperiodic effects of light pre-treatment*. Planta 141:1–7.

King, R. W., D. Vince-Prue, and P. H. Quail. 1978. *Light requirement, phytochrome and photoperiodic induction of flowering of Pharbitis nil Chois. III. A comparison of spectrophotometric and physiological assay of phytochrome transformation during induction*. Planta 141:15–22.

Kirby, E. J. M., and M. Appleyard. 1980. *Effects of photoperiod on the relation between development and yield per plant of a range of spring barley varieties*. Zeitschrift für Pflanzenzüchtung 85:226–39.

Koon, G. 1974. *Timing carnations in Colorado with lights*. Colorado Flower Growers Bull. no. 284.

Krekule, J. 1979. *Stimulation and inhibition of flowering. Morphological and physiological studies*. Pages 19–57 in *La Physiologie de la floraison,* R. Jacques and P. Champagnat, eds. CNRS no. 285, Paris.

Langton, F. A. 1977. *The responses of early-flowering chrysanthemums to daylength*. Scientia Horticulturae 7:277–89.

Langton, F. A., and K. E. Cockshull. 1979. *Screening chrysanthemums for leaf number in long days*. Glasshouse Crops Research Institute Ann. Rep. 1978:177–86.

Marmé, D. 1977. *Phytochrome: membranes as possible sites of primary action*. Ann. Rev. Plant Physiol. 28:173–98.

McMillan, C. 1974a. *Photoperiodic responses of Xanthium strumarium L. (Compositae) introduced and indigenous in Eastern Asia*. Bot. Mag. (Tokyo) 87:261–69.

———. 1974b. *Photoperiodic responses of Xanthium strumarium in Europe, Asia Minor and Northern Africa*. Can. J. Bot. 52:1779–91.

Nakayama, S., H. A. Borthwick, and S. B. Hendricks. 1960. *Failure of reversible control of flowering in Pharbitis nil*. Bot. Gaz. 121:237–43.

Ogawa, Y., and R. W. King. 1979. *Establishment of photoperiodic sensitivity by benzyladenine and a brief red irradiation in dark grown seedlings of Pharbitis nil Chois*. Plant and Cell Physiol. 20:115–22.

Oota, Y., and N. Nakashima. 1978. *Photoperiodic flowering in Lemna gibba G3: time measurement*. Bot. Mag. (Tokyo) Special Issue 1:177–98.

Pittendrigh, C. S. 1966. *The circadian oscillation in Drosophila pseudobscura: a model for the photoperiodic clock*. Zeitschrift für Pflanzenphysiol. 54:275–307.

Pohjakallio, O. 1953. *On the effect of daylength on the yield of potato*. Physio. Plant. 6:140–49.

Post, K. 1948. *Daylength and flower bud development in chrysanthemum.* Proc. Amer. Soc. Hort. Sci. 51:590–92.

Powell, M. C. 1979. *Observations on the growth of carnation (Dianthus caryophyllus L.) in natural and long days.* Ann. Bot. 43:579–91.

Rombach, J. 1981. Personal communication.

Salisbury, F. B. 1981. *The twilight effect: initiating dark measurement in photoperiodism of Xanthium.* Plant Physiol. (in press).

Schwabe, W. W. 1951. *Factors controling flowering in the chrysanthemum. II. Daylength effects on the further development of inflorescence buds and their experimental reversal and modification.* J. Exper. Bot. 2:223–37.

Shibles, R., I. C. Anderson, and A. H. Gibson. 1975. *Soybean.* Pages 151–89 in *Crop Physiology.* Cambridge Univ. Press, London.

Summerfield, R. J., and H. C. Wien. 1979. *Photoperiodic and air temperature effects on growth and yield of economic legumes.* In *Advances in Legume Science.* Her Majesty's Stationery Office, London.

Tafazoli, E., and D. Vince-Prue. 1978. *A comparison of the effects of long days and exogenous growth regulators on growth and flowering in strawberry, Fragaria x ananassa Duch.* J. Hort. Sci. 53:255–59.

Takimoto, A. 1967. *Studies on light affecting the initiation of endogenous rhythms concerned with photoperiodic responses in Pharbitis nil.* Bot. Mag. (Tokyo) 80:241–47.

———. 1969. *Pharbitis nil Chois.* Pages 90–115 in *The Induction of Flowering.* Macmillan of Australia, Melbourne.

Takimoto, A., and K. C. Hamner. 1964. *Effect of temperature and pre-conditioning on photoperiodic response of Pharbitis nil.* Plant Physiol. 39:1024–30.

———. 1965. *Effect of far-red light and its interaction with red light in the photoperiodic response of Pharbitis nil.* Plant Physiol. 40:859–64.

Takimoto, A., and K. Ikeda, 1961. *Effect of twilight on photoperiodic induction in some short day plants.* Plant and Cell Physiol. 2:213–29.

Tcha, K. H., R. Jacques, and M. Jacques. 1976. *Spectre d'action de la lumière sur l'induction florale du Calamintha officinalis ssp. nepetoides et rôle du phytochrome.* Compte Rendu de l'Académie des Sciences, Paris, 283 Série D:341–44.

Vince, D., J. Blake, and R. Spencer. 1964. *Some effects of wavelength of the supplementary light on the photoperiodic behaviour of the long-day plants, carnation and lettuce.* Physiol. Plant. 17:119–25.

Vince-Prue, D. 1975. *Photoperiodism in Plants.* McGraw-Hill, London.

———. 1976. *Phytochrome and photoperiodism.* Pages 347–69 in *Light and Plant Development.* Butterworth, London.

———. 1977. *Photocontrol of stem elongation in light-grown plants of Fuchsia hybrida.* Planta 133:149–56.

———. 1980. *Effect of photoperiod and phytochrome in flowering: Time measurement.* Pages 91–127 in *La Physiologie de la floraison.* R. Jacques and P. Champagnat, eds. CNRS no. 285, Paris.

———. 1981. *Phytochrome and photoperiodic physiology.* In *Biological Time Keeping,* J. Brady ed., Cambridge Univ. Press (forthcoming).

Vince-Prue, D., and K. E. Cockshull. 1981. *Photoperiodism and crop production.* Pages 175–97 in *Physiological Processes Limiting Plant Productivity.* Butterworth, London.

Vince-Prue, D., R. W. King, and P. H. Quail. 1978. *Light requirement phytochrome and photoperiodic induction of flowering of Pharbitis nil Chois. II. A critical examination of spectrophotometric assays of phytochrome transformations.* Planta 141:9–14.

Wall, P. C., and P. M. Cartwright. 1974. *Effect of photoperiod, temperature and vernalization on the phenology and spikelet numbers of spring wheats.* Ann. of App. Biol. 76:299–309.

Whalley, D. N. 1977. *The effects of photoperiod on rooting and growth of hardy ornamentals.* A.D.A.S. Quarterly Rev. 25:41–62.

Zehni, M. S., F. S. Saad, and D. G. Morgan. 1970. *Photoperiod and flower bud development in Phaseolus vulgaris.* Nature 227:628–29.

6 Effect of Far-Red Energy on the Photoperiodic Control of Flowering in Wintex Barley (*Hordeum vulgare* L.)

by GERALD F. DEITZER*

ABSTRACT

Flowering in barley (*Hordeum vulgare* L. var. Wintex), a facultative long-day plant, is enhanced by the addition of far-red energy to the extent that plants flower equally well under 12-hr photoperiods with far-red as with continuous light without far-red. The ability to respond to a 6-hr. pulse of far-red varies with circadian periodicity and is maximal during the second half of the photoperiod. The optimal red to far-red ratio for promotion is between 0.2 and 0.6. The response decreases by 50% with ratios of 1.2 and is not significantly greater than the controls above 2.0. When the ratio is held constant at 0.5, flowering increases with increasing fluence, saturating at 5×10^4 $J{\cdot}m^{-2}$ during a period of maximal sensitivity. At minimal sensitivity, the response does not saturate, even with fluences as high as 8.8×10^6 $J{\cdot}m^{-2}$, and there is a 50-fold difference in half-saturation values between periods of low and high sensitivity. Reciprocity does not hold with periods of less than 6 hr, unless the light is pulsed with a frequency greater than 1 min on/1 min off. Monochromatic light promotes flowering at 710 nm, 730 nm and, to a lesser extent, at 750 nm. There is no response to light at 660 nm and 680 nm. It is concluded that this response is mediated by phytochrome, acting through the "high irradiance response." Also, it is suggested that this response is driven by the far-red component of solar daylight in much the same way as the control of extension growth beneath a leaf canopy.

INTRODUCTION

It has been recognized since the early work of Klebs (1913) and Tournois (1914) that daylength is the controlling factor for reproduction in many plant species.

*Smithsonian Institution, Radiation Biology Laboratory, 12441 Parklawn Drive, Rockville, Maryland 20852.

STRATEGIES OF PLANT REPRODUCTION (BARC Symposium number 6—Werner J. Meudt, ed.)
Allanheld, Osmun, Totowa

Garner and Allard (1920), however, working at the USDA facility in Arlington, Virginia (the predecessor to the facility at Beltsville), were the first to clearly articulate the importance of this phenomenon, which they termed "photoperiodism." They also were the first to recognize the critical importance of events taking place during the dark period (Garner and Allard 1931). This work was confirmed by Hamner and Bonner (1938), who found that flowering could be prevented in the short-day plant *Xanthium* with as little as 1 min of light given in the middle of the dark period. The action spectrum for the inhibitory effect in this light break showed that red light was most effective (Parker et al. 1945, 1946). A similar action spectrum was found for the promotion of flowering by a light break in the long-day plants barley (Borthwick, Hendricks, and Parker 1948) and *Hyoscyamus* (Parker, Hendricks, and Borthwick 1950). This suggested that the same photoreceptor was involved in both responses. When it was found that the effect of red light could be reversed by far-red light (Borthwick, Hendricks, and Parker 1952; Borthwick et al. 1952), it became possible to isolate and characterize the pigment responsible that was called phytochrome (Butler et al. 1959).

Borthwick and Parker (1938) were the first to suggest that the high-intensity light portion of the photoperiod was not simply an interruption of inductive dark periods. They found that a single long dark period alone did not promote flowering in the short-day plant Biloxi soybean. Rather, induction was quantitatively dependent on both the number of short-day cycles given and the intensity of light during each cycle. This work was expanded by Hamner (1940), who showed that induction was dependent on both the light and the dark period, as well as on an interaction between the two. Liverman and Bonner (1953), however, concluded that this requirement was necessary simply to provide photosynthate for some key reaction occurring during the dark period.

Subsequent work with low irradiance daylength extensions of the photoperiod was inconsistent with this view, since the most effective light quality was either nonphotosynthetically active far-red light (DeLint 1958; Takimoto 1961) or less photosynthetically efficient mixtures of red and far-red light (Wassink, Stolwijk, and Beemster 1951; Borthwick and Parker 1952; Takimoto 1957; Downs, Piringer, and Wiebe 1959; Piringer and Cathey 1960; Friend, Helson and Fisher 1961; Vince, Blake, and Spencer 1964; Lane, Cathey, and Evans 1965; Vince 1965). Stolwijk and Zeevaart (1955) compared the light required to promote flowering in the long-day plant *Hyoscyamus* given as a light break with that given as a daylength extension and found that, while red was most effective in the former, blue and far-red were most effective in the latter. More recently, Holland and Vince (1971) and Evans (1976) reported that the light required for promotion of flowering in *Lolium* changes during the course of the photoperiod from far-red to red. They suggested that there are two phytochrome mediated responses, a low Pfr (far-red) response and a high Pfr (red) response, that occur sequentially during the photoperiod.

While the high Pfr reaction is easily ascribed to phytochrome action, since it is readily red/far-red reversible, the low Pfr reaction is less clear. The action spectrum for this response (Schneider, Borthwick, and Hendricks 1967) shows a single peak in the far-red at about 710 nm and some activity in the blue at very high irradiances. It is not readily reversible, does not obey the Bunsen-Roscoe law of reciprocity, and is markedly dependent on irradiance. Such responses have been rationalized solely on the basis of phytochrome (Hartmann 1966) by assuming that they are regulated, in some complex way, by the photostationary state of phytochrome and are known as high energy responses (HER) (Hartmann 1966; Borthwick et al. 1969) or, more appropriately, as high irradiance responses (HIR) (Shropshire 1972). Although a number of models have been proposed to explain this phenomenon (for a recent review, see Mancinelli and Rabino 1978), our understanding of such responses remains quite vague. This chapter deals with this kind of response in an effort, if not to solve the problem, at least to bring the question into sharper focus.

LIGHT ENVIRONMENT

Although the mechanism of action of the high irradiance response is unknown, there is general agreement that it proceeds only while there is an input of light energy. This aspect was most strongly emphasized by José and Vince-Prue (1978), who suggested calling this response the "dynamic mode" of phytochrome action.

In a series of papers in 1977, Holmes and Smith described the function of phytochrome in the natural environment. They showed that, while the relative spectral composition of solar daylight remained remarkably constant throughout the day, changes did occur both at low solar zenith angles around twilight, and as a consequence of the differential screening by chlorophyll when light is passed through a leaf. Both result in a relative increase in the amount of far-red when compared to red.

The twilight shift in relative red to far-red energy was found to vary between about 0.6 and 1.2 (Holmes and Smith 1977a) and was reported to be highly variable and dependent on local atmospheric conditions. This shift was also considerably less than that reported by Shropshire (1973) for changes measured in the direct solar beam and that measured as broad band (100 nm) ratios on clear days at various geographical locations (Goldberg and Klein 1977). Such a shift would change the phytochrome photoequilibrium (Pfr/Ptot) only from about 0.54 to about 0.40. It was concluded that, unless the responses were very sensitive, such a slight change in spectral distribution was unlikely to have any effect.

The change in red to far-red ratio that is produced when sunlight is filtered through a leaf canopy, on the other hand, was substantially greater, ranging from 1.2 above the canopy to about 0.2 below (Holmes and Smith 1977b). This resulted in a constant change in the phytochrome photoequilibrium from 0.54 to as low as 0.13 (values as low as 0.04 were recorded below two layers of sugarbeet leaves) (Smith and Holmes 1977). Morgan and Smith (1976) had previously shown that internode elongation in *Chenopodium album* was linearly related to the phytochrome photostationary state. They have recently demonstrated (Morgan, O'Brian, and Smith 1980), that far-red light, when

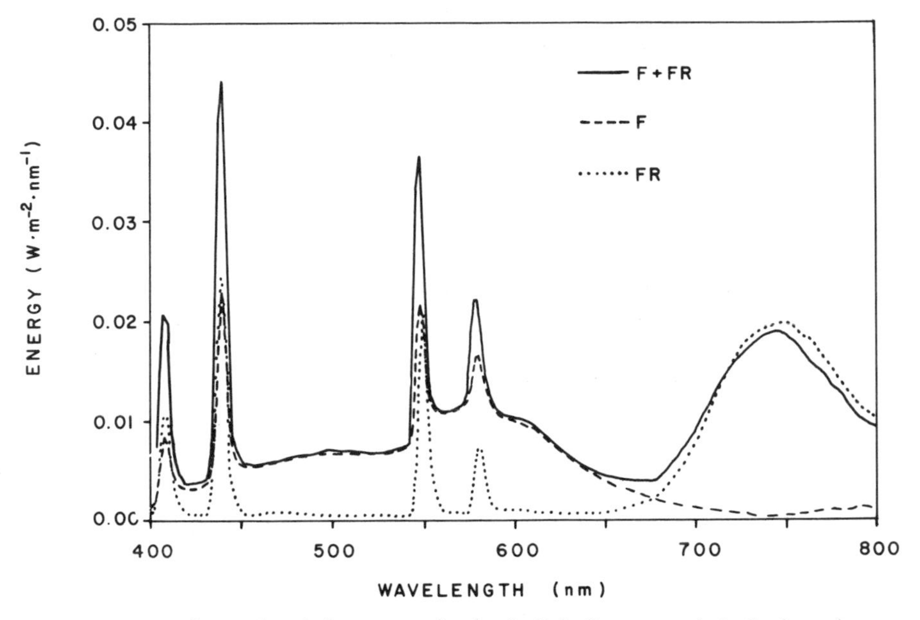

Figure 6.1. Spectral emission curves for the daylight fluorescent (---), the far-red phosphor fluorescent source (..........), and both combined (———). Curves were measured with a Schoeffel GM-100 spectral radiometer calibrated against an NBS standard tungsten source. (After Deitzer et al. 1979.)

added to white light, caused a very rapid increase in stem extension rate in *Sinapis alba*. Based on this information, as well as a large number of other measurements (see Morgan and Smith 1979 for a discussion), they propose that the function of phytochrome in the natural environment is to detect shading by increasing stem elongation in response to a relative increase in far-red energy.

EFFECT OF FAR-RED ENERGY ON FLOWERING

In order to examine the effect of far-red energy during the light portion of the photoperiod in more detail, light sources were chosen to optimize the difference between 660 nm and 730 nm while providing equal photon fluences between 400 nm and 700 nm (PAR). The emission spectra for the sources used throughout the work reported in this paper are presented in Figure 6.1. By combining light from ordinary daylight fluorescent lamps (DF) with that from single far-red–emitting phosphor fluorescent lamps (FR) in a 1:1 ratio, a light environment is produced that has about an order of magnitude lower red to far-red ratio when compared to daylight fluorescent alone (Table 6.1). Both, however, have the same total energy between 400 nm and 700 nm. The phytochrome photoequilibria are also reported in Table 6.1 and show that, although there is a large difference in red to far-red ratio, there is only a

Table 6.1 Measures and Calculated Values for the Photon Fluence Rate (660/730 nm) and Phytochrome Photoequilibria (Pfr/Ptot) Established by the Various Lamp Sources Described in Figure 6.1 for Both Dark- and Light-grown Garry Oat Seedlings

Lamp source	660/730	Pfr/Ptot		
		Calculated[a]	Measured	
			Dark-grown[b]	Light-grown[c]
Daylight fluorescent	5.6	0.73	0.68	0.63
Daylight fluorescent + far-red fluorescent	0.45	0.45	0.38	0.47
Far-red fluorescent	0.02	0.04	0.04	0.05

[a] From equation of Hartman (1966).
[b] Grown in dark for 5 days and exposed to 10 min of the light described.
[c] Grown for 6 days in the light described.

Source: G. F. Deitzer, R. Hayes, and M. Jabben (1979).

difference of about 20% in relative Pfr concentration. This is comparable to the range of difference reported by Holmes and Smith (1977a) for changes in spectral energy distribution around twilight, but is much less than that below a leaf canopy.

Wintex barley (*Hordeum vulgare* L.) is a facultative long-day plant (Altman and Dittmer 1964) that is quantitatively stimulated to flower by increasing daylengths. When grown under 12-hr daylight fluorescent photoperiods, floral initiation takes place between days 17 and 21 (Fig. 6.2). Growth in continuous daylight fluorescent light (24-hr photoperiods) results in initiation between days 6 and 8. This enhancement of flowering is essentially the same when far-red light is added to 12-hr daylight fluorescent photoperiods. An even greater effect is obtained when the far-red is present in the continuous light.

Even though a great deal of care was taken to minimize differences in total light energy available for photosynthesis, there was nevertheless an effect of both photoperiod and far-red light on total dry weight accumulation (Table 6.2). Since floral stage did not correlate easily with dry weight accumulation, apex length was used as a measure of flowering and the correlation is shown in Figure 6.3. The control 12-hr daylight fluorescent photoperiods remain linearly correlated throughout the course of the experiment. The apex length increase under both continuous daylight fluorescent, and 12-hr photoperiods with supplemental far-red, is six to seven times greater than that of the controls. Thus, enhanced flowering results in a stimulation of apical growth that is considerably greater than can be accounted for by increased photosynthetic carbon fixation. This response is biphasic, and the increased rate of elongation

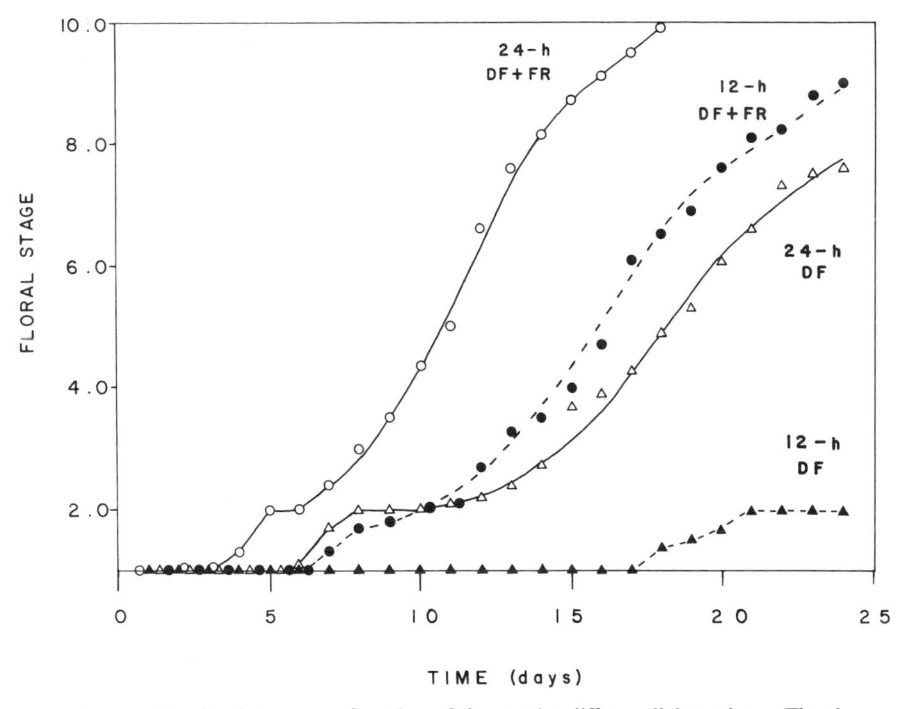

Figure 6.2. Floral stage as a function of time under different light regimes. Floral stage 1 is vegetative, floral stage 2 is the first fully transformed floral apex, and stages 3 to 10 are clearly defined morphological stages leading to the fully mature inflorescence (stage 10). After 8 days pretreatment (4 days dark, 4-day, 12-hr daylight fluorescent photoperiods), plants were grown continuously under either 12-hr daylight fluorescent (▲), 12-hr daylight fluorescent supplemented with far red (△), or 24-hr daylight fluorescent (●) and 24-hr daylight fluorescent supplemented with far-red (0) photoperiods. Each point is the average of five separate determinations of 10 plants each with standard errors of less than 5%. (After Deitzer et al. 1979.)

eventually reverts to the control rate. The stimulation of elongation is roughly additive when far red is given continuously, but is also biphasic. In this case, however, when the morphological development of the inflorescence is complete (stage 10 in Fig. 6.2), further vegetative growth ceases and the apex elongates very rapidly, leading to a much greater slope in the correlation.

Using the same correlation of apex length and dry weight, and varying the number of 24-hr photoperiods with added far-red (Fig. 6.4), it was found that each inductive photoperiod resulted in the same stimulation of the apex, but the duration of the enhanced elongation was dependent on the number of cycles. The elongation rate then reverts to that of the controls. This is true at least for three 24-hr cycles; further inductive cycles all cease elongating at the maximal rate at the same time, but the rate thereafter increases above that of the controls. Therefore, the initial stimulation of apical elongation is saturated

Table 6.2 Dry Weight Accumulation during Growth under Continuous Twelve- or Twenty-four- Hour Photoperiods with or without Supplemental Far-Red

Growth conditions	Dry weight (g/plant) measured on day			
	5	10	15	20
12-hr − far-red	0.032	0.056	0.103	0.150
12-hr + far-red	0.029	0.058	0.122	0.282
24-hr − far-red	0.037	0.074	0.147	0.276
24-hr + far-red	0.041	0.096	0.186	0.343

Note: Total PAR (400 nm − 700 nm) is equal under all conditions and standard errors are less than 5%.

Source: G. F. Deitzer, R. Hayes, and M. Jabben (1979).

by 72 hr of continuous daylight fluorescent light supplemented with far-red energy.

In order to examine the response to far-red light during this 72-hr period more closely, the effect of 72 hr of daylight fluorescent without far-red was compared to the same period with added far-red (Fig. 6.5) and examined after 21 days of further growth under 12-hr photoperiods. The addition of far-red results in about a two fold increase in both apex length and floral stage, but has no effect on dry weight accumulation. If the far-red is added for only 6-hr. at various times during this 72-hr. period, the ability of the plant to respond varies with a circadian rhythm. The maximal response occurs during the second 12 hr. of each 24-hr. period and results in stimulation that is about 50% of that obtained when the far-red is present throughout the 72 hr.

SPECTRAL SENSITIVITY OF THE FAR-RED RESPONSE

The remainder of this chapter deals solely with the response to this 6-hr pulse of far-red, added against a background of white daylight fluorescent light. The dependence of floral stage on the relative red to far-red ratio (660 nm/730 nm) is shown in Figure 6.6. The total energy between 400 nm and 700 nm was kept constant (200 ± 18 $\mu mol \cdot m^{-2} \cdot s^{-1}$) and only the level of far-red (700 nm to 800 nm) was varied. Under these conditions, ratios lower than 0.2 could not be obtained without lowering the energy below 700 nm. Also, the maximal red/far-red ratio (5.6) that could be obtained was set by that of the daylight fluorescent source alone. All plants were treated in the same manner as those in Figure 6.5; the only point tested was the effect of far-red light during a single 6-hr period given between hours 15 and 21. During this period, the response is maximal and invariant. Red to far-red ratios between 0.2 and 0.6 result in maximal stimulaton. The response decreases with increasing ratios from about 0.8 to about 2.0, where there is no increase above that of the controls with a ratio of 5.6. It is of interest to note that the response to a ratio of 1.2, the equivalent of solar daylight, is about 50% of the maximal response.

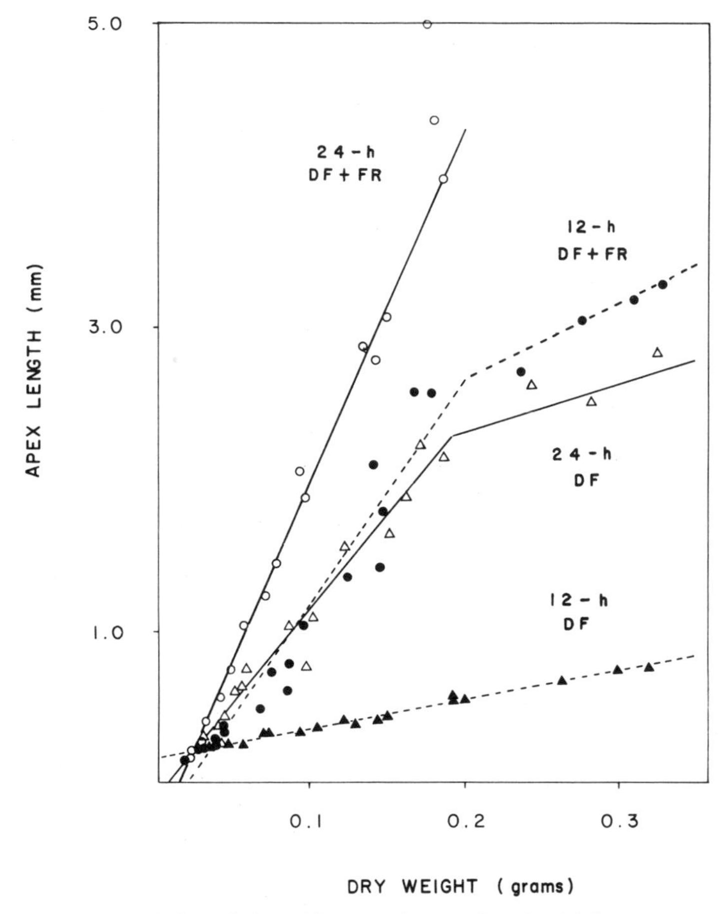

Figure 6.3. Correlation of increase in apex length with increase in dry weight when grown under 12-hr daylight fluorescent (▲), 12-hr daylight fluorescent supplemented with far-red (△), 24-hr daylight fluorescent (●) or 24-hr daylight fluorescent supplemented with far-red (0) photoperiods. Conditions the same as in Figure 6.2. (After Deitzer et al., 1979.)

If the red to far-red ratio is fixed at 0.5 and the total energy is varied during this 6-hr period, the fluence response curve shown in the upper portion of Figure 6.7 is obtained. The data are reported as the fluence of the far-red component (700 nm–800 nm) given during this 6-hr period. The fluence is, of course, much higher when the 400 nm to 700 nm daylight fluorescent component is considered. The flowering response increases exponentially with increasing fluences and saturates at about $5.0 \times 10^4 J \cdot m^{-2}$. This, and subsequent fluence response curves, were generated by increasing the fluence rate for a fixed 6-hr period. When the time of irradiation is varied with a fixed fluence rate, flowering does not respond reciprocally (unpublished data). For example,

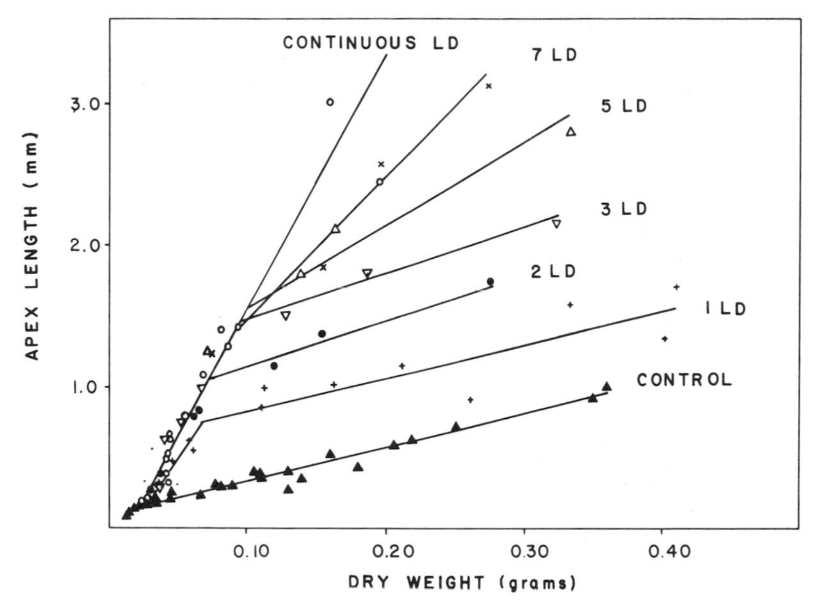

Figure 6.4. Correlation of apex length with dry weight following treatment with varying numbers of 24-hr long-day photoperiods supplemented with far-red light prior to return to 12-hr short-day photoperiods without supplemental far-red light. Control, continuous 12-hr short days (▲), 1 long day (+), 2 long days (●), 3 longs days (▽), 5 longs days (△), 7 longs days (X) and continuous 24-hr long days (0). (After Deitzer et al., 1979.)

a 3-hr period at twice the fluence rate is not significantly above the control level. Attempts to pulse the light with varying fluence rates, and varying on/off duty cycles, over the entire 6-hr period were only partially successful due to the difficulty of rapidly pulsing light at such high fluence rates. Although still preliminary, the data suggest that reciprocity does hold with pulses shorter than 1 min on/1 min off, as predicted by Mancinelli and Rabino (1975).

The data presented in the lower portion of Figure 6.7 show the effect of these increasing fluence rates on the net uptake of CO_2 by photosynthesis. For comparative purposes, these data are plotted on the same basis as the upper curve, but the photon fluence rates between 400 nm and 700 nm were, at the same time, varied from 0 to 500 $\mu mol \cdot m^{-2} \cdot sec^{-1}$) (Fig. 6.8). Unlike the flowering response, photosynthesis is not saturated even at the highest fluence rates (based on whole plant measurements with an IR gas analyzer). Also, the net rate of CO_2 uptake is still below the CO_2 compensation point at fluences sufficient to saturate the flowering response. Hence, although the conditions used in these experiments affect both photosynthesis and flowering, the response is quantitatively very different. It is, therefore, unlikely that photosynthesis plays a significant role in the enhancement of flowering by far-red light.

In order to compare the sensitivity of the flowering response at different

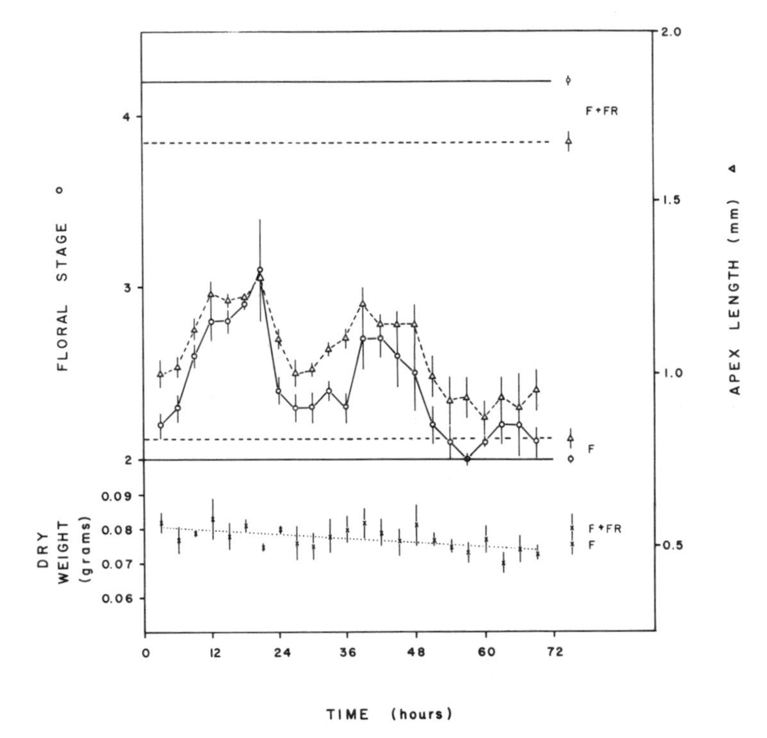

Figure 6.5. Effect on floral stage (0), apex length (△), and dry weight (X) of 6 hr of far-red light added at various times to a continuous 72-hr daylight fluorescent period which was inserted to interrupt 12-hr photoperiods. Flowering was measured 21 days following the onset of treatment. Points are plotted at the center time of each 6-hr treatment. Error bars are the standard errors of the mean of 20–50 replicate determinations per point. (After Deitzer et al., 1979.)

times during the circadian rhythm reported in Figure 6.5 the fluence response relationships were tested at a point of maximal response (between hours 15 and 21) and at a point of minimal response (between hours 3 and 9) (Fig. 6.9). The fluence response curve was the same as that reported in Figure 6.7 for the 6-hr period between hours 15 and 21, but required much higher fluences to produce a response during the 6-hr period between hours 3 and 9. Even far-red fluences as high as 8.8×10^5 J·m^{-2} did not saturate the response during this period. If the saturation level is assumed to be the same at both times and the fluence required to reach 50% saturation is compared, there is about a 50-fold difference ($1.35 \times 10^4 - 6.60 \times 10^5$ J·m^{-2}). Thus, the rhythmic flowering response is a function of a very large change in the sensitivity of the plant to far-red energy during the course of the day.

All the previous fluence response data were obtained by varying the total white light energy with a fixed relative amount of red and far-red energy. Figure 6.10 shows the fluence responses obtained using monochromatic energy

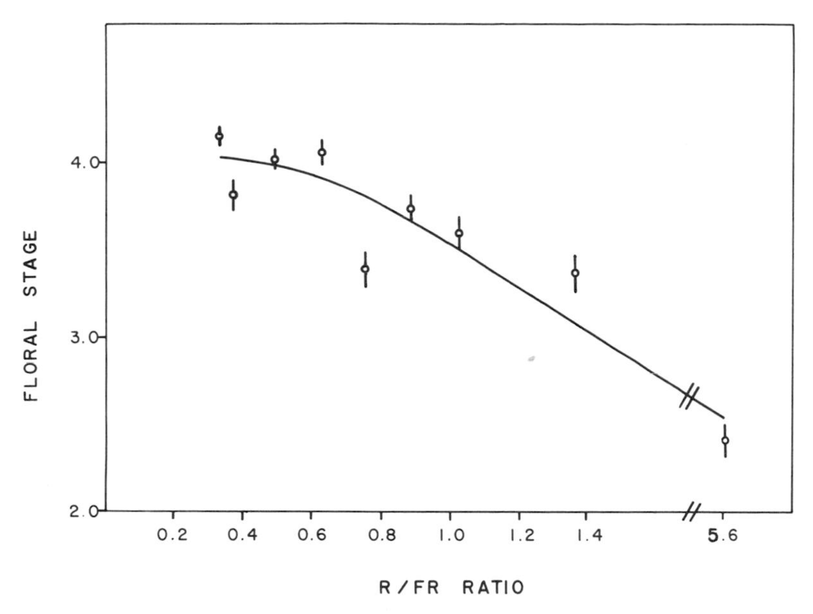

Figure 6.6. Effect of 6 hr of light with various red to far-red ratios on floral stage measured 21 days after the onset of treatment. The 6-hr period was given between hours 15 and 21 with conditions otherwise the same as the controls in Figure 6.5. Error bars are the standard errors of the mean of 50 replicate determinations.

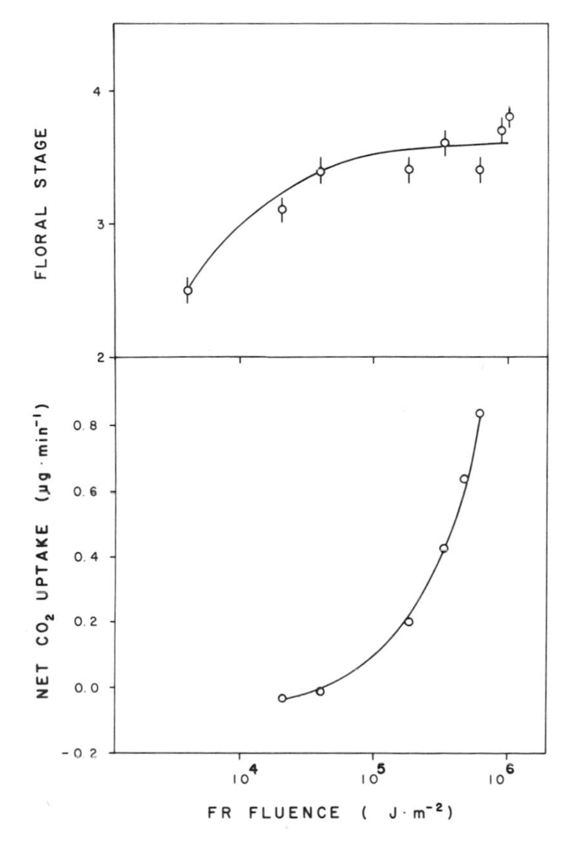

Figure 6.7. Upper curve: Fluence response of floral stage dependence on the far-red fluence given for a 6-hr period between the hours of 15 and 21 with conditions otherwise the same as the controls in Figure 6.5. The red to far-red ratio was held constant at 0.5 and the total energy from 400 nm to 800 given for 6 hours was varied. The data, however, are plotted as a function of only the far-red fluence (700 nm–800 nm). Error bars represent the standard errors of the mean of 50 replicate determinations. *Lower curve:* Fluence response of net CO_2 uptake for light with fluence rates the same as those used in the upper curve. CO_2 was measured by infrared gas analysis for a population of 10 plants under increasing fluence rates of light with a red to far-red ratio of 0.5 for a 15-min period at each point. Measurements are based on whole plant determinations under conditions identical to those for plants treated in the upper curve.

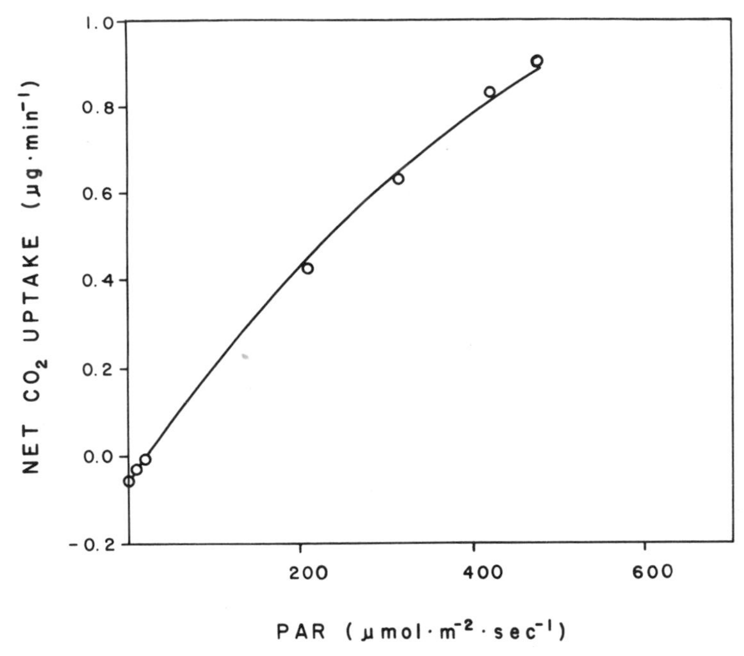

Figure 6.8. Fluence response of net CO_2 uptake for light with fluence rates the same as those reported in the lower half of Figure 6.7, but plotted as a function of the PAR (400 nm–700 nm) fluence rates.

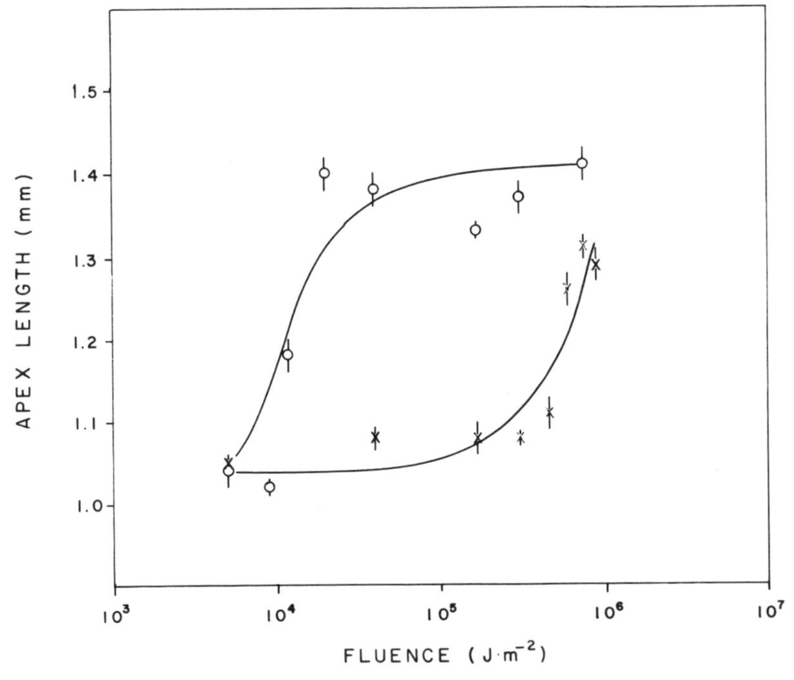

Figure 6.9. Fluence response of apex length to light given for a 6-hr period between the hours of 3 and 9 (X) and 15 and 21 (0) with conditions otherwise the same as the controls in Figure 6.5. Fluences are total far-red fluences with light having a red to far-red ration of 0.5, as in the upper curve of Figure 6.7.

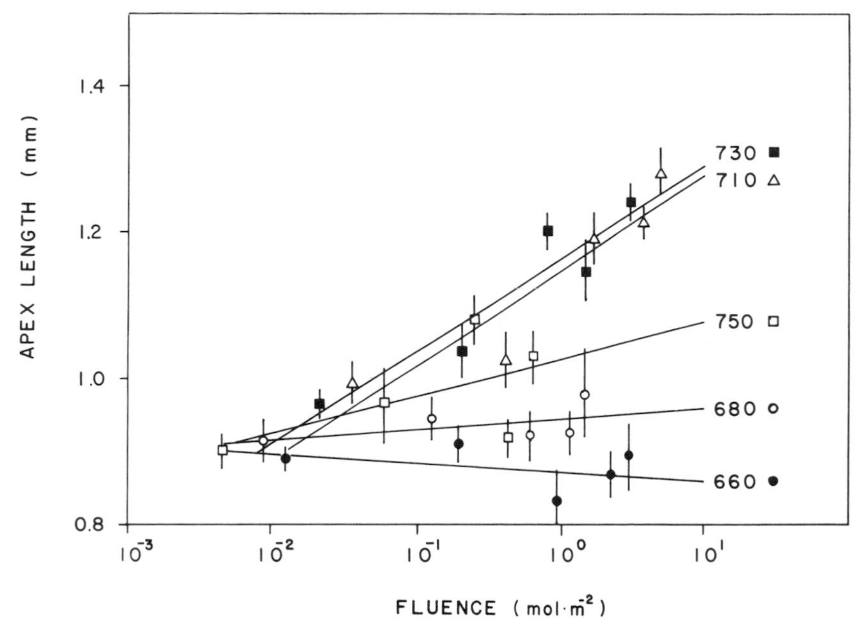

Figure 6.10. Fluence response of apex length to monochromatic light given for a 6-hr period between the hours of 15 and 21 with conditions otherwise the same as the controls in Figure 6.5. Total photon fluence rates of 660 nm (●), 680 nm (0), 710 nm (△), 730 nm (■), and 750 nm (□) were produced by varying a rheostat controlling a 1500 W tungsten filament lamp and were measured with a Gamma Scientific C-3 spectral radiometer calibrated against an NBS standard tungsten source. All wavelengths arc $\lambda_{max} \pm$ 5 nm half bandwidth from Baird-Atomic interference filters. Error bars are the standard errors of the mean of 10 replicate determinations.

during a 6-hr period given between the hours of 15 and 21 (the period of maximal response). Irradiations were carried out in modified Withrow monochromators (Withrow 1957) by filtering light from a 1500 W Tungsten filament source through 10 cm of water containing 0.5% ferrous ammonium sulfate, 2 mm of Schott KG-1 heat-absorbing glass, and Baird-Atomic DIL interference filters with half bandwidths of 10 nm. Only 710 and 730 nm (and to a much less extent, 750 nm) energy showed any response, even with photon fluences as high as 5 mol·m$-^{2}$ (40 J·m^{-2}·sec^{-1} for 6 hr or 8.6 × 10^5 J·m^{-2} total at 710 nm). The slight negative slope at 660 nm is not significant. The response was linear over at least two decades, but did not reach saturation. Due to the relatively weak response and the very high fluence requirements, no attempt was made to further characterize the action spectrum.

CONCLUSIONS

The fluence responses for monochromatic light in Figure 6.10 are generally consistent with the action spectrum of Schneider, Borthwick, and Hendricks

(1967). They, however, reported a single very sharp peak at 710 nm. Since 710 nm and 730 nm light in this study are equally effective, and there is even some response with 750 nm light, the peak in the far-red would have to be much broader. Also, if the response were due solely to the far-red component, and 5×10^4 J·m^{-2} is required for saturation even when the plant is maximally sensitive, then monochromatic light with a 10 nm half-bandwidth would require at least an order of magnitude more energy to saturate the response. Since 8.6×10^5 J·m^{-2} of 710 nm light is insufficient to reach saturation, light energy below 700 nm must contribute substantially to the response. Since there is no measurable response to 660 nm or 680 nm, we assume that the additional energy occurs in the blue region of the spectrum.

Full sunlight on a clear day contains about 8.0 J·m^{-2}·sec^{-1} of far-red energy between 700 nm and 800 nm. If supplied constantly for 6 hr, this would provide about 1.7×10^{-5} J·m^{-2}, which would be sufficient to saturate the response during a period of maximal sensitivity, but not during a period of minimal sensitivity. Such high fluences may be expected only around solar noon which, on the longest day in the Washington, D.C., area, would range between 4.5 and 10.5 hr following sunrise. If time 0 in Figure 6.5 is equivalent to sunrise, the plants would be minimally sensitive during this period. The relative red to far-red ratio during this period is also suboptimal (about 1.2, according to Holmes and Smith 1977a).

It might, therefore, be concluded that such a control mechanism could not function in the natural environment. A number of other factors however must be taken into account. As the solar fluence rate decreases toward twilight, there may be a compensatory shift toward more optimal red to far-red ratios (Holmes and Smith 1977a). The data presented here show that even such a slight change in ratio may result in a twofold increase in the response. While the amount of far-red light around noon is not sufficient to saturate the response during that period, it is sufficient to produce more than 50% of the maximal response. Also, there would be an effect of chlorophyll screening on relative red to far-red ratio, as described by Holmes and Smith (1977b), in such a way that lower leaves would receive much more far-red than the upper leaf. Finally, far-red light also affects the timing mechanism itself, causing the phase of maximal sensitivity to occur much earlier in the photoperiod (Deitzer, Hayes, and Jabben 1982).

We conclude, therefore, that the control of flowering in Wintex barely is mediated by a circadian rhythm in the sensitivity of the plant to the far-red component of solar daylight. The photoreceptor for absorption of the energy is assumed to be phytochrome, acting in some way through what has been described as the "high irradiance response" (Schneider, Borthwick, and Hendricks 1967). The change in sensitivity is suggested to be a function of either the availability, or affinity, of whatever must interact with phytochrome to produce a response.

Finally, it is tempting to further suggest, as DeLint (1960) has done, that the mechanism of phytochrome control in photoperiodism may be the same as that described by Morgan, O'Brian, and Smith (1980) for the control of stem elongation by far-red light. They postulated that the rapid increase in elongation rate may be mediated by a phytochrome-controlled release of gibberellin (Cooke, Saunders, and Kendrick 1975; Evans and Smith 1976). This is a particularly attractive possibility in view of a recent paper by Williams and Morgan (1979) that reported that the capacity of gibberellic acid to induce flowering in sorghum was enhanced by far-red light. Thus, it is conceivable that this phytochrome-mediated event in the control of photoperiodism results in a release of gibberellin that may then promote flowering directly, or lead to the production of some other flower-promoting substance. This could explain the requirement for the action of far-red light over a full 6-hr period and the failure of reciprocity with shorter periods. Therefore, while gibberellin itself probably does not lead to floral induction initiation (Cleland and Zeevaart 1970), it may be responsible for subsequent promotion.

LITERATURE CITED

Altman, P. L., and D. S. Dittmer, eds. 1964. *Biology Data Book*. Fed. Amer. Soc. for Exper. Biol., Bethesda, MD.

Borthwick, H. A., S. B. Hendricks, and M. W. Parker. 1948. *Action spectrum for photoperiodic control of floral initiation of a long day plant, Wintex barley (Hordeum vulgare)*. Bot. Gaz. 110:103–18.

———. 1952. *The reaction controlling floral initiation*. Proc. Nat. Acad. Sci. USA 38:929–34.

Borthwick, H. A., S. B. Hendricks, M. W. Parker, E. H. Toole, and V. K. Toole. 1952. *A reversible photoreaction controlling seed germination*. Proc. Nat. Acad. Sci. USA 38:662–66.

Borthwick, H. A., S. B. Hendricks, M. J. Schneider, R. B. Taylorson, and V. K. Toole. 1969. *The high energy light action controlling plant responses and development*. Proc. Nat. Acad. Sci. USA 64:479–86.

Borthwick, H. A., and M. W. Parker. 1938. *Photoperiodic perception in Biloxi soybeans*. Bot. Gaz. 100:374–87.

———. 1952. *Light in relation to flowering and vegetative development*. Proc. 13th Int. Hort. Cong., London. Pages 801–10.

Butler, W. L., K. H. Norris, H. W. Siegelman, and S. B. Hendricks. 1959. *Detection, assay, and preliminary purification of the pigment controlling photoresponsive development of plants*. Proc. Nat. Acad. Sci. USA 45:1703–8.

Cleland, C. E., and J. A. D. Zeevaart 1970. *Gibberellins in relation to flowering and stem elongation in the long day plant Silene armeria*. Plant Physiol. 46:392–400.

Cooke, R. J., P. F. Saunders, and R. E. Kendrick. 1975. *Red light induced production of GA-like substances in homogenates of etiolated wheat leaves and in suspensions of intact etioplasts*. Planta 125:319–28.

Deitzer, G. F., R. Hayes, and M. Jabben. 1979. *Kinetics and time dependence of the effect of far red light on the photoperiodic induction of flowering in Wintex barley*. Plant Physiol. 64:1015–21.

——— 1982. *Phase shift in the circadian rhythm of floral promotion by far red light in Hordeum vulgare L.* Plant Physiol. in press.

DeLint, P. J. A. L. 1958. *Stem formation in Hyoscyamus niger under short days including supplementary irradiation with near infrared*. Meded. Landbouwhogesch. Wageningen 58:1–5.

———. 1960. *An attempt to analysis of the effect of light on stem elongation and flowering in Hyoscyamus niger L.* Meded. Landbouwhogesch. Wageningen 60:1–59.

Downs, R. J., A. A. Piringer and G. A. Wiebe. 1959. *Effects of photoperiod and kind of supplemental light on growth and reproduction of several varieties of wheat and barley*. Bot. Gaz. 120:170–77.

Evans, A., and H. Smith. 1976. *Localization of phytochrome in etioplasts and its regulation in vitro of GA-levels*. Proc. Nat. Acad. Sci. USA 73:138–42.

Evans, L. T. 1976. *Inflorescence initiation in Lolium temulentum L. XIV. The role of phytochrome in long day induction*. Austral. J. Plant Physiol. 3:207–17.

Friend, D. J. C., V. A. Helson, and J. E. Fisher. 1961. *The influence of the ratio of incandescent to fluorescent light on the flowering response of Marquis wheat grown under controlled conditions*. Can. J. Plant Sci. 41:418–27.

Garner, W. W., and H. A. Allard. 1920. *Effect of the relative length of day and night and other factors of the environment on growth and reproduction in plants*. J. Agric. Res. 18:553–606.

———. 1931. *Effect of abnormally long and short alternations of light and darkness on growth and development of plants*. J. Agric. Res. 42:629–51.

Goldberg, B., and W. H. Klein. 1977. *Variations in the spectral distribution of daylight at various geographical locations on the earth's surface*. Sol. Energy 19:3–13.

Hamner, K. C. 1940. *Inter-relation of light and darkness in photoperiodic induction*. Bot. Gaz. 101:658–87.

Hamner, K. C., and J. Bonner. 1938. *Photoperiodism in relation to hormones as factors in floral initiation and development*. Bot. Gaz. 100:388–431.

Hartmann, K. M. 1966. *A general hypothesis to interpret "high energy phenomena" of photomorphogenesis on the basis of phytochrome*. Photochem. Photobiol. 5:349–66.

Holland, R. W. K., and D. Vince. 1971. *Floral initiation in Lolium temulentum L.: The role of phytochrome in the responses to red and far red light*. Planta 98:232–43.

Holmes, M. G., and H. Smith. 1977a. *The function of phytochrome in the natural environment. I. Characterization of daylight for studies in photomorphogenesis and photoperiodism*. Photochem. Photobiol. 25:533–38.

———. 1977b. *The function of phytochrome in the natural environment. II. The influence of vegetation canopies on the spectral energy distribution of natural daylight*. Photochem. Photobiol. 25:539–46.

José, A. M., and D. Vince-Prue. 1978. *Phytochrome action: a reappraisal*. Photochem. Photobiol. 27:209–16.

Klebs, G. 1913. *Über das Verhältnis der Assenwelt zur Entwicklung der Pflanze*. Sitz. ber. Acad. Wiss. Heidelberg Ser. B. No. 5.

Lane, H., H. M. Cathey, and L. T. Evans. 1965. *The dependence of flowering in several long day plants on the spectral composition of light extending the photoperiod*. Amer. J. Bot. 52:1006–14.

Liverman, J., and J. Bonner. 1953. *The interaction of auxin and light in the growth responses of plants*. Proc. Nat. Acad. Sci. USA 39:905–16.

Mancinelli, A. L., and I. Rabino. 1975. *Photocontrol of anthocyanin synthesis. IV. Dose dependence and reciprocity relationships*. Plant Physiol. 56:351–55.

———. 1978. *The "high irradiance responses" of plant photomorphogenesis*. Bot. Rev. 44:129–80.

Morgan, D. C., T. O'Brian, and H. Smith. 1980. *Rapid photomodulation of stem extension in light-grown Sinapis alba L. Studies on kinetics, site of perception and photoreceptor*. Planta 150:95–101.

Morgan, D. C., and H. Smith. 1976. *Linear relationship between phytochrome photoequilibrium and growth in plants under simulated natural radiation*. Nature 262:210–12.

———. 1979. *A systematic relationship between phytochrome-controlled development and species habitat for plants grown in simulated natural radiation*. Planta 145:253–58.

Parker, M. W., S. B. Hendricks, and H. A. Borthwick. 1950. *Action spectrum for the photoperiodic control of floral initiation of the long day plant Hyoscyamus niger*. Bot. Gaz. 111:242–52.

Parker, M. W., S. B. Hendricks, H. A. Borthwick, and N. J. Scully. 1945. *Action spectrum for the photoperiodic control of floral initiation in Biloxi soybean.* Science 102:152–55.

———. 1946. *Action spectrum for the photoperiodic control of floral initiation of short day plants.* Bot. Gaz. 108:1–26.

Piringer, A. A., and H. M. Cathey. 1960. Effects of photoperiod, kind of supplemental light, and temperature on growth and flowering of petunia plants. Proc. Amer. Soc. Hort. Sci. 76:649–60.

Schneider, M. J., H. A. Borthwick, and S. B. Hendricks. 1967. *Effects of radiation on flowering of Hyoscyamus niger.* Amer. J. Bot. 54:1241–49.

Shropshire, W., Jr. 1972. *Phytochrome, a photochromic sensor. Pages 34–67 in Giese, ed. Photophysiology,* vol. 7. Academic Press, New York.

Shropshire, W., Jr. 1973. *Photoinduced parental control of seed germination and the spectral quality of solar radiation.* Sol Energy 15:99–105.

Smith, H., and M. G. Holmes. 1977. *The function of phytochrome in the natural environment. III. Measurement and calculation of phytochrome photoequilibria.* Photochem. Photobiol. 25:547–50.

Stolwijk, J. A. J., and J. A. D. Zeevaart. 1955. *Wavelength dependence of different reactions governing flowering in Hyoscyamus niger.* Vehr. K. Akad. Wet. 58:386–96.

Takimoto, A. 1957. *Photoperiodic induction in Silene armeria as influenced by various light sources.* Bot. Mag. (Tokyo) 70:312–21.

———. 1961. *On the light controlling flower initiation of Silene armeria.* Plant Cell Physiol. 2:71–75.

Tournois, J. 1914. *Sexualité du Houblon.* Ann. Sci. Nat. Bot. (Paris) 19:49–191.

Vince, D. 1965. *The promoting effect of far red light on flowering in the long day plant Lolium temulentum.* Physiol. Plant. 18:474–82.

Vince, D., J. Blake, and R. Spencer. 1964. *Some effects of wavelength of the supplementary light on the photoperiodic behavior of the long day plants, carnation and lettuce.* Physiol. Plant. 17:119–25.

Wassink, E. C., J. A. J. Stolwijk, and A. B. R. Beemster. 1951. *Dependence of formation and photoperiodic reactions in Brassica rapa var. Cosmos and Lactuca on wavelength and time of irradiation.* Proc. Koninkl. Nederl. Akad. Wet. 54:421–32.

Williams, E. A., and P. Morgan. 1979. *Floral initiation in sorghum hastened by gibberellic acid and far red light.* Planta 145: 269–72.

Withrow, R. B. 1957. *An interference filter monochromator system for the irradiation of biological material.* Plant Physiol. 32:355–60.

7] Molecular Properties of Phytochrome and Their Relationship to Phytochrome Function

by LEE H. PRATT*

ABSTRACT

Light absorbed by the plant chromoprotein phytochrome influences plant reproduction in two obvious ways. First, phytochrome controls germination of light-sensitive seed. Second, phytochrome interacts with the time-measuring system of photoperiodism and thereby influences flowering in photoperiodically sensitive plants. This chapter concentrates on molecular aspects of phytochrome and how they relate to these and other physiological responses. Emphasis is given to reviewing information concerning differences between the inactive form of phytochrome, Pr, and the active form, Pfr. It is presumably one or more of these differences that provide the answer to the question of how Pfr functions.

INTRODUCTION

Light can be a critical factor in plant reproduction in two distinctly different ways. Light in some instances promotes or inhibits seed germination (Frankland 1976). Light also plays a role in the expression of flowering in photoperiodically sensitive plants (Vince-Prue 1975). Both of these roles of light in influencing plant reproduction are discussed elsewhere in this volume (Chapters 5 and 8).

To understand fully how light influences seed germination and flowering, one must of course elucidate the entire sequence of events leading from photoreception to the final response, which may not occur until days or even weeks after reception of the inductive light stimulus. One may study this sequence of events from two distinctly different perspectives: either begin with the final

*Botany Department, University of Georgia, Athens, Georgia 30602.

STRATEGIES OF PLANT REPRODUCTION (BARC Symposium number 6—Werner J. Meudt, ed.)
Allanheld, Osmun, Totowa

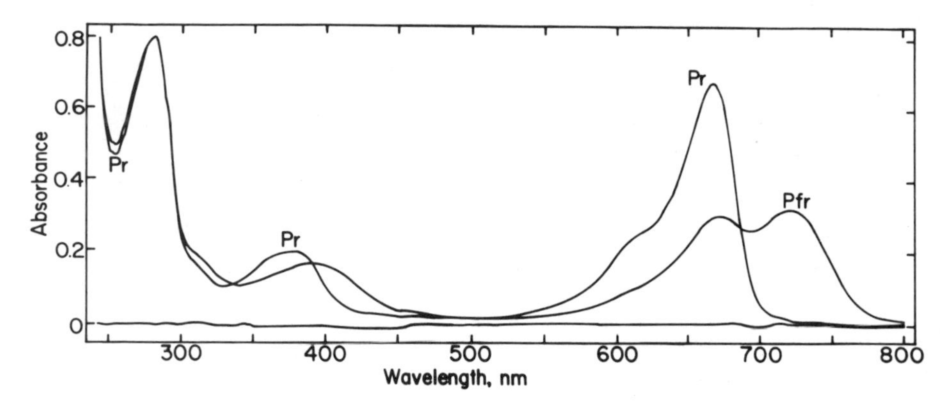

Figure 7.1. Absorption spectra of immunopurified, large oat phytochrome after saturating irradiation with far-red (Pr) or red (Pfr) light. (Data are from Hunt and Pratt, 1979a.)

display (i.e., germination or flowering) and work backwards, or, alternatively, begin with the photoreceptor (i.e., phytochrome) and work forwards. I have chosen to follow this second approach. The subject of this chapter, while possibly seeming far removed from the subject of plant reproduction, is nevertheless an integral part of the overall process.

Phytochrome is a large chromoprotein that exists in solution as a dimer of 120,000-dalton monomers (see Pratt 1982b for review). The molecule is synthesized apparently in a physiologically inactive form (Pr) that absorbs visible light maximally in the red spectral region (Fig. 7.1). Upon absorption of light, Pr transforms to a physiologically active form (Pfr) that absorbs relatively weakly in the red spectral region and has a new absorption band in the far-red (Fig. 7.1). Upon absorption of light, Pfr transforms back to Pr; light therefore produces an endless cycling of the pigment between its inactive and active forms. A number of questions arise from consideration of these simple observations. (a) How does phytochrome get from one form to the other? (b) Do transformation intermediates have any biological activity? (c) Why does Pfr phototransform back to Pr, thereby producing cycling between the two forms? (d) What is the difference between Pr and Pfr that gives Pfr its unique biological activity? The purpose of this chapter will be to address from a molecular perspective these and other questions concerning phytochrome function.

The specific goals of this chapter are three fold: (1) to review the photochemical reactions of phytochrome and relate them to the role of phytochrome in the natural environment; (2) to review the nonphotochemical reactions of phytochrome and relate them to the possible action of phytochrome; and (3) to present recent evidence concerning differences between Pr and Pfr.

Prior to beginning a discussion of phytochrome it is necessary to define two terms that will permit easy reference to the two sizes of phytochrome that have been characterized in vitro. Phytochrome is unusually susceptible to endopeptidases (Gardner et al. 1971) that are present in crude plant extracts (Pike and Briggs 1972b). Consequently, much work with phytochrome has been done with a photoreversible chromopeptide whose size of 60,000 daltons is approximately 50% that of the native molecule (see Pratt 1982b for further discussion). Proteolytically degraded phytochrome of this size will be termed small phytochrome. Phytochrome that is wholly or largely undegraded has a monomer size of about 120,000 daltons and will be referred to as large phytochrome. Recent evidence indicates that even large phytochrome may already have lost about 50 amino acids via a rapid, limited proteolysis (see Pratt 1982b for discussion) and may itself therefore already be partially degraded. Nevertheless, it is evident that, at a minimum, large phytochrome more closely represents the form of phytochrome found in vivo than does small phytochrome.

PHOTOCHEMICAL REACTIONS OF PHYTOCHROME

Phototransformation pathways. Phytochrome phototransformations occur in highly purified preparations (Rice and Briggs 1973a; Hunt and Pratt 1979a; Smith and Daniels 1981). They therefore do not involve the obligatory participation of any cofactor, activator, or reaction partner, as had originally been postulated (Borthwick et al. 1952). The pathways of these phototransformations have been studied by three methods: (1) flash activation analysis; (2) cryogenic spectrophotometry; and (3) intermediate trapping during cycling of phytochrome between its two forms (see Pratt 1979 for review). While the flash activation method has provided the most detailed information about these pathways, the data obtained by all three methods have yielded a single consistent picture of what happens. Upon absorption of light, Pr enters an excited electronic state that decays, possibly via one or more as yet unidentified intermediates, within less than 60 ns at 2°C (Cordonnier, Mathis, and Pratt 1981) to the first detectable transformation intermediate (Linschitz et al. 1966; Shimazaki et al. 1980; Pratt et al. 1982), which has maximum absorbance in a difference spectrum at about 700 nm. This intermediate in turn decays to a relatively bleached intermediate form by at least two reactions that occur in the microsecond time range (Linschitz et al. 1966; Shimazaki et al. 1980; Cordonnier, Mathis, and Pratt 1981). Finally, this bleached intermediate decays to Pfr by multiple reactions in the millisecond to second time range (Linschitz et al. 1966; Cordonnier, Mathis, and Pratt 1981; Pratt et al. 1982). The reverse pathway has so far been described in detail only for small phytochrome. Small Pfr enters an excited electronic state that decays within less than a few microseconds (Linschitz et al. 1966) to an intermediate with maximum absorbance near 650 nm in a difference spectrum. This intermediate then decays to a relatively bleached form, which in turn decays to Pfr. While some superficial similarities between these two opposing transformation pathways might be noted, it is evident that they do not involve any common intermediates.

Can a phytochrome phototransformation intermediate be biologically active? The answer, at least in principle, is yes. As Briggs and Fork (1969a,b) demonstrated originally, and Kendrick and Spruit (1972, 1973) and Spruit (1982) have confirmed, continuous irradiation of phytochrome leads to the formation of a substantial steady state concentration of one or more transformation intermediates in the Pr to Pfr pathway. Kendrick and Spruit (1972) have demonstrated further that as much as 30% of the total phytochrome pool in vivo can be present in intermediate form in conditions found in nature. The three most important environmental factors that control this intermediate level are temperature (since the intermediate reactions are temperature dependent while the initial photochemical events are not), incident photon fluence rate, and relative balance between red (which drives Pr to Pfr) and far-red (which drives Pfr to Pr) photon fluence rates (Kendrick and Spruit 1973). Transformation intermediate concentration is thus a function of both fluence rate and wavelength, two features of the high irradiance responses (Mancinelli and Rabino 1978).

Phytochrome photoequilibria. Phytochrome phototransformations are initiated by self-absorbed light. Because phytochrome absorption spectra overlap (Fig. 7.1), irradiation of phytochrome therefore produces a photostationary equilibrium between Pr and Pfr that is a function of wavelength. This equilibrium relationship can be measured only with purified phytochrome in vitro under well-defined conditions (Butler, Hendricks, and Siegelman 1964b). The original measurement of the amount of Pfr present at photoequilibrium in red light (Pfr^R_∞) was made long before the problem of proteolysis (see above) was appreciated. Consequently, Pfr^R_∞ was almost certainly measured first with the artifactual, small form of phytochrome (Pratt 1975; see Pratt 1979 for discussion). Nevertheless, the value obtained (0.8) was used for so long that it is still widely used now, even though it is evident that there is no longer any firm theoretical or empirical basis to justify its continued use.

The first remeasurement of Pfr^R_∞ with large phytochrome gave a value of 0.75 (Pratt 1975). Because Pratt also obtained the value of 0.8 for small phytochrome, it was possible to conclude that the difference between 0.75 and 0.8 was significant and did not reflect some methodological difference between his measurements and those of Butler, Hendricks, and Siegelman (1964b). More recently, Yamamoto and Smith (1981b) reported a value of 0.84. Unfortunately, they did not repeat the measurement with small oat phytochrome to exclude the problem of a potential methodological difference. Nevertheless, it is evident that there is no independent justification for the continued use of 0.8 for Pfr^R_∞. One should use a value obtained for large phytochrome or, since there is at present disagreement over what that value is and since there is also the possibility that it is variable, one should perhaps use no value at all. Instead, it might be more appropriate to express results on the basis of relative Pfr values.

Given Pfr^R_∞, absorption spectra for phytochrome after saturating red and far-red irradiations (Fig. 7.1), and the reasonable assumption that the phototransformation quantum yields are not wavelength dependent, Pfr_∞ may be calculated as a function of wavelength. $Pfr^R_\infty = 0.75$ is used here because this value was obtained for large phytochrome preparations most comparable to that

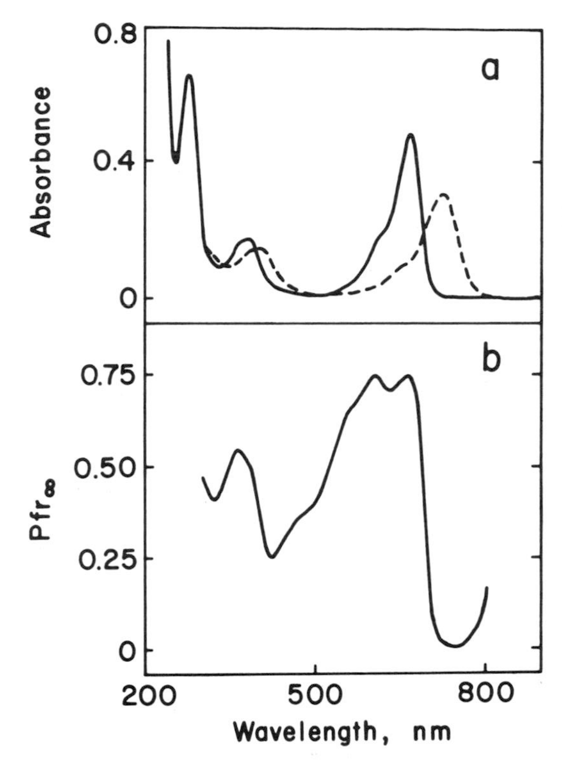

Figure 7.2. (a) Corrected extinction spectra for large oat phytochrome. These spectra were calculated for equimolar solutions of Pr (solid line) and Pfr (dashed line), as described elsewhere (Pratt 1979), based upon the assumption that Pfr_∞ at 665 nm = 0.75 and Pfr_∞ at 724 nm = 0.02. *(b)* Pfr_∞ as a function of wavelength. Pfr_∞ was calculated from the extinction spectra in (a) as described by Pratt (1979). Both the spectra and Pfr_∞ as a function of wavelength were calculated by a microcomputer utilizing spectral data that were obtained with an Hitachi 557 spectrophotometer and were stored in digital form at the time they were obtained.

used for these calculations. The first step is to calculate absorption spectra for equimolar concentrations of Pr and Pfr, uncontaminated by the contributions made by the opposite form that is inevitably present at photoequilibrium (Fig. 7.2a). Given these calculated spectra, Pfr_∞ as a function of wavelength may then be determined (Fig. 7.2b). Pfr_∞ is relatively large throughout the visible spectrum but drops rapidly at wavelengths greater than about 680 nm. Because the calculations were made with spectra stored in a microcomputer, it was possible to extend the determination farther into the far-red. Beyond about 745 nm, Pfr_∞ increases rather than decreases, as might have been anticipated (e.g., Schäfer, Schmidt, and Mohr 1973). This apparent increase in Pfr_∞ at long wavelength has, however, been observed only with a single spectrophotometer (an Hitachi 557 interfaced with a Hewlett-Packard 9845 computer), and the possibility must be considered that it may be an instrumental artifact. It will be of interest to test this observation with other spectrophotometers since, if valid, it will have important implications with respect to the interpretation of physiological data obtained with long wavelength monochromatic light.

This brief discussion of phytochrome photoequilibria has so far emphasized its obvious wavelength dependence. One must not, however, overlook the fact that Pfr_∞ will also be fluence rate dependent at fluence rates found in nature

(Kendrick and Spruit 1973). Pfr_{∞} becomes fluence rate dependent because of the relatively large proportion of the total phytochrome pool that exists as intermediates at high fluence rates and because the steady state level of these intermediates is a function of fluence rate (Kendrick and Spruit 1973).

Phytochrome as a sensor of mutual shading. Over the past several years Professor H. Smith and his colleagues have been examining the role of phytochrome in the natural environment (see H. Smith 1981 for review). Their work, as well as that of others, has led to answers to two questions concerning phytochrome properties: (1) Why is the absorption spectrum of Pr so similar to that of chlorophyll? (2) Why is the photoconversion of Pr to Pfr photoreversible?

Holmes and Smith (1975) have demonstrated that phytochrome photoequilibria, as might be expected, are affected markedly by lighting conditions found in nature, especially when comparing equilibria obtained in open-field situations to those obtained under plant canopies (Table 7.1). Because Pr absorbance is so similar to chlorophyll absorbance, light that has been filtered by a chlorophyll-containing plant canopy will be deficient in those wavelengths that are absorbed preferentially by Pr. The consequence, of course, is that the Pfr to Pr transformation predominates under these conditions and Pfr_{∞} becomes smaller. Furthermore, if phytochrome phototransformation were not photoreversible, then photostationary equilibria could not be established. Phytochrome could then no longer detect mutual shading provided by other green plants.

Can these changes in phytochrome photoequilibria (Table 7.1) lead to changes in plant development? The answer is yes (e.g., Morgan and Smith 1978). Furthermore, one can ascribe to this function of phytochrome (i.e., a detector of mutual shading) a role in plant reproduction. Seed that would give rise to seedlings that would fare poorly under a dense plant canopy (where Pfr_{∞} values would be low) are among those that require light for germination. If these seeds required a sufficiently high Pfr level for germination, then they would not germinate until after the plant canopy were gone (e.g., in the fall or after a fire) or until the seed were removed to a more favorable environment (e.g., an open field). The survival value of such a mechanism is, of course, obvious.

NONPHOTOCHEMICAL REACTIONS OF PHYTOCHROME

Phytochrome undergoes at least three distinct nonphotochemical reactions. Two of these, reversion and destruction, have been reasonably well characterized, at least for phytochrome as it is found in etiolated tissue. The third reaction, which leads to Pfr action, has not been identified. Consequently, the discussion here will deal only with reversion and destruction.

Phytochrome reversion. Pfr reverts by a nonphotochemical reaction to Pr both in vitro (e.g., Pike and Briggs 1972a) and in vivo (e.g., Butler and Lane 1965; see Fig. 7.3). This reversion of Pfr to Pr has long been thought to play an

Table 7.1 Proportion of Phytochrome Present as Pfr at Photoequilibrium (Pfr_∞) as a Function of Different Lighting Conditions

Lighting	Pfr_∞
Midday daylight	73–77
Under a wheat canopy	41
Under a sugarbeet canopy	7–12
Incandescent	68
Fluorescent (unspecified type)	94
Red	99
Far-red	5

Note: Data recalculated to reflect a relative value for Pfr_∞ of 100% at photoequilibrium under 665 nm irradiation.

Source: Holmes and Smith (1975).

important role in phytochrome-mediated physiology (e.g., Borthwick et al. 1954). It is still not clear, however, how important this role is (see Vince-Prue 1975 for discussion).

Not all phytochrome in vivo undergoes measurable reversion. In some plants (for example grasses) phytochrome reversion is not measured in vivo even though phytochrome extracted from the same plant exhibits reversion in vitro (Pike and Briggs 1972a). Furthermore, even if reversion does occur in vivo, only a fraction of the Pfr present is observed to do so (Fig. 7.3). The significance of these two Pfr pools, only one of which undergoes measurable reversion, is not evident. What is evident is that given these experimental observations it is impossible to relate with any confidence physiological observations to what we know about phytochrome reversion (see Vince-Prue 1975 for discussion). An apparent inhibitor of reversion (Manabe and Furuya 1971; Shimazaki and Furuya 1975) might be responsible for producing the apparently nonreverting pool of Pfr in vivo, which approaches 100% of the total pool in the grasses.

Phytochrome reversion in vitro exhibits complex kinetics that may indicate the presence of two pools of phytochrome, each of which reverts by a different rate constant (Pike and Briggs 1972a). Reversion in vitro is also affected by a number of factors. Reversion is enhanced by reductants, such as reduced pyridine nucleotides and dithionite (Mumford and Jenner 1971; Pike and Briggs 1972a) and by low concentrations of metal ions (Negbi, Hopkins, and Briggs 1975; Pratt and Cundiff 1975). While reversion of small phytochrome is enhanced by low pH (Anderson, Jenner, and Mumford 1969), the same does not seem to be true for large phytochrome (Pike and Briggs 1972a). Reversion has a Q_{10} of about 2 (Pike and Briggs 1972a), indicative of its nonphotochemical nature.

The biological role of reversion, if any, would be obvious since the conse-

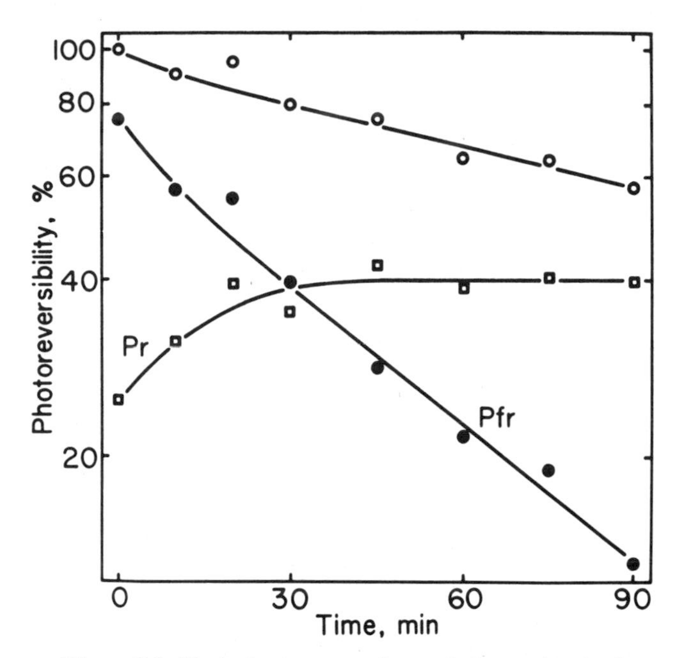

Figure 7.3. Phytochrome reversion and destruction in the hypocotyl hook of etiolated *Sinapis alba* after a saturating 3-min red irradiation at time zero. The decrease in Pfr content (closed circles) is a function of both destruction, as evidenced by the loss in total phytochrome (open circles), and reversion, as evidenced by the increase in Pr level (boxes). Data have been recalculated from Schäfer et al. (1973) to reflect an initial Pfr level of 75% after red irradiation (Pratt 1975).

quence of reversion is to deplete from the cell the biologically active form of phytochrome. Since the rate of this depletion can be varied by redox potential and metal ion availability, for example, it is evident that the rate of reversion could itself be under biological control. A major hypothesis to explain the measurement of time by photoperiodically sensitive plants has been the suggestion that plants sense darkness essentially by the absence of Pfr. Thus, when a plant is placed in darkness, Pfr is presumed to revert to Pr, and when the Pfr level falls below some critical value, the plant initiates the measurement of time. Whether phytochrome functions in this way, however, is uncertain, as discussed in detail by Vince-Prue (1975). It has even been argued in one case (Oelze-Karow and Mohr 1976) that reversion does not take place in the pool of phytochrome controlling the response that was being studied. For the moment it is perhaps best to conclude that we know too little about reversion to evaluate properly its possible role in phytochrome-mediated physiology.

Phytochrome destruction. Phytochrome destruction may be defined as the enhanced rate of loss of phytochrome photoreversibility in vivo that follows

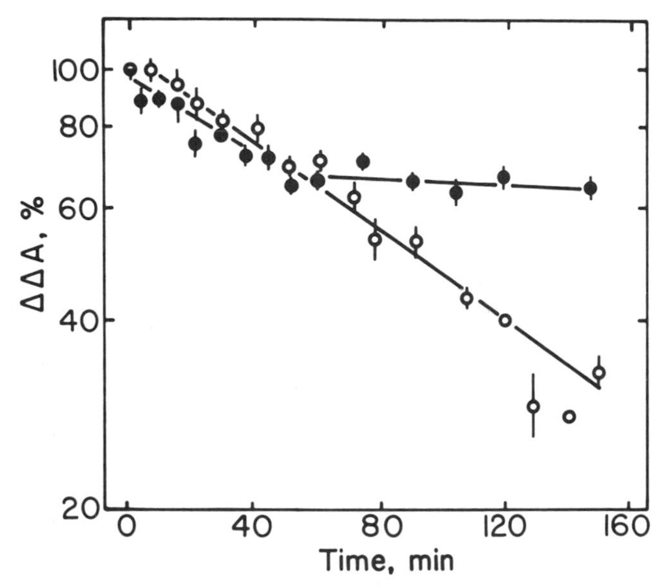

Figure 7.4. Phytochrome destruction as both Pr and Pfr in etiolated oat shoots that were irradiated at zero time with 4-min red light (Pfr destruction; open circles) or with 4-min red light followed by 3-min far-red light (Pr destruction, closed circles). (Data are taken from Stone and Pratt, 1979.)

phototransformation of Pr to Pfr (Butler, Lane, and Siegelman 1963; Butler and Lane 1965; see Fig. 7.3). It has been widely assumed that destruction is specific for Pfr even though a control experiment necessary to establish this point has typically not been done. This control is the measurement of phytochrome levels in tissue irradiated sequentially by first red and then by far-red light. When this control has been done, enhanced rates of phytochrome loss have been observed for the "cycled" Pr as well as for Pfr (Chorney and Gordon 1966; Dooskin and Mancinelli 1968). More recent and more detailed examination of phytochrome destruction as Pr has indicated that for etiolated oats, about 30% of the total phytochrome pool is lost by this process (Mackenzie, Briggs, and Pratt 1978a; Stone and Pratt 1979; see Fig. 7.4).

Both Pr and Pfr destruction result in the loss of antigenically detectable phytochrome (Pratt, Kidd, and Coleman 1974; Stone and Pratt 1979), indicating that both processes lead to a general degradation of the protein moiety. Both Pr and Pfr destruction occur at the same rate (until Pr destruction ceases, see Fig. 7.4), both are equally influenced by growth of oats under conditions that permit ethylene accumulation, and both exhibit comparable sensitivities to 2-mercaptoethanol and to azide (Stone and Pratt 1979). It thus appears likely that both Pr and Pfr destruction occur by the same mechanism. Models to explain phytochrome action that do not explicitly acknowledge this lack of specificity for Pfr are therefore limited in applicability (Schäfer 1975; Johnson and Tasker 1979). Since phytochrome destruction is not always specific for its form, it may well be that it is instead specific for the intracellular localization of phytochrome. For example, destruction may occur only when phytochrome is either "sequestered" (Mackenzie et al 1975; Mackenzie, Briggs, and Pratt 1978b; and see below) or bound to some subcellular organelle or membrane fraction (as may be the case for phytochrome pelletability that is induced in vivo; see below for discussion).

The characteristics of destruction indicate that it is an energy-dependent, enzymatically mediated process. Destruction is sensitive to inhibitors of respiration and to anaerobiosis (Butler and Lane 1965), has a Q_{10} near 3 (Pratt and Briggs 1966), is inhibited by metal complexing agents (Furuya, Hopkins, and Hillman 1965), and by phenylmethylsulfonyl fluoride and 2-mercaptoethanol (Pike and Briggs 1972a). Induction of the mechanism required for phytochrome destruction, as opposed to the expression of this mechanism, is sensitive to inhibitors of protein synthesis (Kidd and Pratt 1973) and possibly also to ethylene (Stone and Pratt 1978).

As for reversion, the biological significance of destruction, if any, would be obvious since destruction results in the rapid depletion from the cell of the active form of phytochrome. As in the case of reversion, destruction should lead to a rapid decrease in Pfr level after the onset of darkness. In addition, except in the case of very brief night periods, plants should contain each morning a large pool of Pr, most of which is probably newly synthesized and may thus be considered "naive," that could serve to signal the onset of day. Whether this potential function of phytochrome as a sensor of daybreak (in addition to its role as a sensor of mutual shading as summarized above) is of biological significance is yet undetermined.

Destruction and reversion in light-grown plants. Two methodological advances have recently made possible the measurement of phytochrome in light-grown plants. Jabben and Deitzer (1978a) have taken advantage of the fact that plants treated with the herbicide Sandoz 9789 are essentially achlorophyllous, even when grown in white light. Chlorophyll-derived interference with spectral assay of phytochrome (see Pratt 1982a for review) is thus eliminated. Of course, a limitation inherent to this method is that the metabolic status of the plants cannot be the same as that of a green plant. Nevertheless, there is evidence that phytochrome control of at least some photomorphogenic responses is not affected by the herbicide treatment (Jabben and Deitzer 1979). In contrast, Hunt and Pratt (1979b) have developed a radioimmunoassay for phytochrome that is insensitive to the presence of chlorophyll and is therefore capable of quantitating phytochrome in crude extracts of green plants (Hunt and Pratt 1980). This method also has a limitation, however, in that it quantitates only that phytochrome that can be extracted in soluble form.

Application of both methods has confirmed that, as a consequence of destruction, light-grown plants do have as expected only about 1% as much phytochrome as do etiolated plants (Jabben and Deitzer 1978b; Hunt and Pratt 1979b, 1980). Phytochrome destruction in light-grown plants, as assayed by both methods, does appear to proceed at a rate comparable to that observed

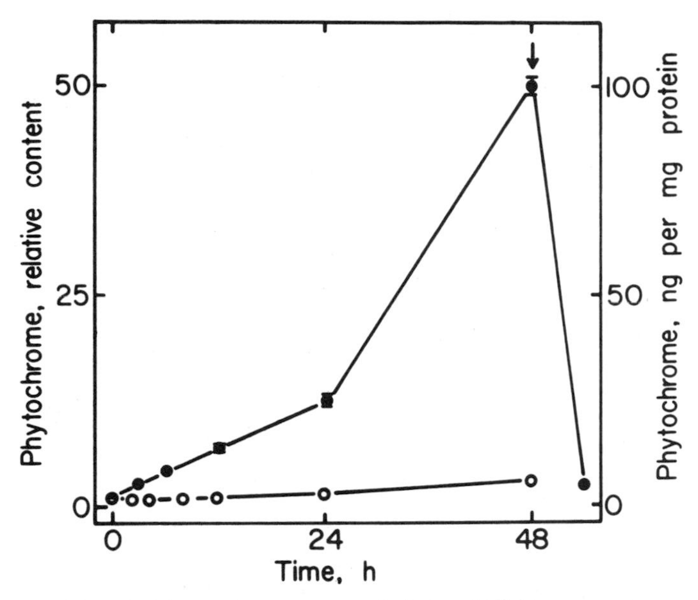

Figure 7.5. Phytochrome content in 6-day-old light-grown oat shoots as a function of time during an extended dark period. Phytochrome content was measured by radioimmunoassay of crude extracts from green oats incubated at 25°C (closed circles) or by spectrophotometry of herbicide-treated, achlorophyllous oats incubated at 20°C (open circles). Radioimmunoassay data are from Hunt and Pratt (1980) and are expressed as ng phytochrome per mg protein; spectrophotometric data are from Jabben and Deitzer (1978b) and are expressed relative to the phytochrome content at zero time. Both sets of data are assigned an arbitrary relative value of 1 at zero time to permit direct comparison. After 48-hr, plants to be assayed by radioimmunoassay were placed in white light.

for etiolated, but otherwise comparable, tissues (Hunt and Pratt 1980; Heim, Jabben, and Schäfer 1981). The observed rate of apparent phytochrome synthesis, however, is dependent upon the method used. Spectrophotometric assay of herbicide-treated plants indicates that the phytochrome level at the end of a 12-hr dark period is about the same as it was at the onset of that 12-hr dark period while after an extended 48-hr dark period it is only about two-fold higher (Jabben and Deitzer 1978b; see Fig. 7.5). By contrast, radioimmunoassay of crude extracts of green plants indicates that phytochrome levels increase three fold during a 12-hr night period and about 50-fold during an extended 48-hr dark period (Hunt and Pratt 1980; see Fig. 7.5). While several explanations for this disagreement are possible (Hunt and Pratt 1980), a simple one would be that the bleached plants obviously have a limited energy supply in the absence of photosynthesis and do not expend energy on the synthesis of phytochrome as they would if they were green and capable of performing photosynthesis.

Why might green plants accumulate high levels of phytochrome in darkness? The answer may be no more than that the plants become more sensitive to light by synthesizing more phytochrome. The more phytochrome the plant has, the smaller would be the proportion that would have to be converted to Pfr to obtain a given response. In turn, this means that for a given response less light would be required. The plant would thus be more sensitive to light at daybreak than at sunset.

DIFFERENCES BETWEEN Pr AND Pfr

It is self-evident that one or more differences between Pr and Pfr must be responsible for the unique biological activity of Pfr. It is therefore crucial that we understand how Pr and Pfr differ.

Biochemical differences. Initial comparisons of Pr to Pfr, which almost certainly were made with small phytochrome, indicated that Pfr is more sensitive to a variety of agents including urea, *p*-chloromercuribenzoate, and trypsin (Butler, Siegelman, and Miller 1964a) as well as to glutaraldehyde (Roux 1972). Nevertheless, other attempts to find differences by methods such as gel filtration, gel electrophoresis under nondenaturing conditions, and behavior during brushite chromatography failed (Briggs, Zollinger, and Platz 1968). An evaluation of these and other data (see Pratt 1982b for review) leads to the conclusion that while Pr and Pfr differ, they do not do so by much.

The need for highly purified phytochrome for detailed comparative studies of Pr and Pfr and the absence until very recently of methods for obtaining this highly purified phytochrome in large form (Hunt and Pratt 1979a; Smith and Daniels 1981) means that little work has been done with large phytochrome. Comparative study of Pr and Pfr with immunopurified large oat phytochrome has detected some specific differences between Pr and Pfr (Hunt and Pratt 1981). Pfr has one more each immediately reactive cysteine and histidine residues as compared to Pr. One intriguing possibility is that these two reactive amino acids are part of an active site. Furthermore, while Pr and Pfr each have two immediately reactive tyrosine residues, phytochrome can exist as Pr after they have been modified but can no longer exist as Pfr, indicating that these residues play an important and specific role in maintaining phytochrome in its active form. The isoelectric focusing profiles of Pr and Pfr do not differ, however, nor do large Pr and Pfr differ significantly in their immunological characteristics (Pratt 1973; Rice and Briggs 1973b; Cundiff and Pratt 1975). As for small phytochrome, it is evident that large Pr and Pfr do not differ by much, although it is clear that they do possess different chemical reactivities.

Behavioral differences. Pr and Pfr differ in their behavior in several respects. I will refer here to four such differences: (1) immunocytochemically observed sequestering of phytochrome as Pfr, (2) light-induced pelletability of phytochrome, (3) preferential binding of Pfr to Cibacron Blue F3GA, and (4) preferential binding of Pfr to artificial liposomes.

Phytochrome sequestering is the association of phytochrome as Pfr with as yet unidentified, discrete subcellular areas about 1 μm in dimension (Mackenzie et al. 1975; Mackenzie, Briggs, and Pratt 1978b; Epel et al. 1980). This redistribution of phytochrome occurs within seconds at 25°C after a brief actinic red irradiation. Whether sequestering reflects the association of Pfr with receptors, or whether sequestering reflects an association of Pfr with the site at which destruction occurs or even results from self-aggregation of Pfr molecules remains uncertain (Mackenzie et al. 1975).

Light-enhanced phytochrome pelletability occurs for at least two reasons (see Pratt 1978 for discussion). In one case, Pfr binds with particulate subcellular debris in vitro (Marmé, Boisard, and Briggs 1973; see Quail 1975a for review). The best studied example of pelletability that arises in vitro, however, deals with an almost certainly artifactual association of Pfr with 31S ribonucleoprotein particles, which most likely derive from ribosome degradation (Quail 1975b, c). In the other case, Pfr is induced in vivo to behave as a particulate protein in vitro. Although the characteristics of the process leading to the induction of this type of pelletability are indicative of a biologically significant reaction (Pratt and Marmé 1976; Quail 1978; Quail and Briggs 1978), the binding partner(s) for phytochrome has(have) not been identified, and it is not even known whether the binding occurs in vivo (Pratt 1980; Quail and Briggs 1980). The biological significance, if any, of phytochrome pelletability has yet to be established.

Pfr binds preferentially to the dye Cibacron Blue F3GA, especially when the dye is conjugated to dextran (W. Smith 1981). The characteristics of this interaction are consistent with the hypothesis that phytochrome has a site that interacts with a flavin and that the Pfr form of phytochrome has a higher affinity for flavin at this site than does Pr. An interaction between phytochrome and a flavin would, of course, provide phytochrome with greater absorbance in the blue spectral region. One would then have, as suggested by W. Smith (1981), a ready explanation for many physiological interactions between red and blue light, especially with respect to the role of phytochrome in the high irradiance responses (Mancinelli and Rabino 1978).

An increasing number of observations indicate that phytochrome, especially as Pfr, may have a significant hydrophobic binding region. For example, Yamamoto and Smith (1981a) demonstrated that phytochrome binds to alkyl agaroses and that the longer the alkyl chain the more tightly Pfr binds as compared to Pr. Similarly, Kim and Song (1981) found that Pfr binds with greater efficiency to artificial liposomes than does Pr. They obtained preliminary evidence that this binding is hydrophobic. Hahn and Song (1981) propose that phototransformation of Pr to Pfr unmasks a hydrophobic region on the surface of Pfr that is covered in Pr by the chromophore. Such a hydrophobic binding site could facilitate a hypothetical interaction between phytochrome and hydrophobic regions of cellular membranes.

CONCLUSIONS

Future work with phytochrome as a molecule deserves emphasis in two areas. First, what we know about phytochrome, as summarized here, has been learned almost exclusively by studying phytochrome from etiolated tissues. The reasons for this emphasis on phytochrome from etiolated plants were practical ones, deriving from the low quantities found in green plants and from the interference with spectral assays caused by chlorophyll. The recent development of affinity purification methods (Hunt and Pratt 1979a; Smith and Daniels 1981) and a radioimmunoassay (Hunt and Pratt 1979b) overcome to a significant extent these practical limitations. Now seems to be both an appropriate and an ideal time to turn our attention to molecular-level studies of phytochrome in green plants.

Second, biochemical and biophysical investigations of phytochrome, regardless of whether isolated from etiolated or green plants, has been hindered by the difficulties associated with obtaining sufficient quantities of highly purified, large phytochrome. The affinity purification methods referred to above overcome this hindrance. Both yield large phytochrome of exceptionally high purity that is apparently suitable for virtually any physicochemical characterization. In particular, the immunoaffinity approach has already been demonstrated to be applicable to the purification of phytochrome from a range of plants (Hunt and Pratt 1979a; Cordonnier and Pratt 1982). The availability of highly purified phytochrome will make possible an extension of the comparative study of Pr and Pfr. An outcome of this comparative study should ultimately be an identification of the difference(s) between Pr and Pfr that is(are) responsible for the biological activity of Pfr.

ACKNOWLEDGMENT

Work cited here from the author's research program has been supported by grants from the National Science Foundation.

LITERATURE CITED

Anderson, G. R., E. L. Jenner, and F. E. Mumford. 1969. *Temperature and pH studies on phytochrome in vitro*. Biochemistry 8:1182–87.

Borthwick, H. A., S. B. Hendricks, and M. W. Parker. 1952. *The reaction controlling floral initiation*. Proc. Nat. Acad. Sci. USA 38:929–34.

Borthwick, H. A., S. B. Hendricks, E. H. Toole, and V. K. Toole. 1954. *Action of light on lettuce-seed germination*. Bot. Gaz. 115:205–25.

Briggs, W. R., and D. C. Fork. 1969a. *Long-lived intermediates in phytochrome transformation*. I. *In vitro studies*. Plant Physiol. 44:1081–88.

———. 1969b. *Long-lived intermediates in phytochrome transformation.* II. *In vitro and in vivo studies.* Plant Physiol. 44:1089–94.

Briggs, W. R., W. D. Zollinger, and B. B. Platz. 1968. *Some properties of phytochrome isolated from dark-grown oat seedlings (Avena sativa L.).* Plant Physiol. 43:1239–43.

Butler, W. L., and H. C. Lane. 1965. *Dark transformations of phytochrome in vivo. II.* Plant Physiol. 40:13–17.

Butler, W. L., H. C. Lane, and H. W. Siegelman. 1963. *Nonphotochemical transformations of phytochrome in vivo.* Plant Physiol. 38:514–19.

Butler, W. L., H. W. Siegelman, and C. O. Miller. 1964a. *Denaturation of phytochrome.* Biochemistry 3:851–57.

Butler, W. L., S. B. Hendricks, and H. W. Siegelman. 1964b. *Action spectra of phytochrome in vitro.* Photochem. Photobiol. 3:521–28.

Chorney, W., and S. A. Gordon. 1966. *Action spectrum and characteristics of the light-activated disappearance of phytochrome in oat seedlings.* Plant Physiol. 41:891–96.

Cordonnier, M-M., and L. H. Pratt. 1982. *Immunopurification and initial characterization of dicotyledonous phytochrome.* Plant Physiol. 69:360–65.

Cordonnier, M-M., P. Mathis, and L. H. Pratt. 1981. *Phototransformation kinetics of undegraded oat and pea phytochrome initiated by laser flash excitation of the red-absorbing form.* Photochem. Photobiol. 34:733–40.

Cundiff, S. C., and L. H. Pratt. 1975. *Phytochrome characterization by rabbit antiserum against high molecular weight phytochrome.* Plant Physiol. 55:207–11.

Dooskin, R. H., and A. L. Mancinelli. 1968. *Phytochrome decay and coleoptile elongation in Avena following various light treatments.* Bull. Torrey Botan. Club 95:474–87.

Epel, B. L., W. L. Butler, L. H. Pratt, and K. T. Tokuyasu. 1980. *Immunofluorescence localization studies of the Pr and Pfr forms of phytochrome in the coleoptile tips of oats, corn and wheat.* Pages 121–33 in J. DeGreef, ed. *Photoreceptors and Plant Development.* Antwerpen Univ. Press, Antwerp, Belgium.

Frankland, B. 1976. *Phytochrome control of seed germination in relation to the light environment.* Pages 447–91 in H. Smith, ed. *Light and Plant Development.* Butterworth, London.

Furuya, M., W. G. Hopkins, and W. S. Hillman. 1965. *Effects of metal-complexing and sulfhydryl compounds* on *nonphotochemical phytochrome changes in vivo.* Arch. Biochem. Biophys. 112:180–86.

Gardner, G., C. S. Pike, H. V. Rice, and W. R. Briggs. 1971. *"Disaggregation" of phytochrome in vitro—a consequence of proteolysis.* Plant Physiol. 48:686–93.

Hahn, T-R., and P-S. Song. 1981. *Hydrophobic properties of phytochrome as probed by 8-anilinonaphthalene-1-sulfonate fluorescence.* Biochemistry 20:2602–9.

Heim, B., M. Jabben, and E. Schäfer. 1981. *Phytochrome destruction in dark- and light-grown Amaranthus caudatus seedlings.* Photochem. Photobiol. 34:89–93.

Hendricks, S. B., and H. A. Borthwick. 1967. *The function of phytochrome in regulation of plant growth.* Proc. Nat. Acad. Sci. USA 58:2125–30.

Holmes, M. G., and H. Smith. 1975. *The function of phytochrome in plants growing in the natural environment.* Nature 254:512–14.

Hunt, R. E., and L. H. Pratt. 1979a. *Phytochrome immunoaffinity purification.* Plant Physiol. 64:332–36.

———. 1979b. *Phytochrome radioimmunoassay.* Plant Physiol. 64:327–31.

———. 1980. *Radioimmunoassay of phytochrome content in green, light-grown oats.* Plant, Cell and Environment 3:91–95.

———. 1981. *Physicochemical differences between the red- and the far-red-absorbing forms of phytochrome.* Biochemistry 20:941–45.

Jabben, M., and G. F. Deitzer. 1978a. *A method for measuring phytochrome in plants grown in white light.* Photochem. Photobiol. 27:799–802.

———. 1978b. *Spectrophotometric phytochrome measurements in light-grown Avena sativa L.* Planta 143:309–13.

———. 1979. *Effects of the herbicide* San *9789 on photomorphogenic responses.* Plant Physiol. 63:481–85.

Johnson, C. B. and R. Tasker. 1979. *A scheme to account quantitatively for the action of phytochrome in etiolated and light-grown plants.* Plant, Cell and Environment 2:259–65.

Kendrick, R. E., and C. J. P. Spruit. 1972. *Light maintains high levels of phytochrome intermediates.* Nature New Biol. 237:281–82.

———. 1973. *Phytochrome intermediates in vivo.* I. *Effects of temperature, light intensity, wavelength and oxygen on intermediate accumulation.* Photochem. Photobiol. 18:139–44.

Kidd, G. H., and L. H. Pratt. 1973. *Phytochrome destruction: an apparent requirement for protein synthesis in the induction of the destruction mechanism.* Plant Physiol. 52:309-11.

Kim, I-S., and P-S. Song. 1981. *Binding of phytochrome to liposomes.* Biochemistry 20:5482–89.

Linschitz, H., V. Kasche, W. L. Butler, and H. W. Siegelman. 1966. *The kinetics of phytochrome conversion.* J. Biol. Chem. 241:3395–3403.

Mackenzie, J. M., Jr., R. A. Coleman, W. R. Briggs, and L. H. Pratt. 1975. *Reversible redistribution of phytochrome within the cell upon conversion to its physiologically active form.* Proc. Nat. Acad. Sci. USA 72:799–803.

Mackenzie, J. M., Jr., W. R. Briggs, and L. H. Pratt. 1978a. *Phytochrome photoreversibility: empirical test of the hypothesis that it varies as a consequence of compartmentalization.* Planta 141:129–34.

———. 1978b. *Intracellular phytochrome distribution as a function of its molecular form and of its destruction.* Amer. J. Bot. 65:671–76.

Manabe, K., and M. Furuya. 1971. *Factors controlling rates of nonphotochemical transformation of Pisum phytochrome in vitro.* Plant Cell Physiol. 12:95–101.

Mancinelli, A. L., and I. Rabino. 1978. *The "high-irradiance responses" of plant photomorphogenesis.* Bot. Rev. 44:129–80.

Marmé, D., J. Boisard, and W. R. Briggs. 1973. *Binding properties in vitro of phytochrome to a membrane fraction.* Proc. Nat. Acad. Sci. USA 70:3861–65.

Morgan, D. C., and H. Smith. 1978. *The relationship between phytochrome photoequilibrium and development in light grown Chenopodium album* L. Planta 142:187–93.

Mumford, F. E., and E. L. Jenner. 1971. *Catalysis of the phytochrome dark reaction by reducing agents.* Biochemistry 10:98–101.

Negbi, M., D. W. Hopkins, and W. R. Briggs. 1975. *Acceleration of dark reversion of phytochrome in vitro by calcium and magnesium.* Plant Physiol. 56:157–59.

Oelze-Karow, H., and H. Mohr. 1976. *An attempt to localize the threshold reaction in phytochrome-mediated control of lipoxygenase synthesis in the mustard seedling.* Photochem. Photobiol. 23:61–67.

Pike, C. S., and W. R. Briggs. 1972a. *The dark reactions of rye photochrome in vivo and in vitro.* Plant Physiol. 49:514–20.

———. 1972b. *Partial purification and characterization of a phytochrome-degrading neutral protease from etiolated oat shoots.* Plant Physiol. 49:521–30.

Pratt, L. H. 1973. *Comparative immunochemistry of phytochrome.* Plant Physiol. 51:203–9.

———. 1975. *Photochemistry of high molecular weight phytochrome in vitro.* Photochem. Photobiol. 22:33–36.

———. 1978. *Molecular properties of phytochrome.* Photochem. Photobiol. 27:81–105.

———. 1979. *Phytochrome: Function and properties.* Photochem. Photobiol. Rev. 4:59–124.

———. 1980. *Phytochrome pelletability induced by irradiation in vivo. Test for in vitro binding of added [^{35}S] phytochrome*. Plant Physiol. 66:903–7.
———. 1982a. *Assay of Photomorphogenic Photoreceptors*. Encyclopedia of Plant Physiology, New Series (in press).
———. 1982b. *Phytochrome: The protein moiety*. Ann. Rev. Plant Physiol. 33 (in press).
Pratt, L. H., and W. R. Briggs. 1966. *Photochemical and nonphotochemical reactions of phytochrome in vivo*. Plant Physiol. 41:467–74.
Pratt, L. H., and S. C. Cundiff. 1975. *Spectral characterization of high-molecular-weight phytochrome*. Photochem. Photobiol. 21:91–97.
Pratt, L. H., and D. Marmé. 1976. *Red light-enhanced phytochrome pelletability: A re-examination and further characterization*. Plant Physiol. 58:686–92.
Pratt, L. H., G. H. Kidd, and R. A. Coleman. 1974. *An immunochemical characterization of the phytochrome destruction reaction*. Biochim. Biophys. Acta 365:93–107.
Pratt, L. H., Y. Shimazaki, Y. Inoue, and M. Furuya. 1982. *Spectral analysis of phototransformation intermediates in the pathway from the red-absorbing to the far-red-absorbing form of phytochrome*. Photochem. Photobiol. (in press).
Quail, P. H. 1975a. *Interaction of phytochrome with other cellular components*. Photochem. Photobiol. 22:299–301.
———. 1975b. *Particle-bound phytochrome: Association with a ribonucleoprotein fraction from Cucurbita pepo L*. Planta 123:223–34.
———. 1975c. *Particle-bound phytochrome: The nature of the interaction between pigment and particulate fractions*. Planta 123:235–46.
———. 1978. *Irradiation-enhanced phytochrome pelletability in Avena: In vivo development of a potential to pellet and the role of Mg^{2+} in its expression*. Photochem. Photobiol. 27:147–53.
Quail, P. H., and W. R. Briggs. 1978. *Irradiation-enhanced phytochrome pelletability: Requirement for phosphorylative energy in vivo*. Plant Physiol. 62:773–78.
———. 1980. *Phytochrome pelletability induced by irradiation in vivo. Mixing experiments*. Plant Physiol. 66:908–10.
Rice, H. V., and W. R. Briggs. 1973a. *Partial characterization of oat and rye phytochrome*. Plant Physiol. 51:927–38.
———. 1973b. *Immunochemistry of phytochrome*. Plant Physiol. 51:939–45.
Roux, S. J. 1972. *Chemical evidence for conformational differences between the red- and far-red-absorbing forms of oat phytochrome*. Biochemistry 11:1930–36.
Schäfer, E. 1975. *A new approach to explain the "high irradiance responses" of photomorphogenesis on the basis of phytochrome*. J. Math. Biol. 2:41–56.
Schäfer, E., W. Schmidt, and H. Mohr. 1973. *Comparative measurements of phytochrome in cotyledons and hypocotyl hook of mustard (Sinapis alba L.)*. Photochem. Photobiol. 18:331–34.
Shimazaki, Y., and M. Furuya. 1975. *Isolation of a naturally occurring inhibitor for dark Pfr reversion from etiolated Pisum epicotyls*. Plant and Cell Physiol. 16:623–30.
Shimazaki, Y., Y. Inoue, K. T. Yamamoto, and M. Furuya. 1980. *Phototransformation of the red-light-absorbing form of undegraded pea phytochrome by laser flash excitation*. Plant and Cell Physiol. 21:1619–25.
Smith, H. 1981. *The role of phytochrome in the natural environment*. *In* C. Hélène, M. Charlier, and Th. Montenay-Garestier, eds. *Trends in Photobiology*. Plenum Press, New York (forthcoming).
Smith, W. O., Jr. 1981. *Probing the molecular structure of phytochrome with immobilized Cibacron Blue 3GA and blue dextran*. Proc. Nat. Acad. Sci. USA 78:2977–80.
Smith, W. O., Jr., and S. M. Daniels. 1981. *Purification of phytochrome by affinity chromatography on agarose-immobilized Cibacron Blue 3GA*. Plant Physiol. 68:443–46.

Spruit, C. J. P. 1982. *Phytochrome intermediates in vivo. IV. Kinetics of Pfr emergence.* Photochem. Photobiol. 35:117–21.

Stone, H. J., and L. H. Pratt. 1978. *Phytochrome destruction. Apparent inhibition by ethylene.* Plant Physiol. 62:922–23.

———. 1979. *Characterization of the destruction of phytochrome in the red-absorbing form.* Plant Physiol. 63:680–82.

Vince-Prue, D. 1975. *Photoperiodism in Plants.* McGraw-Hill, London.

Yamamoto, K. T., and W. O. Smith, Jr. 1981a. *Alkyl and ω-amino alkyl agaroses as probes of light-induced changes in phytochrome from pea seedlings (Pisum sativum* cv. *Alaska).* Biochim. Biophys. Acta 668:27–34.

———. 1981b. *A re-evaluation of the mole fraction of Pfr at the red-light-induced photostationary state of undegraded rye phytochrome.* Plant Cell Physiol. 22:1159–64.

8] Mechanisms of Photothermal Interactions in Phytochrome Control of Seed Germination

by WILLIAM J. VANDERWOUDE*

ABSTRACT

Nondeep dormant seeds such as Grand Rapids lettuce, *Lactuca sativa*, L., display a complex array of responses to the interaction of light and temperature. These responses were investigated by detailed studies of their kinetics and their dependence on temperature and photon fluence. The findings support a dichromophoric mechanism of phytochrome control of its presumptive, morphogenically active receptor, X. They suggest that Pfr:Pfr-X is always active, whereas the activity of Pr:Pfr-X is influenced by membrane properties near X, and therefore by the thermal environment. Theoretical considerations suggest that membrane control of Pr:Pfr-X activity may be central to many plant responses to light and temperature. These include high irradiance responses as well as photoperiodic and thermoperiodic phenomena.

INTRODUCTION

The ultimate success of plant reproduction requires that seeds germinate only when conditions are favorable for seedling growth and development. The environmental factors of light and temperature are two of the primary cues by which seeds sense apropriate conditions for germination. This chapter examines the mechanisms of photothermal interaction in nondeep dormant seeds. In such seeds germination is readily influenced by light or by short periods of incubation at alternate temperatures (Toole 1973). The phenomenology of photothermal control of nondeep dormant seeds is varied and complex. Nevertheless, our studies (VanDerWoude and Toole 1980; VanDerWoude

*Seed Research Laboratory, Agricultural Research Service, U.S. Department of Agriculture, Beltsville, Maryland 20705.

STRATEGIES OF PLANT REPRODUCTION (BARC Symposium number 6—Werner J. Meudt, ed.)
Allanheld, Osmun, Totowa

1982) of the enhancement of phytochrome-dependent lettuce seed germination by prechilling suggest that a single, fundamental mechanism may underlie the many, seemingly disparate responses of seeds to the photothermal environment. At the core of this mechanism is the influence of membrane properties on the activity of Pr:Pfr-X, one photochemical form of dichromophoric phytochrome bound to its photomorphogenically active receptor, X. The proposed mechanism, which will subsequently be described in detail, provides a plausible explanation not only of photothermally induced germination responses but also of many other enigmatic photomorphogenic phenomena in plants. In particular, it forms the basis for a new explanation of high irradiance responses that is also presented and discussed in relationship to photoperiodism and thermoperiodism.

MEMBRANE CONTROL OF PHYTOCHROME ACTIVITY

General experimental approach. Seeds of the Tip Burn Resistant variety of Grand Rapids lettuce, *Lactuca sativa*, L., were used in our studies. These seeds display partial, primary, nondeep dormancy. If imbibed and incubated in darkness at a constant 20°C, germination is typically about 40%. Such dark germination is under phytochrome control, requires Pfr, and may be reduced by far-red (FR)* irradiation of imbibed seeds.

To discern the influence of temperature treatments on the mechanism of germination photocontrol, it is necessary to minimize the influence of temperature on early events in the germination process, such as imbibition and physiological and metabolic equilibration, and also on the later events of potentiation of germination and growth. The following treatment schedule was used for this purpose: seeds were imbibed at 20°C for 1 hr; a brief FR preirradiation, sufficient to establish a low Pfr/Ptot ratio (ca. 7%), was then given; seeds were then incubated at 20°C for an additional 23 hr to permit destruction of remaining Pfr or its dark reversion to Pr. After this pretreatment seeds were given various thermal treatments of defined temperature and duration, returned to 20°C, specifically irradiated, and scored for germination after 48 hr.

The pretreatment with FR and 23 hr of incubation at 20°C completely removed Pfr as indicated by the following: (a) no Pfr was spectrophotometrically detectable in isolated embryos; and (b) the germination of thermally sensitized seeds could be promoted above that of seeds that subsequently remained in darkness by photon fluences known to establish very low Pfr levels. The influence of thermal treatments on germination sensitivity to defined Pfr levels in seeds produced by subsequent irradiations could therefore be examined.

Thermal sensitization to FR irradiation. FR irradiation (adequate to establish phytochrome photoequilibria) did not influence the germination of Pfr-

*Abbreviations: R, red; FR, far-red; LF, low fluence and VLF, very low fluence.

depleted seeds held at constant 20°C. Germination was promoted by FR, however, if given after several hours of low-temperature incubation. The response to FR was inversely related to the incubation temperature and was maximized by temperatures near 4°C. Germination promoted by FR increased linearly with the duration of incubation at 4°C up to a maximum near 18 hr. Similar kinetics were found for the decay of responses to FR when such prechilled seeds were incubated at 20°C before irradiation.

These findings show that low-temperature incubations act by a time-dependent mechanism to greatly increase germination responses to the low levels of Pfr produced by FR irradiation. Additional studies demonstrated that prechilling produces little change in rates of phytochrome potentiation of germination. They indicate that sensitization to FR by prechilling does not involve an alteration of the metabolic or synthetic processes of potentiation. The mechanism of thermal sensitization to low Pfr levels therefore appears to reside at the level of control of potentiation by phytochrome rather than the processes of potentiation.

The kinetics of onset and decay of sensitivity to FR produced by prechilling were similar to those associated with membrane adaptation to temperature change in other systems. We therefore suggested that sensitization to low Pfr levels by prechilling involved such thermal adaptation, and that this leads ultimately to membrane hyperfluidity when seeds are returned to 20°C. This mechanism is supported by our unpublished studies of thermal sensitization to FR by brief incubations at high temperatures (25° to 30°C). In brief, seeds were maximally sensitized by about 1 hr of incubation at high temperature. Thermal sensitization to low Pfr levels may therefore result from the shift to higher temperatures of seeds that have been equilibrated at or adapted to lower temperatures. Although biophysical studies must yet be performed, the findings suggest that both prechilling and brief high-temperature incubation induce temporary membrane hyperfluidity. The influence of such hyperfluidity on the membrane microenvironment in which phytochrome acts may therefore underlie sensitization to low Pfr levels.

Even though only a brief state of membrane hyperfluidity would be expected to result from brief incubations at high temperature, the sensitivities to FR that are produced decay slowly. The sensitivity decay kinetics are similar to those of prechilled seeds. Sensitization to FR therefore appears to be initiated by the hyperfluidity but not dependent upon its continuance.

Sensitization by ethanol to FR irradiation. Support for the role of membrane phenomena in the development of sensitivity to low Pfr levels comes from studies of the influence of ethanol on germination responses. Ethanol, a substance known to alter membrane properties, is known to increase the dark germination of some seeds (Pecket and Al-Charchafchi 1978; Taylorson and Hendricks 1979). We examined the effects of 1% ethanol on lettuce seeds that had been pretreated as usual to remove all Pfr (unpublished data). Seeds were exposed to ethanol for specific times, returned to water, and examined for sensitivity to FR. All incubations were at 20°C. Ethanol treatment did not influence the dark germination of Pfr-depleted seeds. It did induce sensitivity to FR irradiation, however. The kinetics of ethanol enhancement of responses to FR and subsequent decay after return to water were very similar to those described above for the influence of prechilling.

The parallel responses to FR produced by prechilling and ethanol support the involvement of membrane phenomena. In *Escherichia coli*, the presence of ethanol induces an increased proportion of membrane phospholipids containing two unsaturated fatty acids (Berger, Carty, and Ingram 1980). The changes are similar to those caused by a reduction in growth temperature. Such membrane changes may also underlie the similarity of responses to FR induced in seeds by low temperature and ethanol pretreatments. It is clear that studies of membrane properties and composition, and their alteration by the thermal environment, are yet required for a thorough understanding of these responses.

THE DICHROMOPHORIC MECHANISM OF PHYTOCHROME ACTION

Fluence-response studies. The foregoing studies suggest that thermally induced membrane phenomena influence Pfr control of potentiation processes. Photon fluence versus germination response characteristics of prechilled seeds were examined (VanDerWoude 1982) to further elucidate the manner in which this mechanism may produce high-germination responses to very low Pfr levels. Lettuce seeds were pretreated as usual to deplete them of Pfr, incubated at 4°C for a specific time, returned to 20°C, and irradiated with a measured photon fluence of narrow-band, 660 nm (R) or 730 nm (FR) irradiation. Duration of irradiations was usually 15 sec.

The R fluence-response behavior of prechilled seeds was clearly biphasic. Very low fluences (VLF) promoted the first response component which was maximized by fluences near 2×10^{-7} mol/m^2. The second response component required the same low fluences (LF), in the range of 5×10^{-5} to 2×10^{-4}mol/m^2, that promoted the germination of seeds that remained at constant 20°C. The maximum level of VLF responses increased with the duration of the prechilling period. The fluence ranges that promoted VLF and LF response components remained relatively constant, however. Seeds that had been prechilled for 24 hr were promoted to near maximal germination by the VLF response component. Such seeds required about 10,000-fold less fluence than unchilled controls for one-half-maximal promotion of germination above dark control levels.

FR fluences produced response characteristics similar to that of the VLF response component to R, but required 65-fold higher fluences. This factor is roughly the ratio that relates the extinction coefficients of Pr at 660 nm and 730 nm. The response to FR can therefore be interpreted in terms of phototransformation of Pr to Pfr by FR. No second response component was produced by FR, even at fluences as high as 5×10^{-3}mol/m^2. R and FR promoted similar maximum VLF responses in seeds that had been prechilled for the same period. Also, FR reduced responses of R-irradiated seeds down only to the level of maximum VLF responses. Lastly, the FR fluence that was required to maximize responses (near 1×10^{-5}mol/m^2) was unchanged when responses were allowed to partially decay by incubation of seeds at 20°C before irradiation.

Our unpublished studies show seeds treated by brief incubation at 28°C or by incubation for several hours in 1% ethanol to display biphasic, R-fluence-response behavior similar to that of prechilled seeds. In addition, studies by Small et al. (1979) show thermodormant lettuce seeds to also display such behavior. The similarity of the fluence-response characteristics produced by these diverse treatments suggests that they all increase germination sensitivity to low Pfr levels by the same mechanism.

A molecular interpretation of the responses. The following mechanism, based totally on the action of phytochrome and its presumptive receptor, X, was proposed to explain the observed fluence-response behavior (VanDerWoude 1982). Phytochrome, now known to be a molecular dimer (Hunt and Pratt 1980), has two chromophores and exists in the intertransformable states Pr:Pr, Pr:Pfr, and Pfr:Pfr. The concentration of X is very low relative to that of total phytochrome. The establishment by irradiation of very low amounts of Pr:Pfr and its association with X lead to the VLF response component of sensitized seeds. Saturation of X by Pr:Pfr yields maximum VLF responses. The LF response component requires the establishment of Pfr:Pfr-X primarily by phototransformation of Pr:Pfr-X. Fluence requirements for the formation of Pfr:Pfr-X are therefore much greater than those of Pr:Pfr-X.

The presence of two chromophores on the phytochrome molecule is indicated by two lines of evidence. First, chromopeptides that are produced by extensive proteolytic degradation of phytochrome all have the same amino acid sequence near the chromophore (Lagarias and Rapoport 1980). Second, irradiations that produce low Pfr levels induce the subsequent pelletability of nearly equal amounts of "Pr" and "Pfr" (Pratt and Marmé 1976). This behavior suggests the membrane association of phytochrome dimers, only one subunit of which need be in the Pfr form to elicit pelletability.

Examination of the proposed mechanism. We compared the observed R and FR fluence requirements for VLF and LF response components with calculated estimates of the fluence requirements for Pr:Pfr-X and Pfr:Pfr-X formation. The calculations assumed the in vivo photochemical properties of lettuce seed phytochrome to be similar to those determined (Pratt 1978) for purified oat phytochrome. In vivo spectrophotometric measurements showed the fluence dependence of phytochrome phototransformation in seeds to be similar to that expected for purified oat phytochrome in solution, even though seed coat transmittance was low. Light scattering in embryo tissues apparently increases subcellular fluences at the site of photocontrol of germination in the embryo and compensates for seed coat opacity. For the purpose of examining the proposed mechanism the applied fluences could therefore be roughly equated with photochemically effective, subcellular fluences.

A mathematical basis for estimating fluence requirements for the formation of Pr:Pfr, Pfr:Pfr, and their association with X was developed. It assumed that photochemical properties of chromophores were not greatly changed by phototransformation of one-half of the phytochrome dimer, nor by binding of phytochrome to X. These estimates compared very favorably with observed fluence-response characteristics and strongly supported the proposed mechanism. The comparison suggested the following relationships: (a) the concentration of X is about 10^{-3} that of phytochrome in those cells responsible for photocontrol of germination; (b) VLF response magnitude is related to the logarithm of the concentration of Pr:Pfr-X; (c) LF response magnitude is linearly related to the concentration of Pfr:Pfr-X; and (d) maximum VLF responses result from the saturation of X with Pr:Pfr and the magnitude of such maximum responses reflects the activity of Pr:Pfr-X, not its formation.

The very low concentration of X suggested by these findings indicates that the formation of Pr:Pfr-X and Pfr:Pfr-X cannot be equated with the well-characterized phenomena of in vivo, light-induced pelletability and sequestering of phytochrome (Pratt 1978), since the latter suggest massive association of phytochrome with a subcellular component. The formation of Pr:Pfr-X and Pfr:Pfr-X may first involve activities that underlie these phenomena, however. If so, the extreme sensitivity of VLF responses to light indicate Pr:Pfr to be specifically targeted for the component, possibly a membrane, on which X is located.

Control of Pr:Pfr-X activity. The foregoing findings, when taken together with the indicated role of membrane alteration in thermal and ethanol sensitization of seeds to low Pfr levels, suggests that the activity of Pr:Pfr-X is influenced and controlled by its membrane microenvironment, while that of Pfr:Pfr-X is not. The mechanism of such control of Pr:Pfr-X activity remains to be determined. Nevertheless, a possible mechanism is suggested by our findings. The logarithmic dependence of VLF responses on Pr:Pfr-X formation suggests that, unlike Pfr:Pfr-X, the average effectiveness of an individual Pr:Pfr-X is reduced as the concentration of its population is increased. In support of this, our studies indicated that maximum VLF responses required saturation of X with Pr:Pfr, whereas the corresponding LF response levels of control seeds required that less than 35% of X be occupied by Pfr:Pfr. Such behavior would occur if the morphogenic action of phytochrome resulted from the formation of either stable $(\text{Pfr:Pfr-X})_N$ complexes, where $N \geq 2$, or of $(\text{Pr:Pfr-X})_N$ complexes that exist in a dynamic equilibrium with Pr:Pfr-X and have a stability inversely related to its concentration. Changes in the membrane near Pr:Pfr-X that improve such stability would increase VLF responses.

Although related here to responses to very low fluences or to FR irradiations, Pr:Pfr-X would be abundant under most natural light conditions and even during temporary darkness. Of particular experimental importance, green "safelight" conditions are expected to readily saturate X with Pr:Pfr and were not used in our studies. The possibility therefore exists that the regulation of Pr:Pfr-X activity by membrane phenomena may be central to many thermal, hormonal, and chemical influences on dormant seed germination and other morphogenic phenomena in plants. Additionally, the activity of Pr:Pfr-X may in some systems be a natural component of responses to light in the absence of other stimuli. This would explain the biphasic, fluence-response behavior observed in many photomorphogenic phenomena (Mandoli and Briggs 1981). Also, since saturating FR irradiations produce an estimated level of Pr:Pfr 60-fold greater than that required to saturate X (VanDerWoude 1982), Pr:Pfr-X activity may underlie many of the paradoxical responses to FR as well as the lack of photoinhibition by FR of R-induced responses in some systems.

HIGH-IRRADIANCE RESPONSES (HIR) AND PHOTOPERIODISM

Pr:Pfr-X activity and the HIR. Control of the activity of Pr:Pfr-X may also be central to the involvement of phytochrome in the HIR. Previous speakers in this session have discussed the HIR and possible mechanisms for its action. HIR phenomena occur in response to natural irradiations that are usually prolonged and often of high fluence rate. They are often not fully reversible by FR light, and in many instances they display action spectra maxima at wavelengths that produce low photoequilibria where only Pr:Pfr-X would be abundant. They are maximized by combinations of fluence rate and wavelength that produce photochemical half-lives of about 1 min or less for chromophores in the Pfr form (Mancinelli and Rabino 1978).

The possible function of the short "Pfr" half-life in the HIR may be found in the body of evidence concerning the kinetics of light-induced phytochrome destruction, sequestering, and pelletability (reviewed by Pratt 1978). Phytochrome destruction is apparently initiated unless newly formed "Pfr" is phototransformed back to "Pr" within a few minutes. "Pfr" becomes pelletable or sequestered within a few seconds. On phototransformation back to "Pr," loss of pelletability or the sequestered condition occurs over a period of about 1 hr. Consideration of the dynamic interaction of these phenomena suggests that the amount of phytochrome (all photochemical forms) bound to membranes is inversely related to the photochemical half-life of "Pfr." Therefore, combinations of fluence rate and spectral energy distribution that yield short "Pfr" life-times and maximize the HIR are expected to maximize the amount of phytochrome associated with target membranes.

As discussed above, our evidence suggests that thermal and ethanol treatments modify the membrane microenvironment about Pr:Pfr-X in a manner that increases its activity. Similarly, the mechanism of the HIR may involve light-induced associations of phytochrome with membranes on which X is located that induce time-dependent modifications of specific membrane properties and thereby increase the activity of Pr:Pfr-X. This proposed mechanism offers a plausible explanation for most HIR phenomena.

Pr:Pfr-X activity and photoperiodism. Application of the proposed HIR mechanism to a theoretical consideration of photoperiodic responses suggests that light-dark cycles produce oscillations in the state of the membrane microenvironment about X and in Pr:Pfr-X activity. Such oscillations would be expected to reflect the time dependence of changes in membrane properties during the early daylight hours. Subsequently, the steady-state magnitude of Pr:Pfr-X activity would be a function of fluence rate and spectral energy distribution. With the onset of darkness, decay of Pr:Pfr-X activity would occur with a lag that is dependent upon the final phytochrome photoequilibrium and therefore the end-of-day spectral energy distribution. (For example, an end-of-day FR irradiation would be expected to maximize this lag period since, by transforming most "Pfr" to "Pr," it would minimize the destruction of bound phytochrome. Since bound "Pr" is released slowly, maintenance of high Pr:Pfr-X activity would be prolonged into the dark period.) Pr:Pfr-X activity would subsequently decay gradually during the dark period. The length of the critical night period in long- and short-day plants may reflect the time required for Pr:Pfr-X activity to decay to some critical, threshold level.

The phase relationships of photoperiodic responses may involve parallel influences of natural oscillations in membrane properties near X on Pr:Pfr-X activity. Similarly, thermoperiodically induced oscillations in the microenvironment of X would influence Pr:Pfr-X activity. Control of Pr:Pfr-X activity may therefore be central to the activities of photoperiodic, thermoperiodic, and natural rhythmic phenomena, as well as to their interaction.

CONCLUSIONS

Our findings suggest that the influence of membrane phenomena on Pr:Pfr-X activity is central to many plant responses to the photothermal environment. The proposed mechanisms are based primarily on interpretations of morphogenic responses. The need remains to test and refine these hypotheses by biophysical, biochemical, and continued physiological studies. In addition to the fundamental information to be gained, it is anticipated that Pr:Pfr-X will provide a useful focus in the search for improved practical control of plant growth and development.

LITERATURE CITED

Berger, B., C. E. Carty, and L. O. Ingram. 1980. *Alcohol-induced changes in the phospholipid molecular species of Escherichia coli*. J. Bacteriol. 142:1040–44.

Hunt, R. E., and L. H. Pratt. 1980. *Partial characterization of undegraded oat phytochrome*. Biochemistry 19:390–94.

Lagarias, J. C., and H. Rapoport. 1980. *Chromopeptides from phytochrome. The structure and linkage of the Pr form of the phytochrome chromophore*. J. Amer. Chem. Soc. 102:4821–28.

Mancinelli, A. L., and I. Rabino. 1978. *The "high-irradiance responses" of plant photomorphogeneis*. Bot. Rev. 44:129–80.

Mandoli, D. F., and W. R. Briggs. 1981. *Phytochrome control of two low-irradiance responses in etiolated oat seedlings*. Plant Physiol. 67:733–39.

Pecket, R. C., and F. Al-Charchafchi. 1978. *Dormancy in light-sensitive lettuce seeds*. J. Exp. Bot. 29:167–73.

Pratt, L. H. 1978. *Molecular properties of phytochrome*. Photochem. Photobiol. 27:81–105.

Pratt, L. H., and D. Marmé. 1976. *Red light-enhanced phytochrome pelletability: A reexamination and further characterization*. Plant Physiol. 58:686–92.

Small, J. G. C., C. J. P. Spruit, G. Blaauw-Jansen, and O. H. Blaauw. 1979. *Action spectra for light-induced germination in dormant lettuce seeds I. Red region*. Planta 144:125–31.

Taylorson, R. B., and S. B. Hendricks. 1979. *Overcoming dormancy in seeds with ethanol and other anesthetics*. Planta 145:507–10.

Toole, V. K. 1973. *Effects of light, temperature and their interactions on the germination of seeds*. Seed Sci. Technol 1:339–96.

VanDerWoude, W. J. 1982. *A dichromophoric model for the action of phytochrome: Evidence from photothermal interactions in lettuce seed germination*. Proc. Nat. Acad. Sci. USA (in press).

VanDerWoude, W. J., and V. K. Toole. 1980. *Studies of the mechanism of enhancement of phytochrome-dependent lettuce seed germination by prechilling*. Plant Physiol. 66:220–24.

four

HORMONAL CONTROL SYSTEMS

9] Some Concepts Concerning the Mode of Action of Plant Hormones

by HANS KENDE*

ABSTRACT

Plants regulate a large number of physiological processes with a small number of plant hormones. The mechanism of action of plant hormones has not yet been elucidated. This chapter discusses some phenomena of hormone action that seem important for the understanding of hormonal control mechanisms in plants: modulation of sensitivity to plant hormones, the logarithmic dose-response relationship, and known early responses to plant hormones. Possible analogies to the modulation of sensitivity to animal hormones and to adaptation in sensory perception are pointed out.

INTRODUCTION

In plants, many processes are regulated by a relatively small number of substances termed "plant hormones." There are five known plant hormones or groups of plant hormones: the auxins, the gibberellins, the cytokinins, abscisic acid, and ethylene. These substances often fulfill similar functions in different tissues, and each of them regulates a variety of different processes. For example, auxins, gibberellins and cytokinins all induce cell division in different tissues, and auxin and gibberellins also regulate cell elongation, though probably by different mechanisms. Apart from acting as growth regulators, auxins, gibberellins and cytokinins are also involved in the control of enzyme activities and of morphogenetic events. Some hormone-regulated processes are initiated within minutes, such as induction of cell elongation by auxin and abscisic acid-stimulated closure of stomates. Other hormonally controlled reactions show a lag period of several hours between addition of the hormone and appearance of the response, as is the case with gibberellin-enhanced synthesis of α-amylase,

*MSU-DOE Plant Research Laboratory, Michigan State University, East Lansing, Michigan 48824.

STRATEGIES OF PLANT REPRODUCTION (BARC Symposium number 6—Werner J. Meudt, ed.)
Allanheld, Osmun, Totowa

and still other responses become evident after days or weeks only, e.g., the onset of shoot regeneration under the influence of cytokinins. Detailed accounts of hormone action in plants are given in recent textbooks (Galston, Davies, and Satter 1980; Salisbury and Ross 1978; Wareing and Phillips 1978).

The question whether plant hormones should be classified as such in the sense of the classical definition of a hormone is debatable. They certainly act at very low concentrations, but their role as factors of communication between organs has not always been demonstrated. In some instances, plant hormones are synthesized in one organ or tissue and are translocated to another organ or tissue, where they act. This is the case with cytokinins, for example, which are synthesized in the root and translocated via the xylem to the leaves, where they are involved in controlling the proper balance between catabolic and anabolic metabolism (Kende 1971). In other instances, there is no evidence that the site of synthesis and the site of action of plant hormones are clearly separated from each other. Cytokinins in the root are concentrated in the meristematic portion of the tip (Weiss and Vaadia 1965). It is very likely that they are synthesized there and that they are also involved in the regulation of cell division in the root meristem. The situation appears to be similar with gibberellins in the stem apex. To summarize, in a substantial number of cases, there is no indication that plant hormones play the role of organ-to-organ or tissue-to-tissue messengers.

The question whether plant hormones are indeed hormones is perhaps more serious when one considers their mode of action. If the classification of these substances as "hormones" leads to preconceived notions concerning their mode of action, one may well end up following wrong leads. Indeed, attempts to find analogies between the mechanism of action of plant hormones and that of animal hormones have failed up to now. Instead of trying to prove or disprove on the basis of known facts that plant hormones are analogous to animal hormones, it appears more worthwhile to define areas of knowledge and ignorance concerning their mode of action. This may help us to focus our efforts as we try to describe the mechanism of action of plant hormones.

In a systematic attempt to elucidate the hormonal response in plants, one can start from two points of departure. One can follow the fate of a hormone in order to localize its site of action. Alternatively, one can start from a known biochemical response to a hormone and determine the chain of reactions that leads to this response. At present, neither the site of action of a plant hormone nor a primary hormonal response is known. There are many reports on specific binding of plant hormones to soluble proteins or to particulate cell fractions, but the significance of binding has not yet been established in any of these cases. We also know some very fast responses to hormones, but we do not know how these responses are regulated.

Having established in very broad terms the areas of our ignorance, I would like to dwell on some known phenomena of hormonal response systems in plants. Defining such processes and drawing analogies to similar phenomena in organisms other than plants may point to biochemical mechanisms that we may expect to find when probing into the mode of action of plant hormones.

MODULATION OF SENSITIVITY TO PLANT HORMONES

There are many known instances where the sensitivity of a plant to a given hormone is modulated. Such changes in sensitivity occur during ontogeny of a plant, as a result of environmental stimuli, and can be brought about by other hormones. The logarithmic relationship between hormone concentration and the hormonal response in plants is also a manifestation of changing sensitivity to hormones and will be treated in a separate section.

Many examples of developmentally determined changes in hormone sensitivity have been described, of which three will be given below. Wright (1961) has shown that wheat coleoptiles are most responsive to gibberellin early in their development. As sensitivity to gibberellin declines, responsiveness to cytokinin develops. Finally, the wheat coleoptile gains sensitivity to auxin. Similarly, responsiveness to ethylene often develops as a function of tissue age. This is true in the case of ethylene-mediated leaf abscission where the abscission zone acquires sensitivity to ethylene as it ages (Abeles 1968). In flower tissue of *Impomoea tricolor*, sensitivity to ethylene develops during the last days before opening of the flower (Kende and Hanson 1976). Two days before opening of the flower, added ethylene elicits no noticeable response of any kind. One day later, ethylene is capable of inducing loss of turgor in the tissue but not "autocatalytic" ethylene synthesis. On the day of flower opening, exogenously applied ethylene causes both turgor loss and "autocatalytic" ethylene formation.

Environmental stimuli often affect the responsiveness of the plant to endogenous and applied plant hormones. Dwarf peas, for example, grow normally in the dark but are severely stunted in light. Impaired synthesis or utilization of gibberellins in the light has been suspected as reason for the inhibition of growth (Lockhart 1956). Kende and Lang (1964) have shown that the level of gibberellins is not significantly different in light and in darkness. The sensitivity to gibberellins is greatly reduced by light, however. A similar phenomenon has been observed with spinach plants (Zeevaart 1971). When spinach is switched from short days to long days or treated with gibberellin in short days, growth and flowering are induced. It has been found that photoperiodic induction enhances the sensitivity of spinach to gibberellin.

Hormones are also known to affect responsiveness of the plant to other hormones. In some water plants, e.g., *Callitriche platycarpa,* ethylene potentiates the response to gibberellin (Musgrave, Jackson, and Ling 1972). These authors have shown that the growth response to ethylene in air is identical to that caused by submergence of the plant. They have suggested that responsiveness to ethylene may be the mechanism by which aquatic plants adjust to the depth of the water. When the top of *Callitriche,* which forms a rosette, is floating on the surface of the water, ethylene escapes into the air, and the concentration of ethylene in the tissue is low. When the rosette becomes submerged, diffusion of ethylene from the tissue is slowed, and the internal ethylene concentration increases. As a consequence, the plant becomes more sensitive to its endogenous gibberellin, and growth of the internodes within the rosette is enhanced until the top of the plant reaches the water surface again. In the case of leaf abscission, responsiveness of the abscission layer to ethylene appears to be regulated, at least in part, by auxin (Abeles and Rubinstein 1964; Rubinstein and Leopold 1963). If the level of auxin in the abscission zone is kept high following excision of the petiole, development of sensitivity to ethylene and abscission are retarded, even though auxin stimulates ethylene synthesis. Delayed application of auxin leads to accelerated abscission because the tissue has become responsive to the ethylene it forms.

Modulation of the sensitivity to plant hormones can be achieved through modifications at their site of action. The following alterations of the hormone receptor would result in lowered hormone sensitivity: (a) reduced affinity of the hormone to its receptor, which may be caused by the synthesis of an altered receptor, through covalent modification of an existing receptor or through interaction of the receptor with regulatory molecules such as other hormones or inhibitors; (b) reduced number of binding sites, perhaps the result of changes in receptor turn-over or of competitive binding of some regulatory molecule (e.g., of an inhibitor) to the hormonal binding site; and (c) impaired effectiveness of the receptor in mediating the primary response.

There are no documented cases where reduced sensitivity to a plant hormone can be traced to a modification of its receptor site. Nevertheless, two leads appear promising. In dark-grown dwarf peas, radioactive gibberellins A_1 and A_5 (GA_1 and GA_5) accumulate in the apical, GA-sensitive part of the stem (Musgrave, Kays, and Kende 1969). This may be due to the presence of GA binding sites in the target tissue and may be analogous to the accumulation of ^{3}H-estrogen in the uterus (Jenson and Jacobson 1962). In light-grown dwarf peas, which are inhibited in growth and have a lowered sensitivity to GAs, accumulation of radioactive GA_1 and GA_5 in the apical portion of the stem is greatly reduced (Musgrave, Kays, and Kende 1969). These data are consistent with the hypothesis that reduced responsiveness to GA in illuminated dwarf peas is a result of impaired interaction between GAs and their receptor sites. Efforts to isolate GA receptors from dwarf peas have failed thus far, so this hypothesis has not yet been tested directly. In their search for auxin receptors in corn coleoptiles, Ray, Dohrman, and Hertel (1977) have found a soluble substance (supernatant factor) that reduces the affinity of a specific auxin binding site on the endoplasmic reticulum to 1-naphthaleneacetic acid. This supernatant factor has been identified as a mixture of 6-methoxy- and 6,7-dimethoxy-2-benzoxazolinone (Venis and Watson 1978). Both of these compounds have been found to inhibit growth of oat coleoptile sections. While no regulatory function can yet be assigned to benzoxazolinones, they, or compounds of similar biological activity, may very well participate in regulating auxin activity in plants.

Modulation of the responsiveness to plant hormones may also be achieved through a modification of a biochemical reaction that lies on the pathway between the primary response and the ultimate physiological effect. Conditions or factors that reduce sensitivity to a hormone may do so by imposing a limitation on any such biochemical reaction. Rate-limiting steps may result from lowered enzyme activities, reduced availability of substrates or cofactors, or the action of inhibitors.

In summary, changes in responsiveness to hormones are important regulatory mechanisms that permit the plant to react to environmental perturbations at constant hormone levels. In our attempts to elucidate the mechanism of action of plant hormones, such changes in hormone sensitivity may be extremely useful as handles to manipulate either some property of the receptor site or some biochemical response to the hormone. This, in turn, may become invaluable in assessing the significance of any hormone binding site that has been observed in vitro or any biochemical response that is suspected to be part of the hormonal response system.

MODULATION OF SENSITIVITY TO ANIMAL HORMONES

Modulation of sensitivity to hormones is not unique to plants. It has also been observed with animal hormones and neurotransmitters. This phenomenon is referred to as refractoriness or desensitization and has recently been reviewed (Lefkowitz, Wessels, and Stadel 1980; Tell, Haour, and Saez 1978). Examples cited below are described in these two reviews. In general, desensitization has been observed with hormones and other effectors that bind to membrane-bound receptors at the cell surface. Best investigated is refractoriness to hormones that act through adenylate cyclase-coupled systems. In those cases, chronic exposure of cells to high concentrations of a hormone leads to loss of sensitivity to that hormone. For example, treatment of fat cells with lipolytic hormones, e.g., epinephrine, results in a rapid rise of the cAMP titer, but the concentration of cAMP returns to near normal upon prolonged exposure to the hormone. Cases are also known where one hormone affects the sensitivity of the cell to another hormone. Thyroid hormones increase the number of glucagon binding sites, and estrogen reduces the sensitivity of Leydig cells to gonadotropin.

The modulation of sensitivity to animal hormones is achieved by a number of mechanisms. In the case of hormones that act via cAMP as second messenger, there is evidence for alteration of hormone receptor sites, for the formation of inhibitors that reduce the activity of adenylate cyclase, and for increased activity of phosphodiesterase which causes accelerated breakdown of cAMP. Desensitization to animal hormones is viewed as a protective mechanism that maintains homeostasis of cells that are chronically or repeatedly exposed to high levels of a hormone. We do not know whether there are any similarities between the mechanisms that regulate sensitivity to animal hormones and those that govern sensitivity to plant hormones. Since plant hormones do not seem to act via the adenylate cyclase system, cAMP-related mechanisms are probably not relevant for plants. Nonetheless, other mechanisms, e.g., alteration of receptor sites, are biochemical processes of general significance that may also operate in plants.

THE DOSE-RESPONSE RELATIONSHIP

The concentration range over which plant hormones elicit an increasing response covers usually 3 to 4 orders of magnitude and, in some instances, even 5 to 6 orders of magnitude (Kende 1971; Kende and Gardner 1976). Within a certain concentration range, the response is linearly related to the logarithm of the hormone concentration, meaning that low hormone doses are relatively more effective than high ones. This phenomenon constitutes a special case where the sensitivity to a hormone is modulated, and it deserves separate treatment. To give one example, in barley aleurone layers, there is a nearly linear relationship between the logarithm of the gibberellic acid concentration in the range of 1.4×10^{-10}M to 1.4×10^{-6}M and the level of hormonally induced α-amylase (Jones and Varner 1967). In contrast, estrogen-induced synthesis of a specific uterine protein exhibits a saturation curve that covers only two decades, from 10^{-10}M to 10^{-8}M (Katzenellenbogen and Gorski 1972). It has been pointed out that this feature of the hormonal response in plants resembles the response to sensory stimulants (Kende 1971; Kende and Gardner 1976). For example, in odor perception, an animal can distinguish differences in the concentration of olfactants over a concentration range of four to five decades (Schneider 1966). In sensory physiology, the extended dose-response curve has usually been interpreted in terms of adaptation. In the broadest terms, adaptation describes the ability of an organism to decrease its sensitivity toward a stimulant as the concentration of this stimulant is raised.

There is a well-documented case of adaptation to a plant hormone in the fungus *Phycomyces blakesleeanus*. When a barrier is placed close to the growing zone of a sporangiophore of *Phycomyces*, the sporangiophore grows away from the barrier. Cohen et al. (1975) have suggested that the sporangiophore emits a growth-promoting gas and that the increased concentration of this gas between a closely placed object and the sporangiophore causes the side of the sporangiophore that is proximal to the barrier to elongate faster than the side away from the barrier. This asymmetric distribution of growth would lead to the observed avoidance response. Russo, Halloran and Gallori (1977) have tested the above hypothesis and have provided evidence that the avoidance response in *Phycomyces* is indeed mediated by a gas, namely by ethylene. They have also demonstrated that the sporangiophore of *Phycomyces* shows a similar adaptation to ethylene as to light. Upon addition of ethylene, the growth rate of the sporangiophore increases temporarily but returns to the original level after 10–15 min. A second addition of ethylene to adapted sporangiophores is without further growth-promoting effect.

Vanderhoef and Stahl (1975) have found that auxin elicits two sequential growth responses in soybean hypocotyl sections. The first, fast response is transient and can be separated from the second, long-term response by application of inhibitors. We do not know what that biochemical or physiological basis for these two responses is. Nevertheless, one may hypothesize that the transient nature of the first growth response is indicative for adaption to auxin.

In bacteria, adaptation to chemoattractants or repellents and deadaptation are connected to methylation and demethylation of transmembrane receptor proteins (Chelsky and Dahlquist 1980; DeFranco and Koshland 1980; Springer, Goy, and Adler 1979). According to Koshland (1980), binding of sensory stimulants to these transmembrane receptors leads to a transient increase in signal, which is reduced to its original level (adaptation) by methylation of the transmembrane receptor. Therefore, methylation counteracts sensory excitation. Phosphorylation and dephosphorylation of rhodopsin (Shichi and Somers 1978) may, in similar fashion, form the biochemical basis for visual adaptation and de-adaptation (Koshland 1980).

It may be very profitable to look for adaptation to plant hormones. If such a phenomenon can clearly be demonstrated in higher plants, one may test whether or not adaptation and de-adaptation are linked to reversible covalent modification of specific proteins. The finding of proteins that are modified during adaptation would permit localization, isolation, and characterization of probable components of the hormonal response system in plants. The chances are that such proteins would be hormone receptors, a possibility that could be tested in hormone-binding experiments. Testing for reversible covalent modification of proteins may break the current deadlock in the search for plant hormone receptors. This deadlock has arisen because no primary response to any of the plant hormones is known, and hence no assay is available to establish the significance of hormone binding.

The extended range of the dose response and desensitization toward hormones can also be achieved through negative cooperativity, i.e., changes in the affinity of the receptor to the hormone as a function of hormone binding. Negative cooperativity has been found with a number of animal hormones, most notably with insulin, and can result from conformational changes of the receptor or from clustering of receptors (Tell, Haour, and Saez 1978). If a receptor exhibits negative cooperativity, the dissociation constant (K_d) of hormone binding will increase as more hormone becomes bound to the receptor. This will yield nonlinear Scatchard plots with a curve that is steep at low concentrations of bound hormone but becomes less steep as the concentration of the hormone is raised. Such Scatchard plots have been obtained with plant hormones (Kende, unpublished) and may indicate the presence of multiple binding sites with different K_ds or the presence of one type of binding site with changing K_d (negative cooperativity).

THE PRIMARY RESPONSE TO PLANT HORMONES

The question whether plant hormones act at the level of gene activation or at some other point in cellular regulation has frequently been discussed. Based on our current knowledge, we strongly suspect that even one and the same plant hormone may have different modes of action, depending on the target tissue in which it acts. Below, I shall cite some cases where most progress has been made toward the elucidation of the mode of action of plant hormones and where the underlying biochemical responses are well enough defined to make further advances.

In barley aleurone layers, synthesis and secretion of α-amylase and a number of other hydrolases is induced by gibberellic acid (GA_3). Abscisic acid (ABA) inhibits the production of these hydrolases. It has been shown that GA_3 induces the appearance of large amounts translatable mRNA for α-amylase (Higgins, Zwar, and Jacobson 1976; Mozer 1980). This result is consistent with the hypothesis that GA_3 regulates the synthesis of hydrolases in barley aleurone cells at the level of transcription. ABA also induces the appearance of new translatable mRNAs. When given together with GA_3, ABA blocks synthesis of α-amylase but does not prevent the appearance of mRNA for α-amylase. Therefore, ABA inhibits translation of mRNA coding for a α-amylase. This mode of action of ABA in barley aleurone cells can be contrasted with the action of the same hormone in barley leaves. There, ABA causes closure of stomates within minutes of application (Cummins, Kende, and Raschke 1971). Stomatal opening and closure is controlled by rapid changes in the turgor of guard cells. This comparison makes it quite clear that ABA operates through entirely different mechanisms within the same plant, the nature of the response being determined by the target tissue.

The effect of auxin on cell extension and softening of the cell wall constitutes another case where the biochemical basis of hormone action is amenable to experimental analysis. At least during the early phases of the growth response, auxin appears to act by inducing proton excretion into the cell wall (for a review, see Cleland 1980). The question of auxin action is well defined in this case: How does auxin regulate the pumping of protons into the cell wall, and how does acidification lead to softening of the wall?

In summary, three modes of hormone action appear to be likely at this point: regulation of transcription, of translation, and of ion fluxes across cell membranes. At present we do not know the detailed mechanisms by which these events are controlled.

CONCLUSIONS

Reviewing concepts concerning the mode of action of plant hormones is disheartening and encouraging at the same time: disheartening because so little is available in terms of solid information, and encouraging because the field is wide open for major discoveries. On the basis of our current knowledge, we cannot expect to find unified principles for the mode of action of plant hormones, as have been found in animals for polypeptide hormones acting via cAMP or steroid hormones acting via gene activation. It is more likely that plants utilize a very limited number of regulatory substances, referred to as plant hormones, for the control of a wide array of processes via a number of different biochemical mechanisms. The elucidation of these mechanisms is clearly the task facing us.

LITERATURE CITED

Abeles, F. B. 1968. *Role of RNA and protein synthesis in abscission.* Plant Physiol. 43:1577–86.

Abeles, F. B., and B. Rubinstein. 1964. *Regulation of ethylene evolution and leaf abscission by auxin.* Plant Physiol. 39:963–69.

Chelsky, D., and F. W. Dahlquist. 1980. *Structural studies of methyl-accepting chemotaxis proteins of Escherichia coli: Evidence for multiple methylation sites*. Proc. Nat. Acad. Sci. USA 77:2434–38.

Cleland, R. E. 1980. *Auxin and H^+ – excretion: The state of our knowledge*. Pages 71–78 in F. Skoog, ed. *Plant Growth Substances 1979*. Springer-Verlag, Berlin-Heidelberg-New York.

Cohen, R. J., N. Y. Jan, J. Matricon, and M. Delbrück. 1975. *Avoidance response, house response and wind responses of the sporangiophore of Phycomyces*. J. Gen. Physiol. 66:67–95.

Cummins, W. R., H. Kende, and K. Raschke. 1971. *Specificity and reversibility of the rapid stomatal response to abscisic acid*. Planta. 99:347–51.

DeFranco, A. L., and D. E. Koshland, Jr. 1980. *Multiple methylation in processing of sensory signals during bacterial chemotaxis*. Proc. Nat. Acad. Sci. USA 77:2429–33.

Galston, A. W., P. J. Davies, and R. L. Satter. 1980. *The Life of the Green Plant*. Prentice-Hall, Englewood Cliffs, N.J.

Higgins, T. J. V., J. A. Zwar, and J. V. Jacobson. 1976. *Gibberellic acid enhances the level of translatable mRNA for α-amylase in barley aleurone layers*. Nature 260:166–69.

Jenson, E. V., and H. J. Jacobson. 1962. *Basic guides to the mechanism of estrogen action*. Recent Progr. Hormone Res. 18:387–414.

Jones, R. L., and J. E. Varner. 1967. *The bioassay of gibberellins*. Planta. 72:155–61.

Katzenellenbogen, B. S., and J. Gorski. 1972. *Estrogen action in vitro. Induction of the synthesis of a specific uterine protein*. J. Biol. Chem. 247:1299–1305.

Kende, H. 1971. *The cytokinins*. Int. Rev. Cytol. 31:301–38.

Kende, H., and G. Gardner. 1976. *Hormone binding in plants*. Ann. Rev. Plant Physiol. 27:267–90.

Kende, H., and A. D. Hanson. 1976. *Relationship between ethylene evolution and senescence in morning-glory flower tissue*. Plant Physiol. 57:523–27.

Kende, H., and A. Lang. 1964. *Gibberellins and light inhibition of stem growth in peas*. Plant Physiol. 39:435–40.

Koshland, D. E., Jr. 1980. *Biochemistry of sensing and adaptation*. Trends Bioch. Sci. 5:297–302.

Lefkowitz, R. J., M. R. Wessels, and J. M. Stadel. 1980. *Hormones, receptors, and cyclic AMP: Their role in target cell refractoriness*. Curr. Topics Cellular Reg. 17:205–30.

Lockhart, J. A. 1956. *Reversal of the light inhibition of pea stem growth by the gibberellins*. Proc. Nat. Acad. Sci. USA 42:841–48.

Mozer, T. J. 1980. *Control of protein synthesis in barley aleurone layers by the plant hormones gibberellic acid and abscisic acid*. Cell 20:479–85.

Musgrave, A., M. B. Jackson, and E. Ling. 1972. *Callitriche stem elongation is controlled by ethylene and gibberellin*. Nature New Biol. 238:93–96.

Musgrave, A., S. E. Kays, and H. Kende. 1969. *In-vivo binding of radioactive gibberellins in dwarf pea shoots*. Planta. 89:165–77.

Ray, P. M., U. Dohrmann, and R. Hertel. 1977. *Characterization of naphthaleneacetic acid binding to receptor sites on cellular membranes of maize coleoptile tissue*. Plant Physiol. 59:357–64.

Rubinstein, B., and A. C. Leopold. 1963. *Analysis of the auxin control of bean leaf abscission*. Plant Physiol. 38:262–67.

Russo, V. E. A., B. Halloran, and E. Gallori. 1977. *Ethylene is involved in the autochemotropism of Phycomyces*. Planta. 134:61–67.

Salisbury, F. B., and C. W. Ross. 1978. *Plant Physiology*. Wadsworth Publishing Co., Belmont, CA.

Schneider, D. 1966. *Chemical sense communication in insects*. Symp. Soc. Exp. Biol. 20:273–97.

Shichi, H., and R. L. Somers. 1978. *Light-dependent phosphorylation of rhodopsin. Purification and properties of rhodopsin kinase*. J. Biol. Chem. 253:7040–46.

Springer, M. S., M. F. Goy, and J. Adler. 1979. *Protein methylation in behavioural control mechanisms and in signal transduction.* Nature 280:279–84.
Tell, G. P., F. Haour, and J. M. Saez. 1978. *Hormonal regulation of membrane receptors and cell responsiveness: A review.* Metab. Clin. Exp. 27:1566–92.
Vanderhoef, L. N., and C. A. Stahl. 1975. *Separation of two responses to auxin by means of cytokinin inhibition.* Proc. Nat. Acad. Sci. USA 72:1822–25.
Venis, M. A., and P. J. Watson. 1978. *Naturally occurring modifiers of auxin-receptor interaction in corn: Identification as benzoxazolinones.* Planta. 142:103–7.
Wareing, P. F., and I. D. J. Phillips. 1978. *The Control of Growth and Differentiation in Plants.* Pergamon Press, Oxford.
Weiss, C., and Y. Vaadia. 1965. *Kinetin-like activity in root apices of sunflower plants.* Life Sci. 4:1323–26.
Wright, S. T. C. 1961. *A sequential growth response to gibberellic acid, kinetin and indolyl-3-acetic acid in the wheat coleoptile (Triticum vulgare L.)* Nature 190:699–700.
Zeevaart, J. A. D. 1971. *Effects of photoperiod on growth rate and endogenous gibberellins in the long-day rosette plant spinach.* Plant Physiol. 47:821–27.

10] Hormonal Regulation of Flowering and Sex Expression

by CHARLES F. CLELAND* and YOSEF BEN-TAL†

ABSTRACT

The hormonal control of flowering remains a major unsolved problem in plant physiology today. Several different theories on how flowering may be controlled have been suggested. Various attempts have been made to isolate the hormonal substances that control flowering, but with very little success. Each of the known plant hormones has been tested on many different plants for an effect on flowering, and endogenous levels have been measured to look for correlations with flowering. In a few cases one of the hormones has an important influence on flowering, but in no case has one of them been shown to be the critical limiting factor for flowering under non-inductive conditions. Salicylic acid has been shown to induce substantial flowering in many Lemnaceae; it must be present continuously to give the effect. Studies with ^{14}C-salicylic acid indicate that it is taken up rapidly and apparently converted quite quickly to a bound form of salicylic acid. In monoecious and dioecious plants, considerable evidence suggests that sex expression is controlled by the interaction of two or more of the known plant hormones. Since we are not sure how flowering is controlled, it is argued that in future work a variety of different approaches should be taken in studies dealing with the hormonal control of flowering.

INTRODUCTION

Flowering in photoperiodically sensitive plants is a complex process that is controlled by hormonal substances that are produced in the leaves and move via the phloem to the stem apices. Photoinduced leaves produce one or more

*Smithsonian Institution, Radiation Biology Laboratory, 12441 Parklawn Drive, Rockville, Maryland 20852.
†Department of Olive & Viticulture, Institute of Horticulture, The Volcani Center, Bet-Dagan, Israel.

STRATEGIES OF PLANT REPRODUCTION (BARC Symposium number 6—Werner J. Meudt, ed.)
Allanheld, Osmun, Totowa

flower-inducing substances that constitute the flowering stimulus (also called florigen [Chailakhyan 1936]), while vegetative leaves may produce one or more flower-inhibitory substances. When the level of the flowering stimulus at the stem apices exceeds some threshold value, it causes evocation whereby the indeterminate vegetative apex is transformed into a determinate reproductive apex. Evocation in turn leads to initiation of flower formation and development of flower primordia into mature functional flowers.

In most plants the male and female parts of the flower develop fully, producing bisexual or hermaphroditic flowers. Some plants produce separate male and female flowers, however. In these plants sex expression is genetically determined and can often be modified by environmental factors or by certain plant hormones.

POSSIBLE HORMONAL CONTROL SYSTEMS FOR FLOWERING

Most work on flowering has concentrated on the presumed flower-inducing substance(s) that constitute the flowering stimulus. The evidence for its existence comes both from the observation that the photoinductive signal is perceived by the leaves, but flower formation occurs at the stem apices, and also from grafting experiments where transmission of the flowering stimulus from an induced donor plant to a vegetative receptor plant has been demonstrated in a number of cases (Chailakhyan 1936; Zeevaart 1958; Lang 1965). These experiments show that photoinduced leaves produce a flowering stimulus, but they say nothing about the chemical nature of the stimulus. Traditionally, the flowering stimulus was thought of as a unique, organ-specific substance; since other plant hormones are small, water-soluble molecules it has often been assumed that the same was true for the flowering stimulus (Lincoln, Cunningham, and Hamner 1964; Evans 1969). But the failure to isolate such a substance indicates that other possibilities should also be considered.

The first possibility is that the flowering stimulus is present in very small amounts and/or is quite labile, so that successful isolation and demonstration of flower-inducing activity by bioassay is extremely difficult. Zeevaart (1978) has pointed out that, with the exception of abscisic acid (ABA), the other plant hormones were first isolated from a source other than higher plants, and not until a great deal was known about their chemical properties were they found in higher plant extracts. He suggests that it would be much easier to isolate and identify a flower-inducing substance if a source could be found that produced it in large amounts. Since gibberellins (GAs) (Lang 1970), indole-3-acetic acid (IAA) (Gruen 1959), and zeatin (Crafts and Miller 1974) are all produced by one or more species of fungi, it has been hoped that fungi could be found that would produce an active flowering stimulus. But, attempts to do this have so far yielded very little success (Lincoln et al. 1966; Cleland 1978).

A second possibility is that the flowering stimulus is a unique, organ-specific substance but is chemically quite distinct from the known plant hormones. Flowering, tuberization, and dormancy are examples of plant developmental responses that are under photoperiodic control in certain plants with the photoperiodic signal being perceived by the leaves, while the developmental change occurs at some distance from the leaves. ABA has been proposed as the hormonal substance responsible for the control of dormancy (Wareing and Saunders 1971), but in recent years evidence against this idea has accumulated (Alvim, Saunders, and Barros 1979). Thus, the hormonal factors that control these developmental processes are not known and possibly represent a new class of plant hormones that has yet to be discovered.

Efforts have been made to look for lipid-soluble compounds such as steroids, but without much success (Biswas, Paul, and Henderson 1966; Bledsoe and Ross 1978). Soluble proteins or polypeptides have also been suggested as a possibility (Cleland 1978). An examination by gel electrophoresis of proteins synthesized during a single inductive night in leaves of the short-day plant *Xanthium strumarium* L. failed to detect any differences from those synthesized in non-induced leaves (Sherwood, Evans, and Ross 1971). In the short-day plant *Pharbitis nil* Chois, a new protein was found in leaves after photoinduction (Oota and Umemura 1970), but this difference was not confirmed in a subsequent report (Stiles and Davies 1976). Attempts have also been made to examine proteins in the stem apices before and after evocation. In the long-day plant *Sinapis alba* L., one protein was found to disappear and two others to appear in the apical meristem during the transition to flowering (Pierard et al. 1980). These changes were detected by an indirect histoimmunofluorescence technique, however, and the specificity of this technique is open to question.

A third possibility that has been discussed by several authors is that flowering is not controlled by a unique, organ-specific substance, but rather by the interaction of two or more substances, some or all of which may already be well known (Evans 1969; Vince-Prue 1975; Cleland 1978). It has been pointed out that other developmental processes in plants that have been carefully studied are not controlled by organ-specific hormones, but rather by one or more of the known plant hormones (Evans 1969; Cleland 1978). One is thus forced to ask why flowering should be different.

One attractive feature of the idea that flowering is controlled by the interaction of several substances is that it easily accommodates the participation of one or more flower-inhibitory substances. The relative importance of flower inhibitors for the control of flowering may vary considerably from one plant to another, but there is convincing evidence that flower inhibitors are important, at least in certain plants (Evans 1960; Raghavan and Jacobs 1961; Murfet and Reid 1973; Vince-Prue and Guttridge 1973; Lang, Chailakhyan and Frolova, 1977). The difference between vegetative and photoinduced leaves in terms of the hormonal substances they produced may be qualitative, in that vegetative leaves only produce flower inhibitors and photoinduced leaves only produce the flowering stimulus; or it could be quantitative, and with photoinduction there is only a change in the relative amounts of the hormonal substances that are formed (Cleland 1978).

Probably the most convincing demonstration of the involvement of flower inhibitors in the control of flowering comes from work of Lang, Chailakhyan, and Frolova (1977). They showed that flowering in receptor plants of the day-neutral tobacco *Nicotiana tabacum* L. var. Trapezond was greatly inhibited when they were grafted to either of the long-day plants *Nicotiana sylvestris* L. or *Hyoscyamus niger* L. and the donor plant was kept on short days. If the donor plant was kept on long days, then the receptor flowered earlier than did Trapezond/Trapezond control grafts. Similar experiments using the short-day plant *Nicotiana tabacum* L. var. Maryland Mammouth showed stimulation of flowering in the Trapezond receptor when the donor was kept on short days, but failed to demonstrate any significant flower inhibition when the donor was kept on long days. Nevertheless, these results clearly indicated that, at least in these two long-day plants, the leaves are capable of producing translocatable flower-inducing substances on long days and flower-inhibitory substances on short days, and these substances presumably interact at the stem apices to control whether the plants remain vegetative or undergo evocation and flower formation.

Another possibility is that flowering is not controlled primarily by hormonal substances, but rather by the nutrient status of the plant. Sachs (1977) has referred to this as the nutrient diversion hypothesis. It was suggested that assimilate supply limited reproductive development and that environmental parameters regulated the transition from vegetative to reproductive development through their control of assimilate supply to the stem apices (Ramina, Hackett, and Sachs 1979). Evidence has been obtained that increased assimilate supply to the stem apices is important for flowering in several different plants (Bodson 1977; Quedado and Friend 1978; Ramina, Hackett, and Sachs 1979). In most plants this information is lacking, but in view of the rapid development that ensues once evocation has occurred, it is logical to expect that an increased assimilate supply is a normal change when a plant flowers. The crucial question is whether hormonal factors are also involved in controlling flowering. We would argue that, at least in plants that exhibit strict photoperiodic control of flowering, hormonal factors are the primary controlling agents. It is very likely, however, that one of their effects is either to increase source strength in the leaves or sink strength in the stem apices and thus cause an increased movement of assimilates to the stem apices, and without these assimilates flower formation might either be delayed or prevented entirely.

Finally, suggestions have been made that the reason efforts to isolate the flowering stimulus have not been successful is that the flowering stimulus is not a chemical substance, but rather consists of (a) an action potential generated by photoinduced leaves that rapidly moves from leaves to stem apices, or (b) consists of pressure pulses transmitted in the phloem from leaves to the stem apices over a period of about 15 minutes (Zeevaart 1979b). At present there is no direct evidence to support these or any other nonchemical hypothesis concerning the flowering stimulus. Nevertheless, it is hoped that proponents of such ideas will undertake studies aimed at determining if nonchemical factors could have an influence on the flowering process.

EFFORTS TO ISOLATE FLOWERING HORMONES

Numerous efforts have been made to isolate the hormonal factors responsible for the control of flowering. In most cases these efforts have focused solely on trying to obtain substances with flower-inducing activity. Most efforts have involved the direct extraction of leaves and apices with organic solvents such as acetone or methanol (Bonner and Bonner 1948; Lincoln, Cunningham, and Hamner 1964; Carr 1967; Hodson and Hamner 1969; Cleland 1978). One of the most successful reports was by Hodson and Hamner (1969), who prepared aqueous acetone extracts from young leaves and apices of photoinduced *Xanthium*. The extracts gave a good flowering response when tested on either *Xanthium* or the short-day plant *Lemna paucicostata* Hegelm. 6746. Extracts of vegetative plants were inactive. To get activity on *Xanthium* they had to add a small amount of gibberellin A_3 (GA_3), which by itself had no effect on flowering. Unfortunately, efforts to purify the active substance(s) were not successful, and attempts by several other laboratories to duplicate these results have not been successful (Cleland 1978).

Recently, Chailakhyan (1978) has obtained impressive results with extracts of flowering Maryland Mammouth tobacco. The extracts were tested on young plants of the short-day plant *Chenopodium rubrum* L., and they elicited a good flowering response. Comparable extracts of vegetative Maryland Mammouth tobacco gave no flowering when tested on *Chenopodium*. Only incomplete details of the extraction procedure have been published (see Zeevaart 1979b), and thus, at present, it is not possible to try to duplicate these results. Until such time as further progress is reported and details of the extraction procedure are published so that other workers can attempt to duplicate these results, it will be difficult to assess the significance of this work.

Another approach to the isolation of the flowering stimulus has been to make extracts of material derived from the phloem. Since all available evidence indicates that long-distance movement of the flowering stimulus occurs exclusively in the phloem, it is logical to concentrate on phloem material.

In one study aphids were allowed to feed on vegetative or flowering *Xanthium*, and the honeydew they produced was extracted and tested on the long-day plant *Lemna gibba* L. G3 for an effect on flowering (Cleland 1974a). This approach made two basic assumptions. The first was that the flowering stimulus would pass through the aphid digestive tract without being degraded, and since the plant hormones IAA (Maxwell and Painter 1962), GA (Hoad and Bowen 1968), ABA (Bowen and Hoad 1968) and zeatin (Phillips and Cleland 1972) have all been obtained from aphid honeydew in an active form, this seemed like a reasonable assumption. The second assumption was that the hormonal factors that control flowering in *Xanthium* would be active on *L. gibba* G3. Successful transmission of the flowering stimulus in graft unions between plants of different response type and between different species and even different genera has given rise to the concept of a ubiquitous flowering stimulus that is at least functionally similar in different flowering plants. It was hoped that this concept could be extended to plants as far apart taxonomically as *Xanthium* and *L. gibba* G3.

An acidic ethyl acetate fraction was obtained from the honeydew, it was subjected to thin-layer chromatography (TLC), and the different zones from the TLC plate were tested on *L. gibba* G3 for an effect on flowering (Cleland 1974a). At least one zone of flower-inducing activity and two zones of flower-inhibitory activity were found. The flower-inducing substance was identified as salicylic acid (Cleland and Ajami 1974).

Nevertheless, since salicylic acid is present in both vegetative and flowering *Xanthium* and does not induce flowering in *Xanthium,* it is clear that salicylic acid is not the flowering stimulus in *Xanthium*. The zones of flower-inhibitory activity have not been identified, but since they were obtained from both flowering and vegetative plants, it is unlikely that the active flower-inhibitory substances play a crucial role in controlling flowering.

In recent years a technique has been developed that permits the collection of gram dry weight quantities of phloem exudate from the leaves of the short-day plant *Perilla crispa* (Thunb) Tanaka (King and Zeevaart 1974). Purse (see Zeevaart 1979b) tested the effect of the entire exudate on flowering of cultured excised *Perilla* shoot tips. In a few experiments, he obtained considerable flowering with the exudate from induced leaves and no flowering with exudate from vegetative leaves. Unfortunately, in most of his experiments he obtained no flowering, and thus the significance of these results is unclear. In our laboratory we have also collected *Perilla* phloem exudate, but so far no flower-inducing activity has been found.

INFLUENCE OF KNOWN SUBSTANCES ON FLOWERING

Each of the known plant hormones has been studied extensively for its effect on flowering. Since an excellent review on the involvement of the known plant hormones in flowering has appeared (Zeevaart 1978), no attempt will be made here to provide a comprehensive treatment of the subject. Instead, we will discuss the major effects reported for each hormone and consider what role any of these substances may play in the control of flowering. Since there has been considerable work on the influence of salicylic acid and related compounds on flowering in the Lemnaceae, we will also examine this work.

Auxins. With few exceptions, auxin treatment leads to inhibition of flowering. For instance, in both *Xanthium* (Salisbury 1955) and *Pharbitis* (Ogawa 1962) auxins are inhibitory, particularly if applied just before or during the inductive short-day cycle, and the inhibitory effect was shown to be in the leaves. By contrast, the inhibitory effect of auxin is localized at the stem apices for both *Chenopodium* (Seidlová and Khatoon 1976) and the long-day plant *Brassica campestris* L. (Krekule and Seidlová 1977).

In many plants one of the earliest morphological changes that occurs with the onset of flowering is the activation of axillary meristems as they are released from apical dominance. In both *Chenopodium* and *Brassica,* auxin application inhibits flowering and also the activity of the axillary meristems and, thus, promotes apical dominance (Krekule and Seidlová 1977; Seidlová 1980). Since auxins are known to control apical dominance in many plants, these results raise the possibility that a decrease in the level of auxins in the apical meristem is one of the earliest changes that occurs during the transition to flowering.

Efforts have been made to examine endogenous auxin levels in various plants, but there is no consistent pattern to the results (Zeevaart 1978). In most cases the extracts were prepared from leaves or apical buds that included leaves, and since very young leaves generally have high levels of auxins, such results fail to tell what is happening at the apical meristem itself. Even if one assumes that a decrease in the level of auxins at the stem apices is crucial for flower formation, it seems likely that this decrease is in response to the arrival of the flowering stimulus at the stem apices and, thus, is a secondary response. Nevertheless, one hopes that critical measurement of auxin levels in the apical meristems of *Chenopodium, Brassica,* and other plants will be made.

Gibberellins. Many plants that grow vegetatively as a rosette and are induced to flower by cold and/or long-day treatment can be made to flower on strict noninductive conditions by GA treatment (Lang 1965). By contrast, in cold-requiring and long-day plants that do not show the rosette growth habit, GA treatment does not lead to flowering (Chouard 1957; Stoddart 1962; Cleland and Briggs 1969; Zeevaart 1978).

In at least three qualitative short-day plants, GA will induce flowering under strict non-inductive conditions (Nanda et al. 1967; Pharis 1972; Sawhney and Sawhney 1976), but in most short-day plants GA treatment is ineffective in causing flowering (Lang 1965).

Chailakhyan (1979) has attempted to explain the inability of GA to induce flowering in most short-day plants by postulating that flowering is controlled by the interaction of GAs with unknown substances called anthesins. He suggests that short-day plants produce GAs on all daylengths, but anthesins only on short days, while long-day plants produce anthesins on all daylengths but GAs only on long days. This theory fails to take into account nonrosette long-day plants where GA does not induce flowering and the few short-day plants where GA does induce flowering. Also, if this theory is correct, it should be possible to form a graft union between a long-day plant and a short-day plant and obtain flowering, even though both plants were maintained under noninductive conditions. Only a few such grafts have been attempted, but they have generally been negative (Vince-Prue 1975).

In the long-day rosette plants, flowering is accompanied by rapid stem elongation or bolting, and there is also a significant increase in endogenous GAs that starts just before the onset of bolting (Jones and Zeevaart 1980a, 1980b; Metzger and Zeevaart 1980).

Treatment of long-day plants with GA stimulates bolting and usually induces flowering, but in some long-day rosette plants GA does not induce flowering (Jacques 1968; Suge and Rappaport 1968; Peterson and Yeung 1972). Those that do flower show considerable variation in their sensitivity to GA treatment (Lang 1965). In *Silene armeria* L., treatment with the growth retardant AMO 1618, which is known to inhibit GA biosynthesis (Dennis, Upper, and West 1965; Ruddat, Heftmann, and Lang 1965), reduces the level of GAs well below that seen in the short-day control and severely inhibits bolting, but has no effect on flowering (Cleland and Zeevaart 1970). Thus, in *Silene,* and probably most other rosette plants, GAs are critical for bolting but apparently are not crucial for flowering.

In a few cases GA does appear to be important for flowering. GA treatment in the long-short-day plant *Bryophyllum daigremontianum* (R. Hamet et Perrier) Berger causes flowering on short days, but not on long days, and long-day treatment leads to an increase in endogenous GAs (Zeevaart 1969b). Grafting experiments have shown that the GA effect is in the leaf and that GA is necessary in order to get the synthesis of the flowering stimulus (Zeevaart 1969a).

The plant growth retardants AMO 1618 and CCC (also inhibits GA biosynthesis [Dennis, Upper, and West 1965; Ninnemann et al. 1964]) inhibit flowering in *Samolus parviflorus* Raf. (rosette long-day plant), *L. gibba* G3 (nonrosette long-day plant) and *Pharbitis* (short-day plant), and GA will partially or completely reverse this inhibition in each case (Zeevaart 1964; Baldev and Lang 1965; Cleland and Briggs 1969). It is possible that GAs are limiting for flowering in *Samolus* on short days, but this is highly unlikely in view of results with other long-day rosette plants. Nevertheless, it is clear that GAs are not limiting for flowering on non-inductive conditions in *L. gibba* G3 and *Pharbitis,* and in all three plants a certain level of GA is needed in order to produce flowering. In many other plants attempts to inhibit flowering with AMO 1618 or CCC have not been successful (Cathey and Stuart 1961), and thus GAs either are not critical for flowering or they are required in such low amounts that it has not been possible to demonstrate this requirement.

Xanthium is similar to most other short-day plants in that GA will not induce flowering. Nonetheless, the observation by several laboratories (Lincoln, Cunningham, and Hamner 1964; Carr 1967; Hodson and Hamner 1969) that extracts of flowering *Xanthium* were more active or were only active when GA was added to the extracts raises the possibility that a certain low level of GA may be necessary for flowering to occur in this plant.

In some cases GA can strongly inhibit flowering (Hackett and Sachs 1967; Guttridge 1969; Reid, Dalton, and Murfet 1977). In the short-day plant *Fragaria X. ananassa* Duch. (strawberry), flowering seems to be controlled primarily by a translocatable flower-inhibitory, growth-promoting substance(s) produced by the leaves on long days, and since GA treatment on short days promotes growth and inhibits flowering, it has been suggested that GA could be a part of this stimulus (Tafazoli and Vince-Prue 1978).

In *Pisum sativum* L., the G-type plant is a long-day plant and flowering can be inhibited by GA_3 (Murfet and Reid 1973; Proebsting, Davies, and Marx 1978). When ^3H-GA_9 was applied to leaves of G-type plants, it was converted to several compounds of greater polarity (Proebsting and Heftmann 1980). This conversion was blocked by one long day, which is sufficient to induce flowering. Therefore, it has been postulated that on short days flowering is inhibited by a polar GA and on long days the synthesis of this polar GA is inhibited, and consequently flowering is able to occur.

In conclusion, the involvement of GAs in the flowering process varies considerably from one plant to the next. In a few cases GAs appear to have a critical influence on the promotion or inhibition of flowering. In a few other cases they appear to be required for flowering, but are not a critical limiting factor in vegetative plants. In many other cases they can promote or inhibit flowering, but a definite role for endogenous GAs has not been established. Further work concentrating on the endogenous GAs will be required before we can draw a clearer picture of the extent to which endogenous GAs are important for the control of flowering.

Cytokinins. Cytokinins have been shown to induce flowering in only a few plants, and there is no evidence that endogenous cytokinins control flowering in any of these cases (Michniewicz and Kamienska 1965; Gupta and Maheshwari 1970; Venkataraman, Seth, and Maheshwari 1970; T'se et al. 1974; Ogawa and King 1979). In the case of *Pharbitis*, it was concluded that cytokinin treatment increased assimilate transport to the stem apex; and since this would also result in increased transport of the flowering stimulus to the stem apex, it was concluded that the cytokinin effect on flowering was indirect and its primary effect was on assimilate transport (Ogawa and King 1979).

In a few plants cytokinins have been shown to be inhibitory for flowering (Sotta and Miginiac 1975; Fontaine, Dugué, and Miginiac 1977; Miginiac 1978; Seidlová 1980). In the quantitative long-day plant *Scrophularia arguta* Sol., stem segments cultured in the absence of roots flowered well under short days, but if roots were allowed to form, flowering under short days was greatly inhibited (Miginiac 1978). Addition of cytokinin to the medium also inhibited short-day flowering in the absence of roots. Similar results have been obtained for *Chenopodium polyspermum* L., a quantitative short-day plant, and for *Anagallis arvensis* L., a qualitative long-day plant (Sotta and Miginiac 1975; Fontaine, Dugué, and Miginiac 1977). Since roots are well known to export cytokinins to the shoot, the suggestion has been made that flowering in these three plants may be controlled by a balance between flower-inducing substances coming from the leaves and flower-inhibitory substances coming from the roots (cytokinins) (Miginiac 1978). Nevertheless, no effort has been made to measure cytokinin levels in root exudates from these plants to see if there is more cytokinin under non-inductive conditions.

It has been shown for several species that one of the earliest events that is observed after plants receive the photoinductive treatment is a transitory increase in the mitotic index, and any treatment that abolishes this increase also inhibits flowering (Wada 1968; King 1972; Jacqmard et al. 1976; Bernier et al. 1977). For instance, when cytokinins are added directly to the stem apex they induce an increase in mitotic activity similar to that induced by a single long day (Bernier et al. 1977). The long-day treatment, however, also induces flowering, while cytokinin treatment does not. These results have been interpreted to mean that the flowering stimulus in *Sinapis,* and presumably in other plants where an increase in mitotic activity has also been seen, consists of at least two components: one that stimulates mitotic activity, which may be a cytokinin, and a second component that is necessary for actual flower formation (Bernier et al. 1977).

The major problem with this interpretation is that direct evidence for foliar production and transport of cytokinins to stem apices is lacking. Examination of honeydew produced by aphids feeding on flowering or vegetative *Xanthium* showed that the level of cytokinins in the phloem is substantially higher in the flowering plants than in the vegetative plants (Phillips and Cleland 1972). In this case the flowering plants had received a minimum of five short days before aphids were added to the plants, and the plants remained in short days for ten days, during which honeydew was collected. Seemingly contradictory results have been obtained by Henson and Wareing (1974, 1977). They examined extracts of *Xanthium*; after just one short day there was a significant decrease in cytokinin levels in leaves, buds, and root exudate, and with additional short days the decreases became more pronounced. Perhaps these two reports can be reconciled if one assumes that short-day treatment leads to an increased export of cytokinins from leaves (resulting in higher levels in the phloem and lower levels in leaves) and increased turnover of cytokinins at stem apices, which could result in lower levels of cytokinins in extracts of stem apices. What is clearly needed is work on cytokinin levels in relation to flowering, using a labeled cytokinin so that any movement within the plant may be detected.

Abscisic acid. ABA is well known for its ability to inhibit growth and to counteract the effect of other exogenously applied plant hormones (Zeevaart 1979a). With regard to flowering, ABA treatment usually leads to flower inhibition (Evans 1966; El-Antably, Wareing, and Hillman 1967; Venkataraman, Seth, and Maheshwari 1970; Schwabe 1972). In a few cases ABA treatment causes flower formation (El-Antably, Wareing, and Hillman 1967), but later work disputed these findings (Krekule and Horavka 1972; Nakayama and Hashimoto 1973). Examination of ABA levels in spinach showed that ABA was two to three times higher in flowering plants on long days than in vegetative plants on short days (Zeevaart 1974). Other studies have also failed to show a negative correlation between ABA levels and flowering, and thus there is no conclusive evidence that ABA plays a critical role in flowering in any plants.

Ethylene. With few exceptions ethylene has inhibitory effects on flowering (Zeevaart 1978). In *Pharbitis* ethylene inhibited flowering completely, and the critical time for its application was during the second half of the 16-hour dark period. It was shown that ethylene inhibits the induction process in the cotyledons (Suge 1972).

Originally it was thought that auxin caused flowering in pineapple, but it was shown that the auxin treatment caused ethylene production and the ethylene actually was responsible for inducing flowering (Burg and Burg 1966). Zeevaart

(1978) has stated that as far as we know the ethylene effect applies to all bromeliads. But, so far, there is no information to indicate if endogenously produced ethylene is responsible for induction of flowering under natural conditions.

Salicylic acid. The discovery that salicylic acid has flower-inducing activity was made when extracts of honeydew produced by aphids feeding on *Xanthium* were shown to contain a substance that would induce flowering in *L. gibba* G3 on short days (Cleland 1974a). This substance was identified as salicylic acid (Cleland and Ajami 1974). Subsequent work has shown that although salicylic acid will not induce flowering in every Lemnaceae, it does cause substantial flowering in many different members of the family (Khurana and Maheshwari 1978, 1980; Scharfetter, Rottenburg, and Kandeler 1978; Cleland and Tanaka 1979; Watanabe and Takimoto 1979; Cleland, Tanaka, and Feldman 1981) and also in the closely related aquatic plant *Pistia stratiotes* L. (Pieterse 1978). Flowering can also be induced in the terrestrial short-day plant *Impatiens balsamina* L. (Nanda, Kumar, and Sood 1976), but repeated attempts to induce flowering in *Xanthium* with salicylic acid, either alone or in combination with GA and cytokinin, have not been successful (Cleland, unpublished data).

A number of closely related compounds have been tested to determine the specificity of this response (Cleland 1974b; Watanabe and Takimoto 1979). It was found that the effect was quite specific, being limited to salicylic acid, benzoic acid, and a small number of closely related compounds. For induction of flowering in *L. gibba* G3, salicylic acid was more effective than benzoic acid, but in *L. paucicostata* 151 benzoic acid was more effective (Cleland 1974b; Watanabe and Takimoto 1979). There are no published reports of the occurrence of salicylic acid or benzoic acid in the Lemnaceae. In preliminary work, however, significant levels of benzoic acid have been found in several different Lemnaceae, including *L. gibba* G3 and *L. paucicostata* 151, but it is not clear if the level changes in response to photoinduction in either of these plants (A. Takimoto and N. Takahashi, personal communication).

The effect of salicylic acid on flowering in *L. gibba* G3 is strongly daylength-dependent, and the effect of optimal levels of salicylic acid is to cause a shift in the critical daylength from approximately ten hours to approximately eight hours (Cleland and Tanaka 1979). This means that on the appropriate daylengths, salicylic acid will cause significant increases in the flowering percent (FL%), but if the daylength is a little too short or too long, there may be little or no salicylic acid effect. Consequently, we have developed two other systems where control plants are vegetative and salicylic acid treatment gives FL% of about 80. One system involves *Lemna obscura* (Austin) Daubs 7133. This plant does not flower on any daylength or any medium we have tried so far, but the addition of 10 μM salicylic acid to the medium yields rapid and abundant flowering (Cleland, Tanaka, and Feldman 1981).

The other system involves *L. gibba* G3. Most of the work with *L. gibba* G3 has utilized a Hoagland-type medium. We add 30 μM EDTA to this medium and designate it as E medium. On E medium, *L. gibba* G3 shows a qualitative long-day flowering response with a critical daylength of about 10 hr (Cleland and Tanaka 1979). Another medium that is commonly used for work with the Lemnaceae is Hutner's medium. We use it at half strength; compared with E medium it contains 850 μM EDTA and 1.25 mM NH_4NO_3 (Tanaka, Cleland, and Hillman 1979). On half-strength Hutner's medium (0.5 H), *L gibba* G3 shows excellent growth but no flowering under continuous light (Tanaka, Cleland, and Hillman 1979). This effect has been attributed to the presence of ammonium in the medium (Kandeler 1969), but even in ammonium-free half-strength Hutner's medium (NH_4-free 0.5 H) there is virtually no flowering. With the addition of 10 μM salicylic acid, however, the inhibition is rapidly reversed and a very large flowering response is obtained (Tanaka, Cleland, and Hillman 1979).

In any effort to understand better the possible mechanisms by which salicylic acid induces flowering, it is necessary to know how long the salicylic acid must be present in order to give the response. This has been investigated with the *L. gibba* G3 system, and it has been found that salicylic acid must be present continuously in order to obtain the optimal response (Table 10.1). If salicylic acid is present for 6 or 8 days substantial flowering is seen, but if the plants are given salicylic acid for 6 or 8 days and then transferred to control medium for an additional 5 or 3 days, respectively, flowering drops off substantially. Apparently, the effect of salicylic acid on flowering disappears almost immediately when it is removed from the medium. Similar results have been obtained in preliminary experiments dealing with the effect of salicylic acid on *L. obscura* 7133 (Cleland, unpublished data).

Recently we have investigated the action of salicylic acid further by studying the fate of ^{14}C-salicylic acid using the *L. gibba* G3 system on NH_4-free 0.5 H medium. The plants were grown for 5 days on NH_4-free 0.5 H medium. On the sixth day, 10 μM ^{14}C-salicylic acid was added to the medium (1 μCi of ^{14}C-salicylic acid was added to each flask). After 6 or 24 hr the plants were harvested, extracts were prepared, and the different fractions examined for radioactivity (Table 10.2). For the 6-hr labeling period, about 47% of the counts in the methanol extract partitioned into the acidic ethyl acetate fraction, and of these, 88% co-chromatographed with authentic salicylic acid. After 24 hr the percentage of counts in the acidic ethyl acetate fraction had dropped to about 35 and only about 75% co-chromatographed with authentic salicylic acid. So far no attempt has been made to identify the labeled compounds in the other fractions, but it is our assumption that much of the activity in the acidic butanol fraction is due to a bound form of salicylic acid, possibly a glycoside, and the increase in this fraction between 6 and 24 hr results from some of the free salicylic acid present after 6 hr being converted to the bound form over the next 18 hr.

These results indicate that salicylic acid is taken up by *L. gibba* G3 quite rapidly and much of it is presumably converted to a bound form quite quickly. Preliminary results indicate that when *L. gibba* G3 is grown on NH_4-free 0.5 H medium plus 10 μM ^{14}C-salicylic acid for 5 days and then switched to control medium, considerable free salicylic acid is still present 3 days later. Since the effect of salicylic acid appears to disappear very quickly, it is possible that its

Table 10.1 Influence of Duration of Salicylic Acid (SA) Treatment on Ability of 10 μM SA to Reverse the Inhibition of Flowering in *Lemna gibba* G3 due to Ammonium-free Half-strength Hutner's Medium

Number of days SA treatment	Number of days on control medium	FL%	Number VF
0	11	2.0 ± 0.8	614.9 ± 15.0
6	0	28.6 ± 2.9	25.7 ± 0.7
8	0	57.8 ± 2.7	29.7 ± 2.0
11	0	79.6 ± 0.3	33.5 ± 1.5
4	7	5.8 ± 0.8	346.0 ± 8.5
6	5	4.0 ± 0.8	270.4 ± 15.6
7	4	18.5 ± 2.8	204.0 ± 7.0
8	3	35.3 ± 0.6	145.3 ± 2.6
9	2	53.1 ± 0.8	98.0 ± 3.5
0 (LD control)	11	78.7 ± 0.4	49.3 ± 1.9

Note: Long-day control grown under continuous light on E medium, which is a Hoagland-type medium that contains 30 μM EDTA and 1% sucrose. All other plants grown under continuous light on ammonium-free half-strength Hutner's medium, which contains 850 μM EDTA and 1% sucrose.

action is tied to its uptake and that it is acting at the level of the cell membrane. One possibility would be an effect on permeability, and salicylic acid has been shown to alter ion permeability in roots (Glass and Dunlop 1974). More work will be required, however, before we can decide whether salicylic acid is active at the membrane level or whether salicylic acid or some metabolite is having an effect on cell metabolism.

HORMONAL CONTROL OF FLOWER DEVELOPMENT AND SEX EXPRESSION

Most work on flowering has dealt with events in the leaf or at the apex that lead to evocation and formation of a flower primordium. Much less work has been done on environmental and chemical control of the development of flower primordia into mature flowers.

The common perception is that once the minimum number of inductive cycles has been given to a plant, it can be put under non-inductive conditions and the flower primordia will develop fully into mature flowers. A few plants like this are known (Vince-Prue 1975), but it is much more common for flower development also to show a photoperiodic requirement. In *Xanthium* flowering can be induced by a single short day, but the time it takes to produce ripe seed

Table 10.2 Distribution of ^{14}C-Salicylic Acid (^{14}C-SA) in Extracts of *Lemna gibba* G3 Grown under Continuous Light on Ammonium-free Half-strength Hutner's Medium and Exposed to ^{14}C-SA for Six or Twenty-four Hours.

	6 hr		24 hr	
Fraction	cpm/g FW	%	cpm/g FW	%
Methanol extract	17,822	100.0	42,035	100.0
Ethyl acetate, pH 3	8,394	47.1	14,670	34.9
Butanol, pH 3	3,404	19.1	13,830	32.9
Ethyl acetate, pH 3 after acid hydrolysis	1,907	10.7	3,531	8.4
Aqueous phase	2,210	12.4	6,726	16.0

Note: Plants were grown in control medium for 138 or 120 hr, followed by 6 or 24 hr respectively, in media containing 10 μM ^{14}C-SA. The cpm in the methanol extracts represent 0.6% and 1.3% of the radioactivity added to the media for the 6 and 24 exposures, respectively. TLC of the ethyl acetate, pH 3 fraction indicated 88.7% and 80.9% of the counts cochromatographed with authentic SA for the 6- and 24-hr exposures, respectively.

can vary greatly—from 7 or 8 months after one short day to only about 30 days when the plants are left on short days (Naylor 1941).

In many plants there is an absolute photoperiodic requirement for flower development. In *L. gibba* G3, flower development ceases almost immediately when plants are switched from long days to short days (Cleland and Briggs 1967). Although salicylic acid can induce flower initiation in *L. gibba* G3, it has very little effect on flower development (Cleland 1974b). *Caryopteris* x *clandonensis* A. Simmonds is day-neutral with respect to flower formation, but requires short days for flower development. If the long night is interrupted flower development is inhibited, and this light interruption was shown to be red/far-red reversible, indicating phytochrome control (Piringer, Downs, and Borthwick 1963). In *Phaseolus vulgaris* L., flower development also requires short days and is inhibited by ABA and promoted by cytokinins. Examination of extracts showed more growth-inhibitory substances (including ABA) in leaves and buds of plants grown on long days and more cytokinin in xylem sap from plants grown on short days (Morgan and Zehni 1980). These results are suggestive that cytokinins and ABA may be involved in the control of flower development in *Phaseolus,* but in other plants, such as *Xanthium* and *L. gibba* G3, there is no indication that one of the known hormones can substitute for the photoperiodic requirement for flower development. Thus, in these two plants and many others, it appears that the flowering stimulus is requied not only for flower formation, but also for continued development of the flower primordia.

The final stage of flower development is the production of fertile stamens and carpels. Very little work has been done on environmental or chemical control of flower fertility, but at least, in a few cases where it has been studied, it is clear that returning the plant to unfavorable conditions can interfere with or inhibit normal development. In *Glycine max* (L.) Merrill a minimum of two short days are needed to induce flowering, but even with as many as ten short days there was a high percentage of degenerate microspores if the plants were switched back to long days (Nielson 1942). In *Zea mays* L. (corn), short days favor carpel development and lead to the production of sterile pollen (Moss and Heslop-Harrison 1968).

In most plants both the male and female flower parts develop fully, giving rise to bisexual or hermaphroditic flowers. But in other plants, after the flower primordia pass through a bisexual stage where both stamen and carpel primordia are present, only one organ develops to maturity, resulting in either staminate (male) or carpellate (female) flowers. Monoecious plants have separate male and female flowers on the same plant, while in dioecious plants the male and female flowers are on separate male and female plants.

Sex expression is genetically determined, yet it can often be easily modified by environmental and chemical factors. Although exceptions do exist, when most monoecious and dioecious plants are subjected to strong photoinductive conditions, there is an increase in femaleness as compared to when they are grown in marginally inductive conditions. Thus, it seems that treatments that promote initiation and development of flowers tend to promote femaleness and to depress maleness (Heslop-Harrison 1957). For instance, *Xanthium* plants that received one short day produced more male than female flowers, but if plants were given continuous short days just the opposite was true (Naylor 1941). In most *Cucumis sativus* L. (cucumber) cultivars, high temperature and long days promote maleness and low temperature (mainly at night) and short days promote femaleness (Heslop-Harrison 1957).

Considerable investigation of hormonal effects on sex expression has been done. In monoecious plants auxins and ethylene usually promote femaleness, while GAs promote maleness. Treatment of monoecious cucumber with auxin lowers the node of the first female flower (i.e., causes earlier formation of the first female flower) (Laibach and Kribben 1950). The auxin level was found to be higher in gynoecious plants (that produce only female flowers) than in monoecious plants and also higher on short days (which promote femaleness) than on long days (Rudich, Halevy, and Kedar 1972b). The question is whether the auxin effects are due to auxin or to ethylene, since high levels of auxin are known to stimulate ethylene production and ethylene also promotes femaleness (Rudich, Halevy and Kedar 1969; Byers et al. 1972). Comparison of gynoecious and monoecious lines showed more ethylene production by the former (Byers et al. 1972; Rudich, Halevy and Kedar 1972a). In addition, under short days a gynoecious line produced almost twice as much ethylene as under long days, but there was only a slight increase for a monoecious line (Rudich Halevy and Kedar 1972a). Thus, it seems clear that ethylene is involved and promotes femaleness, but the extent to which auxin is acting by itself remains to be determined.

More recent work has shown that in addition to GA, $AgNO_3$ and aminoethoxyvinyl glycine (AVG) also promote development of male flowers in gynoecious cucumbers (Atsmon and Tabbak 1979). AVG and $AgNO_3$ are antiethylene agents, with AVG blocking ethylene synthesis and $AgNO_3$ inhibiting ethylene action. AVG causes a substantial reduction in ethylene production by gynoecious cucumbers, and $AgNO_3$ treatment leads to a significant increase in ethylene production, but in both cases the action of endogenous ethylene on the plants was greatly reduced (Atsmon and Tabbak 1979). Thus, the effect of AVG and $AgNO_3$ can be understood in terms of an effect on ethylene. But GA treatment has no effect on ethylene production and, thus, promotion of maleness by GA is presumably through a basically different mechanism that has yet to be elucidated.

ABA and cytokinins have also been shown to influence sex expression in cucumbers and related plants. ABA promotes maleness in monoecious cucumbers by increasing the node number of the first female flower and promotes femaleness in gynoecious cucumbers by lowering the node of the first female flower (Friedlander, Atsmon, and Galun 1977a). Shoot tips of monoecious plants have a higher ABA content that those of gynoecious plants (Friedlander, Atsmon, and Galun 1977b), but whether ABA has a controlling influence on sex expression is not clear. Cytokinin application to *Luffa cylindrica* Roem, which is a member of the Cucurbitaceae, caused transformation of male inflorescences into bisexual and female flowers and in some cases caused the inflorescences to develop into vegetative shoots (Takahashi, Suge, and Saito 1980), but information on endogenous cytokinin levels is lacking.

In corn the hormonal control of sex expression differs somewhat from that in cucumber in that GAs promote femaleness. Application of GA causes male flowers to become sterile and female flowers to develop in the apical inflorescence, where normally only male flowers develop (Hansen, Bellman, and Sacher 1976). In certain genetic dwarf varieties of corn where endogenous GA levels are very low, male flowers are sometimes seen on the lateral inflorescence, where normally only female flowers develop (Rood, Pharis, and Major 1980). Sex expression in corn can also be influenced by daylength and light intensity, with short days and low light intensity promoting femaleness. Rood, Pharis, and Major (1980) examined endogenous GA levels in corn and showed that during development of the male inflorescence GA levels drop substantially, and at anthesis there is much more GA in the female inflorescence than in the male inflorescence. Giving low light intensity, which causes female flower formation at the apical inflorescence, causes a substantial increase in the level of GA in the apical inflorescence over what is seen when only male flowers are formed in the apical inflorescence. Thus, these results provide strong evidence that endogenous GAs, perhaps in combination with other hormones, are involved in the control of sex expression in corn.

Considerable work has also been done on several dioecious species. In such plants the flower primordia are sexually uncommitted at the time of initiation, and under appropriate environmental conditions, or with certain chemical treatments, both sexes are able to form flowers of the opposite sex. In *Cannabis sativa* L., auxin and ethylene promote the formation of female flowers on male plants and GA promotes the formation of male flowers on female plants (Heslop-Harrison 1956; Mohan Ram and Jaiswal 1970, 1972), and thus the hormonal effects are similar to those seen in most monoecious plants. In several other dioecious plants the hormonal interactions are somewhat different. In *Mercurialis annua* L., auxin causes maleness and male plants have higher auxin levels than do female plants (Kahlem et al. 1975). By contrast, auxin has no effect on sex expression in *Ricinus communis* L., and GA promotes femaleness (Shifriss 1961).

The influence of cytokinins has been examined in several different dioecious plants. In *Cannabis, Mercurialis*, spinach, and several species of *Vitis*, cytokinin treatment leads to formation of hermaphroditic or female flowers on otherwise male plants (Moore 1970; Negi and Olmo 1972; Kahlem et al. 1975; Chailakhyan and Khryanin 1978a). When *Cannabis* and spinach plants were cut at the root-stem junction and the stem cuttings cultured on nutrient solution, a high percentage of the plants formed female flowers as long as secondary roots were allowed to form (Chailakhyan and Khryanin 1978a, c). When all secondary roots were removed, a high percentage of the plants formed male flowers, but addition of cytokinins to the nutrient solution could substitute for the roots and shift the balance back to predominantly female. Experiments comparing cytokinin levels in root exudate from male and female plants were not done, and thus one can only speculate that cytokinins produced by roots could have an important influence on sex expression in *Cannabis* and spinach.

The level of endogenous cytokinins was measured in differentiating apices and whole plants of *Mercurialis* using combined gas chromatography–mass spectrometry (Dauphin-Guerin, Teller, and Durand 1980). Zeatin was found in female plants but not in male plants, while zeatin nucleotide was found only in male plants. It was concluded that free zeatin has an important regulatory influence on sex expression in this dioecious plant.

CONCLUSIONS

It is clear that in many monoecious and dioecious plants one or more of the plant hormones can exert a controlling influence on whether the flower primordia develop into male or female flowers. The hormonal control of flower initiation and the development of flower primordia into mature flowers, however, remains a mystery that has resisted the best efforts of many researchers for more than 40 years.

Several of the plant hormones and other substances such as salicylic acid can induce flowering in a variety of plants, but there is no convincing evidence that any of these substances, by itself, acts to control flowering. There is reasonable evidence, however, that in some cases, one or more of these substances has an important influence on the flowering process.

Various theories on how flowering might be controlled have been discussed, ranging from control by a single organ-specific flower-inducing substance, to control by the interaction of one or more flower-inducing and flower-inhibitory substances, to a nutrient diversion hypothesis that assumes no more than a secondary role for hormonal substances. Obviously, it is impossible to decide between such varied hypotheses at this time. Instead, what is needed is a multifaceted approach to elucidate the control mechanisms for flowering. Thus, efforts should continue toward the possible discovery of new hormonal substances that may control flowering, but we should also undertake more detailed examinations of endogenous changes of known hormones and other substances, such as sugars. It is hoped that from such a wide range of efforts will come the crucial breakthroughs that will enable us finally to understand the chemical basis for the hormonal control of flowering in photoperiodically sensitive plants.

LITERATURE CITED

Alvim, R., P. F. Saunders, and R. S. Barros. 1979. *Abscisic acid and the photoperiodic induction of dormancy in Salix viminalis L.* Plant Physiol. 63:774–77.

Atsmon, D., and C. Tabbak. 1979. *Comparative effects of gibberellin, silver nitrate and aminoethoxyvinvy glycine on sexual tendency and ethylene evolution in the cucumber plant (Cucumis sativus L.).* Plant Cell Physiol. 20:1547–55.

Baldev, B., and A. Lang. 1965. *Control of flower formation by growth retardants and gibberellin in Samolus parviflorus, a long-day plant.* Amer. J. Bot. 52:408–17.

Bernier, G., J. Kinet, A. Jacqmard, A. Havelange, and M. Bodson. 1977. *Cytokinin as a possible component of the floral stimulus in Sinapis alba.* Plant Physiol. 60:282–85.

Biswas, P. K., K. B. Paul, and J. H. M. Henderson. 1966. *Effect of chrysanthemum plant extract on flower initiation in short-day plants.* Physiol. Plant. 19:875–82.

Bledsoe, C. S., and C. W. Ross. 1978. *Metabolism of mevalonic acid in vegetative and induced plants of Xanthium strumarium.* Plant Physiol. 62:683–86.

Bodson, M. 1977. *Changes in the carbohydrate content of the leaf and the apical bud of Sinapis during transition to flowering.* Planta 135:19–23.

Bonner, J., and D. Bonner. 1948. *Note on induction of flowering in Xanthium.* Bot. Gaz. 110:154–56.

Bowen, M. R., and G. V. Hoad. 1968. *Inhibitor content of phloem and xylem sap obtained from willow (Salix viminalis L.) entering dormancy.* Planta 81:64–70.

Burg, S. P., and E. A. Burg. 1966. *Auxin-induced ethylene formation: Its relation to flowering in the pineapple.* Science 152:1269.

Byers, R. E., L. R. Baker, H. M. Sell, R. C. Herner, and D. R. Dilley. 1972. *Ethylene: A natural regulator of sex expression of Cucumis melo L.* Proc. Nat. Acad. Sci. USA 69:717–20.

Carr, D. J. 1967. *The relationship between florigen and the flower hormones.* Ann. N.Y. Acad. Sci. 144:305–12.

Cathey, H. M., and N. W. Stuart. 1961. *Comparative plant growth retarding activity of AMO-1618, Phosfon, and CCC.* Bot. Gaz. 123:51–57.

Chailakhyan, M. Kh. 1936. *On the hormonal theory of plant development.* C. R. (Dokl.) Acad. Sci. USSR 12:443–47.

———. 1979. *Genetic and hormonal regulation of growth, flowering, and sex expression in plants.* Amer. J. Bot. 66:717–36.

Chailakhyan, M. Kh. and V. N. Khryanin. 1978a. *The influence of growth regulators absorbed by the root on sex expression in hemp plants.* Planta 138:181–84.

———. 1978b. *The role of roots in sex expression in hemp plants.* Planta 138:185–87.

———. 1978c. *Effect of growth regulators and role of roots in sex expression in spinach.* Planta 142:207–10.

Chouard, P. 1957. *Diversité des mécanismes des dormances, de la vernalisation et du photopériodisme, reveleé notamment par l'action de l'acide gibberéllique*. Mem. Soc. Bot. France 1956/1957:51–64.

Cleland, C. F. 1974a. *Isolation of flower-inducing and flower-inhibitory factors from aphid honeydew*. Plant Physiol. 54:889–903.

———. 1974b. *The influence of salicylic acid on flowering and growth in the long-day plant Lemma gibba G3*. Pages 553–57 in R. L. Bieleski, A. R. Ferguson, M. M. Cresswell, eds. *Mechanisms of Regulation of Plant Growth*. Bull. 12, Roy. Soc. New Zealand, Wellington.

———. 1978. *The flowering enigma*. BioScience 28:265–69.

Cleland, C. F., and A. Ajami. 1974. *Identification of the flower-inducing factor isolated from aphid honeydew as being salicylic acid*. Plant Physiol. 54:904–6.

Cleland, C. F., and W. R. Briggs. 1967. *Flowering responses of the long-day plant Lemna gibba G3*. Plant Physiol. 42:1553–61.

———. 1969. *Gibberellin and CCC effects on flowering and growth in the long-day plant Lemna gibba G3*. Plant Physiol. 44:503–7.

Cleland, C. F., and O. Tanaka. 1979. *Effect of daylength on the ability of salicylic acid to induce flowering in the long-day plant Lemna gibba G3 and the short-day plant Lemna pauciostata 6746*. Plant Physiol. 64:421–24.

Cleland, C. F., O. Tanaka, and L. J. Feldman. 1981. *Influence of plant growth substances and salicylic acid on flowering and growth in the Lemnaceae (Duckweeds)*. Aquatic Bot. (in press).

Cleland, C. F., and J. A. D. Zeevaart. 1970. *Gibberellins in relation to flowering and stem elongation in the long-day plant Silene armeria*. Plant Physiol. 46:392–400.

Crafts, C. B., and C. O. Miller. 1974. *Detection and identification of cytokinins produced by mycorrhizal fungi*. Plant Physiol. 54:586–88.

Dauphin-Guerin, B., G. Teller, and B. Durand. 1980. *Different endogenous cytokinins between male and female Mercurialis annua L*. Planta 148:124–29.

Dennis, D. T., C. D. Upper, and C. A. West. 1965. *An enzymic site of inhibition of gibberellin biosynthesis by AMO 1618 and other plant growth retardants*. Plant Physiol. 40:948–52.

El-Antably, H. M. M., P. F. Wareing, and J. Hillman. 1967. *Some physiological responses to d.1. abscisin (dormin)*. Planta 73:74–90.

Evans, L. T. 1960. *Inflorescence initiation in Lolium temulentum L. II. Evidence for inhibitory and promotive photoperiodic processes involving transmissible products*. Austral. J. Biol. Sci. 13:429–40.

———. 1966. *Abscisin II: Inhibitory effect on flower induction in a long-day plant*. Science 151:107–8.

———. 1969. *The nature of flower induction*. Pages 457–80 in L. T. Evans, ed. *The Induction of Flowering*. Cornell Univ. Press, Ithaca, NY.

Fontaine, D., N. Dugué, and E. Miginiac. 1977. *Inhibition de la floraison par la zéatine chez l'Anagallis arvensis L. soumis au traitement photopériodique inductif*. C. R. Acad. Sci. Paris. Ser. D. 284:1413–16.

Friedlander, M., D. Atsmon, and E. Galun. 1977a. *Sexual differentiation in cucumber: The effects of abscisic acid and other growth regulators on various sex genotypes*. Plant Cell Physiol. 18:261–69.

———. 1977b. *Sexual differentiation in cucumber: Abscisic acid and gibberellic acid contents of various sex genotypes*. Plant Cell Physiol. 18:681–91.

Glass, A. D. M., and J. Dunlop. 1974. *Influence of phenolic acids on ion uptake. IV. Depolarization of membrane potentials*. Plant Physiol. 54:855–58.

Gruen, H. E. 1959. *Auxins and fungi*. Ann. Rev. Plant Physiol. 10:405–40.

Gupta, S., and S. C. Maheshwari. 1970. *Growth and flowering of Lemna paucicostata. II. Role of growth regulators*. Plant Cell Physiol. 11:97–106.

Guttridge, C. G. 1969. *Fragaria*. Pages 247–67 in L. T. Evans, ed. *The Induction of Flowering*. Cornell Univ. Press, Ithaca, N.Y.

Hackett, W. P., and R. M. Sachs. 1967. *Chemical control of flowering in Bougainvillea "San Diego Red."* Proc. Amer. Soc. Hort. Sci. 90:361–64.

Hansen, D. J., S. K. Bellman, and R. M. Sacher. 1976. *Gibberellic acid-controlled sex expression of corn tassels*. Crop Sci. 16:371–74.

Henson, I. E., and P. F. Wareing. 1974. *Cytokinins in Xanthium strumarium: A rapid response to short day treatment*. Physiol. Plant. 32:185–87.

———. 1977. *Cytokinins in Xanthium strumarium L.: Some aspects of the photoperiodic control of endogenous levels*. New Phytol. 78:35–45.

Heslop-Harrison, J. 1956. *Auxin and sexuality in Cannabis sativa*. Physiol. Plant. 9:588–97.

———. 1957. *The experimental modification of sex expression in flowering plants*. Biol. Rev. 32:38–90.

Hoad, G. V., and M. R. Bowen. 1968. *Evidence for gibberellin-like substnaces in phloem exudates of higher plants*. Planta 82:22–32.

Hodson, H. K., and K. C. Hamner. 1969. *Floral inducing extracts from Xanthium*. Science 167:384–85.

Jacqmard, A., M. V. S. Raju, J. M. Kinet, and G. Bernier. 1976. *The early action of the floral stimulus on mitotic activity and DNA synthesis in the apical meristem of Xanthium strumarium*. Amer. J. Bot. 63:166–74.

Jacques, 1968. *Diversité et caractéristiques des processus de l'induction florale chez deux Chénopodicees*. C. R. Acad. Sci. Paris. Ser. D. 267:1592–95.

Jones, M. G., and J. A. D. Zeevaart. 1980a. *Gibberellins and the photoperiodic control of stem elongation in the long-day plant Agrostemma githago L.* Planta 149:269–73.

———. 1980b. *The effect of photoperiod on the levels of seven endogenous gibberellins in the long-day plant Agrostemma githago L.* Planta 149:274–79.

Kahlem, G., A. Champault, J. P. Louis, M. Bazin, A. Chabin, M. Delaigue, B. Dauphin, and R. Durand. 1975. *Détermination génétique et régulation hormonale de la différenciation sexuelle chez Mercurialis annua L.* Physiol. Vég. 13:763–79.

Kandeler, R. 1969. *Hemmung der Blütenbildung von Lemna gibba durch Ammonium*. Planta 84:279–91.

Khurana, J. P., and S. C. Maheshwari. 1978. *Induction of flowering in Lemna paucicostata by salicylic acid*. Plant Sci. Lett. 12:127–31.

———. 1980. *Some effects of salicylic acid on growth and flowering in Spirodela polyrrhiza* SP_{20}. Plant Cell Physiol. 21:923–27.

King, R. W. 1972. *Timing in Chenopodium rubrum of export of the floral stimulus from the cotyledons and its action at the shoot apex*. Can. J. Bot. 50:697–702.

King, R. W., and J. A. D. Zeevaart. 1974. *Enhancement of phloem exudation from cut petioles by chelating agents*. Plant Physiol. 53:96–103.

Krekule, J., and B. Horavka. 1972. *The response of short day plant Chenopodium rubrum L. to abscisic acid and gibberellic acid treatment applied at two levels of photoperiodic induction*. Biol. Plant. 14:254–59.

Krekule, J., and F. Seidlová. 1977. *Brassica campestris as a model for studying the effects of exogenous growth substances on flowering in long-day plants*. Biol. Plant. 19:462–68.

Laibach, F., and F. J. Kribben. 1950. *Der Einfluss von Wuchstoff auf die Bildung mannlicher und weiblicher Blüten bei einer monozischen Pflanze (Cucumis sativus L.)*. Ber. Deut. Bot. Ges. 62:53–55.

Lang, A. 1965. *Physiology of flower initiation*. Pages 1380–1536 in W. Ruhland, ed. *Handbuch der Pflanzenphysiologie XV/1*. Springer-Verlag, Berlin.

———. 1970. *Gibberellins; structure and metabolism*. Ann. Rev. Plant Physiol. 21:537–70.

Lang, A., M. Kh. Chailakhyan, and I. A. Frolova. 1977. *Promotion and inhibition of flower formation in a dayneutral plant in grafts with a short-day plant and a long-day plant*. Proc. Nat. Acad. Sci. USA 74:2412–16.

Lincoln, R. G., A. Cunningham, B. H. Carpenter, J. Alexander, and D. L. Mayfield. 1966. *Florigenic acid from fungal culture*. Plant Physiol. 41:1079–80.

Lincoln, R. G., A. Cunningham, and K. C. Hamner. 1964. *Evidence for a florigenic acid*. Nature 202:559–61.

Maxwell, F. G., and R. H. Painter. 1962. *Auxin in honeydew of Toxoptera graminum, Therioaphis maculata and macrosiphum pisi and their relation to degree of tolerance in host plants*. Ann. Entomol. Soc. Amer. 55:229–33.

Metzger, J. D., and J. A. D. Zeevaart. 1980. *Effect of photoperiod on the levels of endogenous gibberellins in spinach as measured by combined gas chromatography-selected ion current monitoring*. Plant Physiol. 66:844–46.

Michniewicz, M., and A. Kamienska. 1965. *Flower formation induced by kinetin and Vitamin E treatment in long-day plant (Arabidopsis thaliana) grown in short day*. Naturwissenschaften 52:623.

Miginiac, E. 1978. *Some aspects of regulation of flowering: Role of correlative factors in photoperiodic plants*. Bot. Mag. Tokyo Special Issue 1:159–73.

Mohan Ram, H. Y., and V. S. Jaiswal. 1970. *Induction of female flowers on male plants of Cannabis sativa L. by 2-chloroethanephosphonic acid*. Experientia 26:214–16.

———. 1972. *Induction of male flowers on female plants of Cannabis sativa by gibberellins and its inhibition by abscisic acid*. Planta 105:263–66.

Moore, J. N. 1970. *Cytokinin-induced sex conversion in male cones of Vitis species*. J. Amer. Soc. Hort. Sci. 95:387–93.

Morgan, D. G., and M. S. Zehni. 1980. *The effects of light breaks in the dark period on flower-bud development in varieties of Phaseolus vulgaris L*. Ann. Bot. 46:37–42.

Moss, G. I., and J. Heslop-Harrison. 1968. *Photoperiod and pollen sterility in maize*. Ann. Bot. 32:833–46.

Murfet, I. C., and J. B. Reid. 1973. *Flowering in Pisum: Evidence that gene Sn controls a graft-transmissible inhibitor*. Austral. J. Biol. Sci. 26:675–77.

Nakayama, S., and T. Hashimoto. 1973. *Effects of abscisic acid on flowering in Pharbitis nil*. Plant Cell Physiol. 14:419–22.

Nanda, K. K., H. N. Krishnamoorthy, T. A. Anuradha, and K. Lal. 1967. *Floral induction by gibberellic acid in Impatiens balsamina, a qualitative short-day plant*. Planta 76:367–70.

Nanda, K. K., S. Kumar, and V. Sood. 1976. *Effects of gibberellic acid and some phenols on flowering of Impatiens balsamina, a qualitative short-day plant*. Physiol. Plant. 38:53–56.

Naylor, F. L. 1941. *Effect of length of induction period on floral development of Xanthium pennsylvanicum*. Bot. Gaz. 103:146–54.

Negi, S. S., and H. P. Olmo. 1972. *Certain embryological and biochemical aspects of cytokinin SD 8339 in converting sex of a male Vitis vinifera (sylvestris)*. Amer. J. Bot. 59:851–57.

Nielson, C. S. 1942. *Effects of photoperiod on microsporogenesis in Biloxi soya bean*. Bot. Gaz. 104:99–106.

Ninnemann, H., J. A. D. Zeevaart, H. Kende, and A. Lang. 1964. *The plant growth retardant CCC as inhibitor of gibberellin biosynthesis in Fusarium moniliforme*. Planta 61:229–35.

Ogawa, Y. 1962. *Über die photoperiodische Empfindlichkeit der Keim Pflanzen von Pharbitis nil Chois. mit besonderer Berucksichtigung auf den Wuchsstoffgehalt der Kotyledonen*. Bot. Mag. Tokyo 75:92–101.

Ogawa, Y., and R. W. King. 1979. *Indirect action of benzyladenine and other chemicals on flowering of Pharbitis nil Chois*. Plant Physiol. 63:643–49.

Oota, Y., and K. Umemura. 1970. *Specific RNA produced in photoperiodically induced cotyledons of Pharbitis nil seedlings*. Pages 224-240 in G. Bernier, ed. *Cellular and Molecular Aspects of Floral Induction*. Longmans, London.

Phillips, D. A., and C. F. Cleland. 1972. *Cytokinin activity from the phloem sap of Xanthium strumarium L.* Planta 102:173–78.

Peterson, R. L., and E. C. Yeung. 1972. *Effect of two gibberellins on species of the rosette plant Hieracium*. Bot. Gaz. 133:190–98.

Pharis, R. P. 1972. *Flowering of Chrysanthemum under non-inductive long days by gibberellins and N^6-benzyladenine*. Planta 105:205–12.

Pierard, D., A. Jacqmard, G. Bernier, and J. Salmon. 1980. *Appearance and disappearance of proteins in the shoot apical meristem of Sinapis alba in transition to flowering*. Planta 150:397–405.

Pieterse, A. H. 1978. *Experimental control of flowering in Pistia stratiotes L.* Plant Cell Physiol. 19:1091–93.

Piringer, A. A., R. J. Downs, and H. A. Borthwick. 1963. *Photocontrol of growth and flowering in Caryopteris*. Amer. J. Bot. 50:86–90.

Proebsting, W. M., P. J. Davies, and G. A. Marx. 1978. *Photoperiod-induced changes in gibberellin metabolism in relation to apical growth and senescence in genetic lines of peas (Pisum sativum L.)*. Planta 141:231–38.

Proebsting, W. M., and E. Heftmann. 1980. *The relationship of (^{3}H) GA_9 metabolism to photoperiod-induced flowering in Pisum sativum L.* Z. Pflanzenphysiol. 98:305–9.

Quedado, R., and D. J. C. Friend. 1978. *Participation of photosynthesis in floral induction of the long-day plant Anagallis arvensis L.* Plant Physiol. 62:802–6.

Raghavan, V., and W. P. Jacobs. 1961. *Studies on the floral histogenesis and physiology of Perilla. II. Floral induction in cultured apical buds of P. frutescens*. Amer. J. Bot. 48:751–60.

Ramina, A., W. P. Hackett, and R. M. Sachs. 1979. *Flowering in Bougainvillea. A function of assimilate supply and nutrient diversion*. Plant Physiol. 64:810–13.

Reid, J. B., P. J. Dalton, and I. C. Murfet. 1977. *Flowering in Pisum: Does gibberellic acid directly influence the flowering process?* Austral. J. Plant Physiol. 4:479–83.

Rood, S. B., R. P. Pharis, and D. J. Major. 1980. *Changes of endogenous gibberellin-like substances with sex reversal of the apical inflorescence of corn*. Plant Physiol. 66:793–96.

Ruddat, M., E. Heftmann, and A. Lang. 1965. *Chemical evidence for the mode of action of AMO 1618, a plant growth retardant*. Naturwissenschaften 52:267.

Rudich, J., A. H. Halevy, and N. Kedar. 1969. *Increase in femaleness of three cucurbits by treatment with ethrel, an ethylene-releasing compound*. Planta 86:69–76.

———. 1972a. *Ethylene evolution from cucumber plants as related to sex expression*. Plant Physiol. 49:998–99.

———. 1972b. *The level of phytohormones in monoecious and gynoecious cucumbers as affected by photoperiod and ethephon*. Plant Physiol. 50:585–90.

Sachs, R. M. 1977. *Nutrient diversion: An hypothesis to explain the chemical control of flowering*. Hort. Sci. 12:220–22.

Salisbury, F. B. 1955. *The dual role of auxin in flowering*. Plant Physiol. 30:327–34.

Sawhney, S., and N. Sawhney. 1976. *Floral induction by gibberellic acid in Zinnia elegans Jacq. under non-inductive long days*. Planta 131:207–8.

Scharfetter, E., Th. Rottenburg, and R. Kandeler. 1978. *Die Wirkung von EDDHA and Salicylsäure auf Blütenbildung und vegetative Entwicklung von Spirodela punctata*. Z. Pflanzenphysiol. 87:445–54.

Schwabe, W. W. 1972. *Flower inhibition in Kalanchoe blossfeldiana. Bioassay of an endogenous inhibitor and inhibition by (±) abscisic acid and xanthoxin*. Planta 103:18–23.

Seidlová, F. 1980. *Sequential steps of transition to flowering in Chenopodium rubrum L.* Physiol. Vég. 18:477–87.

Seidlová, F., and S. Khatoon. 1976. *Effects of indol-3yl acetic acid on floral induction and apical differentiation in Chenopodium rubrum L.* Ann. Bot. 40:37–42.

Sherwood, S. B., J. O. Evans, and C. Ross. 1971. *Gel electrophoresis studies of proteins from leaves of photoperiodically induced and vegetative cocklebur plants.* Plant Cell Physiol. 12:111–16.

Shifriss, O. 1961. *Gibberellins as sex regulator in Ricinus communis.* Science 133:2061–62.

Sotta, B., and E. Miginiac. 1975. *Influence des racines et d'une cytokinine sur le développement floral d'une plante de jours courts, le Chenopodium polyspermum L.* C.R. Acad. Sci. Paris Ser. D. 281:37–40.

Stiles, J. I., Jr., and P. J. Davies. 1976. *Qualitative analysis by isoelectric focusing of the protein content of Pharbitis nil apices and cotyledons during floral induction.* Plant Cell Physiol. 17:855–57.

Stoddart, J. L. 1962. *Effect of gibberellin on a non-flowering genotype of red clover.* Nature 194:1063–64.

Suge, H. 1972. *Inhibition of photoperiodic floral induction in Pharbitis nil by ethylene.* Plant Cell Physiol. 13:1031–38.

Suge, H., and L. Rappaport. 1968. *Role of gibberellins in stem elongation and flowering in radish.* Plant Physiol. 43:1208–14.

Tafazoli, E., and D. Vince-Prue. 1978. *A comparison of the effects of long days and exogenous growth regulators on growth and flowering in strawberry, Fragaria x ananassa Duch.* J. Hort. Sci. 53:255–59.

Takahashi, H., H. Suge, and T. Saito. 1980. *Sex expression as affected by N^6-benzylaminopurine in staminate inflorescence of Luffa cylindrica.* Plant Cell Physiol. 21:525–36.

Tanaka, O., C. F. Cleland, and W. S. Hillman. 1979. *Inhibition of flowering in the long-day plant Lemna gibba G3 by Hunter's medium and its reversal by salicylic acid.* Plant Cell Physiol. 20:839–46.

T'se, A. T. Y., A. Ramina, W. P. Hackett, and R. M. Sachs. 1974. *Enhanced inflorescence development in Bougainvillea "San Diego Red" by removal of young leaves and cytokinin treatment.* Plant Physiol. 54:404–7.

Venkataraman, R., P. N Seth, and S. C. Maheshwari. 1970. *Studies on growth and flowering of a short-day plant, Wolffia microscopica. I. General aspects and induction of flowering by cytokinins.* Z. Pflanzenphysiol. 62:316–27.

Vince-Prue, D. 1975. *Photoperiodism in Plants.* McGraw-Hill, London.

Vince-Prue, D., and C. G. Guttridge. 1973. *Floral initiation in strawberry: Spectral evidence for the regulation of flowering by long-day inhibition.* Planta 110:165–72.

Wada, K. 1968. *Studies on the flower initiation in Pharbitis seedlings, with special reference to the early stimulation of cell division at the flower-induced shoot apex.* Bot. Mag. Tokyo 81:46–47.

Wareing, P. F., and P. F. Saunders. 1971. *Hormones and dormancy.* Ann. Rev. Plant Physiol. 22:261–88.

Watanabe, K., and A. Takimoto. 1979. *Flower-inducing effects of benzoic acid and some related compounds in Lemna paucicostata 151.* Plant Cell Physiol. 20:847–50.

Zeevaart, J. A. D. 1958. *Flower formation as studied by grafting.* Meded. Landbouwhogesch. Wageningen. 58:1–88.

———. 1964. *Effects of the growth retardant CCC on floral initiation and growth in Pharbitis nil.* Plant Physiol. 39:402–8.

———. 1969a. *The leaf as the site of gibberellin action in flower formation in Bryophyllum daigremontianum.* Planta 84:339–47.

———. 1969b. *Gibberellin-like substances in Bryophyllum daigremontianum and the distribution and persistence of applied gibberellin A_3.* Planta 86:124–33.

———. 1974. *Levels of (+)*-abscisic acid and xanthoxin in spinach under different environmental conditions. Plant Physiol. 53:644–48.

———. 1978. *Phytohormones and flower formation*. Pages 291–327 in D. S. Letham, P. B. Goodwin, T. J. V. Higgins, eds. *Phytohormones and Related Compounds—A Comprehensive Treatise*, vol. 2. Elsevier, Amsterdam.

———. 1979a. *Chemical and biological aspects of abscisic acid*. Pages 99–114 in N. B. Mandava, ed. *Plant Growth Substances*. ACS Symposium Series No. 111. American Chemical Society.

———. 1979b. *Perception, nature and complexity of transmitted signals*. Pages 60–90 in: *La Physiologie de la floraison,* R. Jacques and P. Champagnat, eds. CNRS No. 285, Paris.

11] Hormonal Control of Stolon and Tuber Development, Especially in the Potato Plant

by P. F. WAREING*

ABSTRACT

The development of an axillary bud of the potato plant into a stolon is promoted by high gibberellin levels in association with a degree of apical dominance. Stolons can be induced experimentally by the application of IAA and GA_3 to a decapitated plant. Natural or experimentally produced stolons can be converted into orthotropic leafy shoots by application of cytokinin to the stolon tip. In stolons released from apical dominance, the conversion to leafy shoots is promoted by the presence of roots or, in derooted plants, by the basal application of cytokinin. Gibberellin inhibits and, under certain conditions, cytokinin promotes tuber initiation. Abscisic acid (ABA) promotes tuberization in leafless cuttings of short-day–induced plants of *Solanum andigena*. Evidence is presented which suggests that ABA supplied by the leaves is one of the components of the tuberization stimulus, and interacts with a second factor. The nature of this "second factor" is obscure, but it appears to involve other substances in addition to cytokinins. The control of tuber development thus appears to involve interactions between gibberellins, ABA, cytokinins and probably other unknown factors.

INTRODUCTION

Stolons and rhizomes are formed by a wide range of species.† Their morphology is diverse, but the most characteristic features of stolons and rhizomes are:

*Department of Botany and Microbiology, The University College of Wales Aberystwyth, Aberystwyth, Dyfed SY23 3DA, UK.

†A rhizome is defined as an underground stem, whereas the term stolon is frequently applied where an elongated shoot grows plagiotropically at the soil surface. These terms are not applied rigorously, however, and the rhizomes of potato plants are commonly called stolons.

STRATEGIES OF PLANT REPRODUCTION (BARC Symposium number 6—Werner J. Meudt, ed.)
Allanheld, Osmun, Totowa

(a) elongated internodes and suppressed development of leaves, which are frequently represented by cataphylls; (b) a plagiotropic habit. In many species, as in the potato, the stolons are subterranean, being produced from the basal nodes of the shoot, while in others (e.g., strawberry) the stolons or runners are formed above ground. If they are subterranean, the main axis of the rhizome may remain permanently plagiotropic, with only lateral buds forming the leafy aerial shoots (e.g., *Pteridium*), or the terminal shoot apex may ultimately become orthotropic and produce leafy aerial shoots (e.g., *Agropyron*).

Tubers are also formed by a wide range of species (Gregory 1965) and may represent either swollen rhizomes (e.g., *Helianthus tuberosus*) or swollen roots (e.g., *Dahlia, Manihot*). The potato is unusual in forming tubers at the tips of stolons.

An understanding of the developmental physiology of stolen development and tuberization is important not only for its general biological interest but also from a practical viewpoint, since many important weeds are stoloniferous (e.g., *Agropyron*) and better understanding of their physiology is required as a basis for developing control measures, while several types of tuber constitute important food crops, e.g., potato, sweet potato (*Ipomoea batatas*), and yams (*Dioscorea*). It has become apparent in recent years that endogenous hormones play an essential role in regulating stolon and tuber development, and this offers the possibility of controlling these processes by exogenous growth regulators.

STOLON DEVELOPMENT

In the potato plant, stolons are formed from basal, subterranean buds of shoots developing from tubers. Cultivars of *S. tuberosum* normally produce stolons under both long and short photoperiods, but some populations of *S. andigena* do so only under short days. Short days increase stolon production even in cultivars and clones, which are able to produce them under long days. The formation of stolons appears to be stimulated by factors supplied by the mother tuber, the removal of which leads to delayed stolon development (P. F. Wareing unpublished).

Stolon development appears to be promoted by darkness (Kumar and Wareing 1972), so that their normal restriction to subterranean nodes may be partly due to the inhibitory effects of light on aerial axillary buds. There also appears to be an inherent polarity of potato shoots, leading to the production of stolons from basal buds.

Subjection to a degree of apical dominance is also necessary for a bud to develop as a stolon. If a potato plant is decapitated and all lateral buds are removed from the aerial nodes, the stolons rapidly become orthotropic and emerge above ground as aerial, leafy shoots (Booth 1959). The importance of apical dominance can also be demonstrated in 2- or 3- node cuttings, the upper buds of which grow as orthotropic leafy shoots, whereas in certain clones of *S. andigena* the lowermost bud develops as a stolon. If the upper buds are removed, however, the lowest one developes as an orthotropic leafy shoot (Kumar & Wareing 1972).

Studies on the hormonal control of stolen development in the potato plant began with the observations of Booth (1959) on *S. andigena*. It was found that if plants were decapitated and indole-3-acetic acid (IAA) applied to the stem stump, the uppermost lateral buds were partially inhibited, but with gibberellic acid (GA_3) they grew as orthotropic shoots with somewhat elongated internodes. If IAA and GA_3 were applied together, the uppermost lateral bud developed as a plagiotropic, stolonlike shoot, with elongated internodes, greatly reduced leaf development, and an apical hook. The application of GA_3 to intact (nondecapitated) shoots of *S. andigena* also promotes the development of stolons from axillary buds that are subject to apical dominance (Kumar and Wareing 1972). If, after inducing stolon development in decapitated plants by application of IAA and GA_3, the supply of both hormones, or of IAA alone, is withdrawn, the stolon tip turns up in a few days and develops into an upright leafy shoot. Thus, it would appear that the role of IAA in stolon development is to exert a degree of apical dominance, whereas GA_3 is presumably necessary for the greater internode elongation in stolons than in leafy shoots. Again, the suppression of leaf development and the plagiotropic habit occur only with combined IAA and GA_3 applications.

Since correlative inhibition of lateral buds appears to be regulated by an interaction between IAA and cytokinins, what is the effect of cytokinins on stolon development? Application of exogenous kinetin or benzyladenine (BA) to the stump of a decapitated stem has no effect, whether applied alone or in combination with IAA or GA_3. Application of kinetin, BA, or zeatin to a stolon tip causes its rapid conversion to an orthotropic, leafy shoot (Kumar and Wareing 1972). This effect occurs both with natural stolons and with stolons produced experimentally by application of IAA and GA to a decapitated shoot. Thus, as in correlative inhibition of lateral buds, IAA and cytokinin appear to exert opposite and mutually antagonistic effects on stolon development.

Since roots are generally regarded as a source of supply of endogenous cytokinins for the shoot (van Staden 1979), the question arose as to whether roots are necessary for the conversion of a stolon to a leafy shoot. This question was investigated first with natural stolons produced on rooted cuttings (Kumar and Wareing 1972; Woolley and Wareing 1972). It was found that conversion of the stolons to orthotropic leafy shoots occurred only when the plants were decapitated (without application of IAA) and that when the roots were removed such conversion occurred only when BA was supplied to the base of the cuttings. Thus, roots or basally applied cytokinin was necessary for the conversion of a stolen into a leafy shoot following decapitation.

If cuttings are taken from plants with stolons produced at the uppermost nodes by application of IAA and GA, conversion of stolons to leafy shoots will occur following withdrawal of IAA after some delay, if only mineral nutrients are supplied through the base of the cuttings. Even more rapid conversion of experimental stolons could be obtained by supplying both BA and high levels of mineral nutrients through the base of the cuttings, whereas without BA, conversion did not occur during the 10-day period of the experiment (Clifford E. LaMotte, personal communication). Thus, basally supplied BA appears to promote stolon conversion in rootless, decapitated cuttings to which IAA is *not* applied; however, if IAA is applied, conversion of a stolon to a leafy shoot cannot be induced by supplying cytokinin to the base of a cutting, and can only be achieved if cytokinin is *applied directly to the stolon tip*. In a further experiment, zeatin was applied to the stolen at various distances from the stolon tip, and it was found that no conversion occurred when the cytokinin was applied at more than 1 cm from the tip.

How does IAA applied to the stump of the stem prevent the effect of applied cytokinin on stolon conversion except when the cytokinin is applied directly to the tip? The IAA must either be acting directly (i.e., by transport of IAA from the point of application at the decapitated stump to the stolon tip), or indirectly, possibly by affecting the movement and distribution of the applied cytokinin.

In order to study the possible movement of IAA into the stolon, 2 – ^{14}C IAA was applied to the stump of decapitated plants, and its accumulation within the stolon was determined and analyzed (Knox and Wareing, unpublished). After 12 hours, 1.4% of the toal ^{14}C applied was present in the distal half of the stolon, but further analysis showed that 90% of the ^{14}C was present as polar metabolites of IAA, and free IAA represented only a very small proportion of the total ^{14}C applied. Such IAA as was present in the stolon tip had presumably been transported in the phloem, since transport of IAA in segments of stolons appears to be strictly basipetal (Clifford E. LaMotte, personal communication). Thus, it seems unlikely that IAA is acting directly in regulating stolon development, but further studies are necessary to confirm this conclusion.

If IAA is acting indirectly, then one possibility is that the presence of IAA in the stem in some way prevents the movement of cytokinins to the stolon tip. There is some evidence consistent with this hypothesis. When ^{14}C-benzyladenine was applied below experimentally produced stolons at the uppermost node, with and without IAA applied to the decapitated stump, significantly lower amounts of ^{14}C were found in the stolon tips with IAA applied than without (Woolley and Wareing 1972). More recently, we have repeated this type of experiment, but applying ^{14}C-zeatin riboside (ZR) instead of ^{14}C-BA, and similar results were obtained (Knox and Wareing unpublished). There is active metabolism of exogenous ZR in stem and stolon tissue, however, and only 20% of the radioactivity at the stolon tips was present as ^{14}C-ZR.

The mechanism whereby IAA affects the movement of cytokinin to the stolon tip remains obscure. One possibility is that the cytokinin applied to the stem is diverted to the point of application of the IAA by a "hormone-directed transport" effect, and when ^{14}C-ZR was applied there was, indeed, greater accumulation of ^{14}C in the decapitated stump when IAA was applied there, but this was not the case when ^{14}C-BA was used (Woolley and Wareing 1972).

Another possibility is that ^{14}C-BA or ^{14}C-ZR are accumulated at the stolon tip in response to "sink" activity and that the greater accumulation in the absence of IAA applied to the stem stump is the *result* of the release from inhibition on withdrawal of IAA. This latter hypothesis would seem to imply that leaf development in a stolon subjected to apical dominance is actively *inhibited* by IAA and that the failure of leaf development is not due to lack of cytokinin.

Although there is evidence that the conversion of stolons to leafy shoots can be promoted by cytokinins from other parts of the plant, the observation that such conversion can occur even in the absence of roots, provided that mineral nutrients are supplied, indicates that the supply of cytokinins from the roots is not essential for such conversion. Indeed, other experiments have indicated that de-rooted shoots of *S. andigena* can continue growth for several weeks if they are supplied with mineral nutrients, and that if such shoots are decapitated the lateral buds are capable of continued growth for a further period of 30 days (Wang and Wareing 1979). This observation raises the question as to whether the roots are the sole source of cytokinins in the plant.

It is generally assumed that leaves are dependent upon cytokinins produced by the roots and that the senescence of detached leaves results from the fact that they are deprived of such cytokinin supply. Nevertheless, the senescence of detached sunflower and *Xanthium* leaves is arrested if they are supplied with inorganic nutrients, and indeed the endogenous cytokinin levels of such leaves appears to increase (Salama and Wareing 1979), suggesting that they may be capable of cytokinin production when supplied with nutrients. If so, then leaves could be a source of cytokinin supply to the stolons of rootless stem cuttings of *S. andigena,* and IAA may regulate the supply of such leaf-produced cytokinins to the stolons. The question of whether shoot apical meristems can synthesize cytokinins or are dependent upon supply from elsewhere in the plant is unresolved (Wareing et al. 1977), but there is evidence that the buds of sprouting potato tubers are dependent upon the supply of cytokinins from the tubers (van Staden and Brown 1979).

Gibberellin appears to be necessary for rhizome development in quack grass *(Agropyron repens)* (Rogan and Smith 1976). In excised rhizomes, GA_3 applied to the basal end sustains the continued growth of the terminal apex as a rhizome, whereas in the absence of GA_3 the terminal bud turns upward and develops as an aerial shoot. It is suggested that a continuous supply of gibberellin from the parent plant may be essential to maintain the characteristic morphology and growth habit of the rhizome.

In the strawberry plant (*Fragaria* x *ananassa* Duch.), long days promote runner formation, but runners can be induced to form under short days by application of GA_3 (Thompson and Guttridge 1959). Exogenous GA_3 only partly replaces the effects of long days (Guttridge 1970), however, so that the involvement of gibberellin in the long-day promotion of runnering is not unequivocally established.

HORMONAL CONTROL OF TUBERIZATION

Introduction. Several theories have been suggested to account for tuber formation in the potato plant, but two main theories have been put forward: (a) that tuber initiation is promoted by high levels of carbohydrate, especially sugars, and (b) that tuberization involves a specific tuber-inducing stimulus. It has been suggested that tuber initiation is associated with those conditions which lead to a high concentration of soluble carbohydrates at the stolon tips (Borah and Milthorpe 1962); that is, under conditions in which the balance between production and utilization of assimilate is such that a surplus of assimilate occurs. The importance of high sugar levels for tuber initiation has been shown by studies of stem segments and isolated stolons in aseptic culture. There have been several reports that tuber initiation occurs only in such cultures when the sucrose concentration exceeds about 5% (Mes and Menge 1954; Okazawa 1955; Palmer and Smith 1969).

That tuberization may involve a specific stimulus was suggested by several earlier workers, but the work of Gregory (1956) first provided strong experimental evidence for the existence of a specific tuber-forming stimulus. He found that cuttings from short-day–induced parent plants rapidly formed tubers from axillary buds on the stem, whereas similar cuttings from non-induced (long-day) plants did not do so. When induced stem pieces were grafted onto non-induced ones, the latter formed tubers, indicating the existence of a graft-transmissible stimulus. Gregory concluded that a tuber-forming stimulus arises in the plant under specific conditions of temperature and photoperiod, and that the stimulus occurs throughout the plant and is not restricted to the underground parts. These results were effectively confirmed by Chapman (1958) and Kumar and Wareing (1973a,b).

There is circumstantial evidence that the tuberization stimulus may be formed or stored in seed tubers. Thus, Madec and Perennec (1959, 1962) found that whereas rooted cuttings (i.e., without the parent tuber attached) of certain varieties of potato will form only under short-day–inducing conditions, if the mother tuber is allowed to remain attached to the plant tubers may be formed even under long-day conditions.

Although the nature of the tuberization stimulus is still not fully resolved, it is now well established that several classes of growth hormone have profound effects upon tuber initiation, as will be summarized below.

Gibberellins. Whereas stolon development appears to be promoted by high gibberellin levels, tuberization has repeatedly been reported to be inhibited by application of exogenous gibberellins, not only in potato (Harmey, Crowley, and Clinch 1966; Lovell and Booth 1967; Tizio 1971), but also in *Dahlia* (Biran, Gur, and Halevy 1972), *Helianthus tuberosus* (Courduroux 1964), and *Begonia evansia* (Okagami, Esashi, and Nagao 1977).

Cytokinins. Cytokinins have been reported to promote tuberization in potato in certain types of test. Palmer and Smith (1969) carried out experiments with isolated stolons of *S. tuberosum*. Stolons from sprouting tubers were first cultured aseptically on the nutrient medium of Murashige and Skoog (1962) but without cytokinin, and the tips were then transferred to fresh medium containing inorganic nutrients and 6% sucrose. Under these conditions tuber formation occurred when kinetin was added to the medium, but not in its absence. Similar results were obtained with isolated stolons by Mauk and Langille (1978), using zeatin riboside (ZR) instead of kinetin. There was a twofold increase in the levels of endogenous ZR in the below-ground parts of potato plants (including roots and stolons) on transfer from long days to short days.

In sterile cultures of *Ullucus tuberosus,* which normally require short days for tuberization, benzyladenine (BA) was found to promote tuber formation under long days (Asahira and Nitsch 1967). Kinetin stimulates tuber formation in *Helianthus tuberosus* (Courduroux 1967) and bulbil formation in *Dioscorea bulbifera* (Udebo 1971).

The suggestion that cytokinins are necessary for tuberization is, of course, entirely consistent with the fact that this process involves active cell division. Nevertheless, the tests of Palmer and Smith and of Mauk and Langille were carried out with stolons of *S. tuberosum,* cultivars of which do not have an obligate requirement for short days for tuberization. We have tested the ability of cytokinins to stimulate tuberization with a clone of *S. andigena,* which forms tubers only under short days. When one-node leafless stem cuttings are taken from plants growing under long days and cultured under aseptic conditions, they fail to form tubers in response to kinetin or benzyladenine. Moreover, when stolons were taken from parent plants growing under long days and short days, respectively, and after sterilization were placed on a nutrient medium (Murashige and Skoog, plus 6% sucrose) with the addition of 10^{-6}M benzyladenine (BA), tuberization occurred in the stolons taken from short-day plants, but not in those from long-day plants. This matter is further discussed below.

Abscisic acid. Okazawa (1959, 1960) observed that tuber initiation was accompanied by a decline in the levels of endogenous gibberellins and an increase in endogenous inhibitor activity in stolons of potato. He suggested that tuber initiation is regulated by an interaction between endogenous gibberellins and inhibitors.

The discovery of abscisic acid (ABA) as a natural growth inhibitor raised the question as to its possible role in tuber initiation. There have been conflicting reports of the effects of exogenous ABA on tuberization. El-Antably, Wareing, and Hillman (1967) reported that although direct application of ABA to the tips of stolons growing under long days did not induce tuber initiation, application to the leaves did lead to an increase in the number of tubers formed. On the other hand, Smith and Rappaport (1969) were unable to induce tuber formation by spraying plants with ABA or by treating the stolon tips in aseptic culture. In other experiments, ABA was reported to inhibit tuber initiation in isolated stolons supplied with kinetin (Palmer and Smith 1969). Subsequently, other workers have reported that application of ABA promotes tuberization in whole potato plants, in one-node stem cuttings (Menzel 1980), and when applied to stolon tips (Krauss and Marschner 1976). Application of ABA also promotes the formation of root tubers in *Dahlia variabilis,* when applied both to whole plants (Biran, Gur, and Halevy 1972) and to isolated, budless leaf cuttings (Biran et al. 1974). Similar effects have been reported for *Helianthus tuberosus* (Charnay and Courduroux 1972).

We have obtained evidence that endogenous ABA may be required for tuber initiation in *S. andigena* (Wareing and Jennings 1980). When one-node, leafy cuttings are taken from plants that have been fully induced under short days, the axillary bud develops directly into a tuber. If the leaf is removed, the axillary bud develops as a strongly growing, erect shoot with extended internodes. If ABA is supplied to leafless, one-node, short-day–induced cuttings, however, the bud develops into a tuber. Thus, ABA replaces the effect of the "short-day" leaf. But ABA does not cause tuber formation in similar stem cuttings from noninduced (long-day) plants, either with or without leaves.

These findings raised the question as to whether endogenous ABA is supplied to the bud from the leaf and is an essential part of the tuber-inducing stimulus produced by short-day leaves. There is evidence that ABA is produced in leaves and transported in the phloem to other parts of the plant (Zeevaart 1977). Endogenous ABA, both free and conjugated, can be extracted from potato leaves, but there appear to be no significant differences between the levels of ABA in extracts of long-day and short-day leaves (Kumar and Wareing 1974; Wareing and Jennings 1980). In order to test whether leaves of *S. andigena* export ABA, leaves were pretreated with EDTA to inhibit callose fomation in the sieve tubes, following the technique of King and Zeevaart (1974), and phloem exudate was collected over a period of 7 hr. Significant amounts of both free and conjugated ABA were collected, but there appeared to be no appreciable differences in the amounts produced by long-day and short-day leaves. In further experiments, ^{14}C-ABA was supplied in solution to the terminal leaflets of one-node cuttings, and the subsequent movement and distribution of ^{14}C within the cuttings was determined. It was found that a relatively high proportion of the ^{14}C taken up by the leaf accumulated in the buds, indicating that it is likely that endogenous ABA exported by the leaf is accumulated by the axillary bud.

From the foregoing evidence it would appear that the leaves of *S. andigena* do export endogenous ABA to the axillary buds of one-node cuttings and that this ABA is an essential part of the "tuberization stimulus" produced by the leaf. Yet since there appears to be no appreciable difference between the amounts of ABA exported by induced and non-induced leaves, if ABA is an essential factor in the "leaf-effect" then a long-day leaf should be effective if grafted onto a short-day–induced stem, and this was found to be the case. Indeed, a tomato leaf was also found to be effective in inducing tuber development when grafted onto an induced stem cutting of *S. andigena*.

Since tuber initiation involves a switch in the pattern of cell division in the sub-apical region of the stolen tip (Booth 1963; Cutter 1978) from polarized divisions leading to internode extension to nonpolarized divisions resulting in tuber initiation, it would seem probable that ABA promotes tuber formation by inhibiting stolen extension growth, possibly by antagonizing the effects of endogenous gibberellins. Some support for this interpretation is provided by the observation that other treatments that inhibit stolen extension growth, including solutions of high osmotic activity (Wareing and Jennings 1980), and growth retardants, such as CCC, also promote tuberization in induced cuttings.

Growth retardants. Since tuber initiation is inhibited by gibberellic acid (GA_3), it might be expected that growth retardants and inhibitors that antagonize the effects of GA_3 in other processes might promote tuberization. Indeed, there have been several reports that application of growth retardants, such as chlormequat (CCC) and daminozide (B9) increase the number and yield of tubers. Dyson (1965) and Humphries and Dyson (1967) reported that daminozide increased total tuber number in potatoes, and Gifford and Moorby (1967) reported that application of CCC accelerated tuber initiation. Tizio (1969) found that CCC promoted tuberization in sections of potato sprouts grown in vitro. Biran, Gur, and Halevy (1972) reported that daminozide and ethephon promoted tuberization in Dahlia.

The nature of the tuber-inducing stimulus. The foregoing evidence strongly suggests that endogenous ABA constitutes an essential part of the tuber-promoting effect of leaves in *S. andigena*. Although there appear to be no great differences between long-day and short-day leaves with respect to ABA production, there must be other differences between them since a short-day leaf will induce tuberization when grafted onto a long-day stem, and a long-day leaf will not do so. Since gibberellins inhibit tuber initiation, it is possible that the differences between induced and non-induced leaves lies in amounts of gibberellins they export. There is indeed evidence that endogenous gibberellin levels are appreciably lower in short-day than in long-day leaves of *S. andigena* (Railton and Wareing 1973). Moreover, several scientists have argued that tuberization is regulated by the relative levels of gibberellin and ABA or other endogenous inhibitors (Okazawa 1959, 1960; Dimalla, van Staden, and Smith 1977; Menzel 1980). Thus, one might postulate that the promotion of tuberization by short-day leaves depends primarily upon the reduction of the inhibitory effects of long-day leaves arising from the higher gibberellin levels the latter produce. This hypothesis appears to be untenable for the following reasons:

1. If the sole difference between long-day and short-day leaves lay in the levels of exported gibberellin, then a long-day leaf should not be effective when grafted onto a short-day stem, since the higher gibberellin level which it is postulated to produce will inhibit tuberization.
2. Since a long-day leaf will induce tuber formation when grafted onto a short-day stem, but not when grafted onto a long-day stem, it is clear that short-day induction results in a change in the stem tissues, the most probable hypothesis being that some tuber-promoting substance accumulates in the stem under short days. (The observation that short-day induction *promotes* the growth of the axillary bud of one-node cuttings is not consistent with the alternative hypothesis that gibberellin levels are lower in short-day stems).
3. Even high concentrations of ABA will not induce tuberization in non-induced cuttings.

These observations appear to lead unequivocally to the conclusion that even if the lower levels of gibberellin produced by short-day leaves create more favorable conditions for tuber initiation, short-day induction must also involve the increased production of some other factor that has a positive effect on tuberization.

In view of the reports that exogenous cytokinins promote tuber initiation in isolated stolons of *S. tuberosum* and that levels of endogenous cytokinin (identified as ZR) increase on transfer to short days, the question arises as to whether the tuber-inducing effects of short-day leaves is due to the greater production of ZR under short days than under long days. Since, as stated above, exogenous cytokinins, including kinetin, BA, and ZR, have not been found to promote tuber initiation in one-node stem cuttings from non-induced plants of *S. andigena* (with or without exogenous ABA), and grafted short-day leaves will do so, there is doubt as to whether cytokinin is the only or main component of the tuber-inducing stimulus arising in short-day leaves. Tizio (1973) has argued that the effects of kinetin and BA on tuber formation is due to a nonspecific inhibition of sprout and root growth, and that cytokinins are not specific factors for tuber formation.

We have attempted to extract tuber-inducing factors from shoots of short-day–induced plants of *S. andigena,* using single-node leafless cuttings of non-induced plants as the test material. We have extracted mainly stem tissue, since the tuberization stimulus appears to accumulate in stems, and we have tested acidic, neutral, and basic ethyl acetate- and butanol-soluble fractions. The tests were carried out under aseptic conditions, using the nutrient medium of Murashige and Skoog, with the addition of 10^{-6}M ABA. Large numbers of tests have been made, and although some indications of tuber-promoting activity have been obtained from time to time, the effects have not been consistent. We have not yet extensively tested the water-soluble residues after partition against ethyl acetate and butanol, since these contain material that is toxic in the bioassay, but more polar fractions need to be tested.

Thus, the tuber-inducing stimulus appears to be as elusive as the hypothetical "flower hormone." It is clear, however, that the regulation of tuber development involves interactions between several hormonal factors, including gibberellins, cytokinins, ABA, and a possible unknown further factor. Even if tuberization should prove to involve a new type of regulator, the latter may not necessarily be specific for tuberization; thus Nitsch (1965) observed that tubers can be induced in *Helianthus tuberosus* if leaves of *H. annuus,* which does not itself form tubers, are grafted onto it and are maintained under short days.

HORMONES AND ENVIRONMENTAL CONTROL OF STOLON DEVELOPMENT AND TUBERIZATION

Stolon development and tuber initiation are markedly affected by various environmental factors, including photoperiod, temperature, and nitrogen nutrition, and it is pertinent to consider how far these responses to environmental factors are mediated through modulation of endogenous hormones.

Tuber initiation is normally preceded by a phase of stolon growth. In most cultivars of *S. tuberosum,* stolon development commences quite early in the growth cycle, but in some clones of *S. andigena* it occurs only under short days and is delayed until the onset of short days in the autumn. On the other hand, other clones of *S. andigena* produce stolons under long days but require short days for tuber initiation, so that under long days stolon development appears to continue indefinitely without tuber initiation. With all these types of response, tuber initiation is normally delayed until some stolon growth has occurred, and this would appear to be of adaptive advantage in ensuring that it results in a degree of spacing of the tubers.

A similar sequence of stolon development and tuber initiation occurs in stem cuttings of clones of *S. andigena,* which require short days for both stolon development and tuber initiation (Ewing and Wareing 1978). If 2- or 3-node cuttings are taken from plants that have been exposed to a small number of short days, the basal bud rapidly grows out strongly as an erect shoot with extended internodes, whereas in cuttings from long-day plants the bud remains inhibited. As cuttings are taken from parent plants which have received progressively increasing numbers of short days, the basal bud shows corresponding changes and develops as (a) plagiotropic stolons without tubers, (b) stolons terminating in tubers, and (c) sessile tubers, with progressive short-day induction.

It is difficult to reconcile this sequence of changes in stolon development and tuber initiation in response to increasing short-day induction with current information regarding the hormonal regulation of these processes. Studies with exogenous gibberellins indicate that stolon development is promoted by high gibberellin levels, whereas studies on endogenous gibberellins indicate that short days result in reduced gibberellin levels in the leaves (Okazawa 1959, 1960; Railton and Wareing 1973). There is no information on how gibberellin levels are affected by transfer from long days to short days in other parts of the plant, and it is possible that the decrease in gibberellin levels in the leaves is due to export to other parts of the plant, where the gibberellin levels may rise.

As indicated above, increasing short-day induction leads first to stolon development and then to tuber initiation. If stolon development is promoted by increased gibberellin levels under short days, how is it that continued short-day induction promotes tuber initiation, which is *inhibited* by application of exogenous gibberellin? It is possible that progressive short-day induction causes an initial rise in endogenous gibberellin levels, followed by a decline, but there is no evidence of any such change. Alternatively, it may be that short days lead also to increases in other types of hormone favorable to tuber initiation (cytokinins, ABA) but at a slower rate, until they reach a threshold necessary to counteract the effects of the endogenous gibberellins and cause a switch in the pattern of stolon development. As we have seen, there is evidence for a rapid increase in cytokinin levels in the stolons and roots on transfer from long days to short days, although the effect is not great. There is no evidence that ABA levels increase with short days, but it may be the ratio of gibberellin to ABA levels that regulates the capacity of the stolon apices to respond to cytokinins or some other tuber-inducing signal.

Menzel (1980) studied the effects of high temperatures (32° day/28°C night and 32°C day/18°C night) on growth and tuberization in *S. tuberosum*. The responses to high temperatures and gibberellin are similar, in that they promote shoot growth and suppress tuber initiation. The inhibitory effect of high temperature on tuber production was almost completely reversed by application of CCC and partly reversed by ABA. Single-leaf cuttings responded in the same way to temperature and applied growth regulators as did whole plants, so that it appeared that in the whole plant the growth regulators act directly on the stolon tip, rather than through a general influence on shoot growth. Menzel suggests that temperature exerts its influence by alerting the balance between the levels of endogenous gibberellins and inhibitors.

Similar conclusions were drawn by Krauss (1978) in a study of the effects of nitrogen nutrition on tuberization and ABA content in *S. tuberosum*. Interrupting the nitrogen supply of plants grown in water culture promotes tuberization and causes an increase in ABA content in all parts of the plant except the stolons, where the ABA level remains constant. Withdrawal of nitrogen also results in an increase in endogenous cytokinin levels in the roots, stolon tips, and shoots (Sattlemacher and Marschner 1978a,b), but the authors conclude that despite the close correlation between tuberization and cytokinin activity, cytokinins are not directly responsible for the onset of tuber initiation, although they play an important role in tuber growth.

CONCLUSIONS

Stolon development and tuber initiation in the potato plant provide excellent model systems for the study of the control of development in the whole plant, since they are readily modified by both environmental factors and exogenous growth hormones. Development requires both spatial and temporal coordination (Wareing 1977), and both aspects appear to be involved in the regulation of stolon and tuber development. The limitation of stolons to the basal part of the shoot constitutes an aspect of spatial coordination, as does the apical dominance required for their development. It has been argued (Wareing 1977) that IAA is the primary hormone involved in spatial coordination, and it appears to play an essential role in the apical dominance effects in stolon development, although its mode of action in this phenomenon still remains obscure.

On the other hand, the responses of the potato plant to environmental factors, such as photoperiod, may be regarded as aspects of temporal control, in which modulation of gibberellin and cytokinin levels appear to play an important role. The regulation of stolon and tuber development also offers excellent examples of processes regulated by interactions between two or more types of hormone. Thus, in studying these processes in the potato plant we are brought up against some of the central problems of hormonal control within the whole plant.

LITERATURE CITED

Asahira, T., and J. P. Nitsch. 1967. *Tuberisation in vitro: Ullucus tuberosus et Dioscorea.* Bull. Soc. Bot. France 115:345–52.

Biran, I., I. Gur, and A. H. Halevy. 1972. *The relationship between exogenous inhibitors and endogenous levels of ethylene and tuberisation of dahlias.* Physiol. Plant. 27:226–30.

Biran, I., B. Leshem, I. Gur, and A. H. Halevy. 1974. *Further studies on the relationships between growth regulators and tuberization of Dahlias*. Physiol. Plant. 31:23–28.
Booth, A. 1959. *Some factors concerned in the growth of stolons in potato*. J. Linn. Soc. 56:166–69.
———. 1963. *The role of growth substances in the development of stolons*. Pages 99–113 in *The Growth of the Potato,* J. D. Ivins and F. L. Milthorpe, eds. Butterworth, London.
Borah, M. H., and F. Milthorpe. 1962. *Growth of the potato as influenced by temperature*. Indian J. Pl. Physiol. 5:53–72.
Chapman, H. W. 1958. *Tuberisation in the potato plant*. Physiol. Plant., 11:215–24.
Charnay, D., and J. C. Courduroux. 1972. *Acide abscissique et tubérisation in vitro de bourgeons de Topinambour (Helianthus tuberosus L. var D19)*. C.R. Acad. Sci. (Paris) 275:2351–54.
Courduroux, J. C. 1964. *Inhibition de croissance et tubérisation*. C.R. Acad. Sci. (Paris) 259:4346–49.
———. 1967. *Étude du méchanisme physiologique de la tubérisation chez le Topinambour (Helianthus tuberosus L.)*. Ann. Sci. Nat. Bot. 8:215–355.
Cutter, E. G. 1978. *Structure and development of the potato plant*. Pages 70–152 in *The Potato Crop,* P. M. Harris ed., Chapman & Hall, London.
Dimalla, G. C., J. van Staden, and A. R. Smith 1977. *A comparison of the endogenous hormone levels in stolons and sprouts of the potato, Solanum tuberosum*. Physiol. Plant. 41:167–71.
Dyson, P. W. 1965. *Effects of gibberellic acid and 2-chloroethyl-trimethyl ammonium chloride on potato growth and development*. J. Sci. Fd. and Agric. 16:542–49.
El-Antably, H. M. M., P. F. Wareing, and J. Hillman, 1967. *Some physiological responses to D. L. Abscisin (dormin)*. Planta 73:74–90.
Ewing, E. E., and P. F. Wareing. 1978. *Shoot, stolon and tuber formation on potato (Solanum tuberosum L.) cuttings in response to photoperiod*. Plant Physiol. 61:348–53.
Gifford, R. M., and J. Moorby. 1967. *The effect of CCC on the initiation of potato tubers*. Eur. Potato J. 10:235–38.
Gregory, L. E. 1956. *Some factors for tuberisation in the potato plant Amer. J. Bot. 43:281–88.*
——. *1965. Physiology of tuberisation in plants (tubers and tuberous roots).* Pages 1328–54 in *Encyclopedia of Plant Physiology,* vol. 15. W. Ruhland, ed. Springer-Verlag, Berlin.
Guttridge, C. G. 1970. *Interaction of photoperiod, chilling and exogenous gibberellic acid on growth of strawberry petioles*. Ann. Bot. 34:349–64.
Harmey, M. A., M. P. Crowley, P. E. M. Clinch. 1966. *The effect of growth regulators on tuberisation of cultured stem pieces of Solanum tuberosum*. Eur. Potato J. 9:146–51.
Humphries, E. C., and P. W. Dyson. 1967. *Effect of a growth inhibitor, N-dimethylaminosuccinamic acid (B9) on potato plants in the field*. Eur. Potato J. 10:116–26.
King, R. W., and J. A. D. Zeezaart. 1974. *Enhancement of phloem exudation from cut petioles by chelating agents*. Plant Physiol. 53:96-103.
Krauss, A. 1978. *Tuberisation and abscisic acid content in Solanum tuberosum as affected by nitrogen nutrition*. Potato Res. 21:183–93.
Krauss, A., and H. Marschner. 1976. *Einfluss von Stickstoffernährung und Wuchstoffapplikation auf die Knollen-induktion bei kartoffelpflanze*. Z. Pflanzenern. Bodenk. 2:143–55.
Kumar, D., and P. F. Wareing. 1972. *Factors controlling stolon development in the potato plant*. New Phytol. 71:639–48.
———. 1973a. *Studies on tuberisation in Solanum andigena. I. Evidence for the existence and movement of a specific tuberisation stimulus*. New Phytol. 73:283–87.

———. 1973b. *Studies on tuberisation of Solanum andigena. II. Growth hormones and tuberisation.* New Phytol. 73:833–40.
Lovell, P. H., and A. Booth. 1967. *Effect of gibberellic acid on growth, tuber formation and carbohydrate distribution in Solanum tuberosum.* New Phytol. 68:1175–85.
Madec, P., and P. Perennec. 1959. *Le rôle respectif du feuillage et du tubercule-mere dans la tubérisation de la Pomme de terre.* Eur. Potato J. 2:22–49.
———. 1962. *Les relations entre l'induction de la tubérisation et la croissance chez la plante de Pomme de terre (Solanum tuberosum L.)* Ann. Physiol. 4:5–84.
Mauk, C. S., and A. R. Langille. 1978. *Physiology of tuberisation in Solanum tuberosum L. Cis-zeatin riboside in the potato plant: Its identification and changes in endogenous levels as influenced by temperature and photoperiod.* Plant Physiol. 62:438–42.
Menzel, C. M. 1980. *Tuberisation in potato at high temperatures: Responses to gibberellin and growth inhibitors.* Ann. Bot. 46:259–65.
Mes, M. G., and I. Menge. 1954. *Potato shoot and tuber cultures in vitro.* Physiol. Plant. 7:637ff.
Murashinge, T., and F. Skoog. 1962. *A revised medium for rapid growth and bioassays with tobacco tissue cultures.* Physiol. Plant. 15:473–96.
Nitsch, J. P. 1965. *Existence d'un stimulus photoperiodique non specifique capable de provoquer la tubérisation chez Helianthus tuberosus L.* Bull. Soc. Bot. France 112:7–8.
Okagami, N., Y. Esashi, and M. Nagao. 1977. *Gibberellin-induced inhibition and promotion of sprouting in aerial tubers of Begonia evansiana and in relation to photoperiodic treatment and tuber storage.* Planta (Berlin), 136:1–6.
Okazawa, Y. 1955. *Physiological studies on the mechanism of tuberisation of potato plants.* Proc. Crop. Sci. Soc. Japan 23:247–48 (in Japanese).
———. 1959. *Studies on the occurrence of natural gibberellin and its effect on the tuber formation in the potato plant.* Proc. Crop Sci. Soc. Japan. 28:129–33 (in Japanese).
———. 1960. *Studies on the relation between the tuber formation of potato and its natural gibberellin.* J. Crop Sci. Soc. Japan 29:121–24 (in Japanese).
Palmer, C. E., and O. E. Smith. 1969. *Cytokinin and tuber initiation in the potato Solanum tuberosum L.* Plant & Cell Physiol. 10:657–65.
Railton, I. D., and P. F. Wareing. 1973. *Effects of daylength on endogenous gibberellins in Solanum andigena. I. Changes in levels of free acidic gibberellinlike substances.* Physiol. Plant. 28:88–94.
Rogan, P. G., and D. L. Smith. 1976. *Experimental control of bud inhibition in rhizomes of Agropyron repens (L.) Beauv.* Z. Pflanzenphysiol. 78:113–21.
Salama, A.M.S.El-Din A., and P. F. Wareing. 1979. *Effects of mineral nutrition on endogenous cytokinins in plants of sunflower (Helianthus annuus L.)* J. Exp. Bot. 30:971–81.
Sattelmacher, B., and H. Marschner. 1978a. *Relation between nitrogen nutrition, cytokinin activity and tuberisation in Solanum tuberosum L.* Physiol. Plant. 44:65–69.
———. 1978b. *Cytokinin activity in stolons and tubers of Solanum tuberosum during the period of tuberisation.* Physiol. Plant. 44:69–72.
Smith, O. E., and L. Rappaport. 1969. *Gibberellins, inhibitors and tuber formation in the potato, Solanum tuberosum L.* Am. Potato. J. 46:185–91.
Thompson, P. A., and C. G. Guttridge. 1959. *Effect of gibberellic acid on the initiation of flowers and runners in the strawberry.* Nature 184:72–73.
Tizio, R. 1969. *Action du CCC [chlorure de (2-chloroethyl-trimethyl-ammonium)] sur la tubérisation de la pomme de terre.* Eur. Potato J. 12:3–7.
———. 1971. *Action et rôle probable de certaine gibberellines (A1, A3, A4, A5, A7, A9 et A13) sur la croissance des stolons et la tubérisation de la Pomme de terre (Solanum tuberosum L.).* Potato Res. 14:193–204.
Tizio, R., and M. M. Biain. 1973. *Are cytokinins the specific factors for tuber formation in the potato plant.* Phyton 31:3–13.

Udebo, A. E. 1971. *Effect of external supply of growth substances on axillary proliferation and development in Dioscorea bulbifera*. Ann. Bot. 35:159–63.

van Staden, J., and J. E. Davy. 1979. *The synthesis, transport and metabolism of endogenous cytokinins*. Plant, Cell and Environment 2:93–106.

van Staden, J., and N.A.C. Brown. 1979. *Investigations into the possibility that potato buds synthesize cytokinins*. J. Exp. Bot. 30:391–97.

Wang, T. L., and P. F. Wareing. 1979. *Cytokinins and apical dominance in Solanum andigena: Lateral shoot growth and endogenous cytokinin levels in the absence of roots*. New Phytol. 82:19–28.

Wareing, P. F. 1977. *Growth substances and integration in the whole plant*. Soc. for Exp. Biol. Symp. 31:337–65.

Wareing, P. F., and A.M.V. Jennings. 1980. *The hormonal control of tuberisation in potato*. Pages 293–300 in *Plant Growth Substances 1979,* F. Skoog, ed. Springer-Verlag, Berlin.

Wareing, P. F., R. Horgan, I. E. Henson, and W. H. Davis. 1977. *Cytokinin relations in the whole plant*. Pages 147–53 in *Plant Growth Regulation,* P. E. Pilet, ed. Springer-Verlag, Berlin, Heidelberg.

Woolley, D. J., and P. F. Wareing. 1972. *The role of roots, cytokinins and apical dominance in the control of lateral shoot form in Solanum andigena*. Planta (Berlin) 105:33–42.

Zeevaart, J.A.D. 1977. *Sites of abscisic acid synthesis and metabolism in Ricinus communis L*. Plant Physiol. 59:788–91.

12] Hormonal Control of Plant Embryogeny and Synthesis of Embryo-Specific Proteins

IAN SUSSEX*

SHORT COMMUNICATION

The plant embryo has to complete a minimum of two developmental programs in order to produce the post-embryonic plant. The first of these is a morphogenetic program in which all of the embryonic organs are initiated, and the second is a germination program in which the pattern of growth is switched from that characteristic of embryogeny to that characteristic of post-embryonic development. In primitive land plants such as the mosses and ferns, these two programs occur consecutively without interruption, and development is continuous through embryogeny into post-embryonic stages. In the evolutionary more advanced gymnosperms and angiosperms, however, development is interrupted by the formation of a developmentally arrested seed. Thus, in seed plants there must be a third developmental program not present in the primitive land plants that leads to developmental arrest and seed dormancy. This dormancy program is not an obligate requirement for development to proceed, because embryos of a number of species can be removed from the seed and germinated precociously without first becoming developmentally arrested. This observation raises the question of how during seed development the germination program is held in abeyance while the dormancy program is completed. That is, what prevents the onset of precocious germination of the embryo in normal seed development? In fact, there are a number of seed plants where in normal development embryos germinate without first becoming developmentally arrested. This occurs in mangroves (Sussex 1975) and in the viviparous mutants of maize (Robichaud, Wong, and Sussex 1980). Since the ancestors of all these plants are known to have had developmentally

*Department of Biology, Yale University, New Haven, CT 06520.

STRATEGIES OF PLANT REPRODUCTION (BARC Symposium number 6—Werner J. Meudt, ed.)
Allanheld, Osmun, Totowa

arrested seeds, mangroves and the viviparous maize mutants must have lost the dormancy program, or the controls for this program, secondarily.

Several lines of evidence suggest that the plant hormone abscisic acid (ABA) is responsible for blocking the germination program and is also involved in the program leading to seed dormancy. In developing seeds ABA is initially present in low concentrations and increases rapidly about mid-stage of seed development (King 1976; Quebedeaux, Sweetser, and Rowell 1976), and in *Phaseolus vulgaris* most of the ABA has been shown to be present in the embryo itself (Hsu 1979). Young embryos of *Brassica napus* or of wheat removed from the seed and cultured on a medium lacking ABA germinate precociously within a few days, but fail to germinate as long as they are cultured on a medium containing ABA (Crouch 1979; Triplett and Quatrano 1982). Ihle and Dure (1970) showed that in cotton embryos removed from the seed and cultured aseptically, ABA blocked the translation of mRNA that was synthesized about midpoint in embryogeny and which coded for two proteins that were synthesized during germination. From observations of this kind it can be inferred that at about midpoint in embryogeny the embryo develops the ability to germinate, but that the expression of this ability is blocked by the ABA that accumulates to high levels in the embryo at this time, and that the way in which ABA blocks precocious germination is by preventing translation of the germination-related mRNAs that have been transcribed during embryogeny.

Studies on the importance of ABA in the dormancy pathway have concentrated on its role as a regulator of proteins that are synthesized in the embryo during late developmental stages and which are not synthesized in the post-embryonic plant. These embryo-specific proteins include the major storage protein in the seed of *Phaseolus vulgaris,* phaseolin, the 12S storage protein in *Brassica napus* seeds, the lectin wheat germ agglutinin, and a group of about six proteins of unknown function in the cotton embryo. In *Phaseolus vulgaris* embryos, little phaseolin is accumulated before day 20, but it accumulates rapidly in the next week. This is the time of rapid ABA accumulation in the embryo. Immature bean embryos removed from the seed and cultured aseptically with ABA were shown to accumulate phaseolin more rapidly than control embryos, and this increased accumulation rate depended on an increased rate of phaseolin synthesis (Sussex and Dale 1979). A similar dependence on ABA to stimulate the rate of storage protein synthesis in cultured embryos of *Brassica napus* was shown by Crouch and Sussex (1981); and in immature wheat embryos removed from the grain and grown in sterile culture, Triplett and Quatrano (1982) showed that the agglutinin was not synthesized in the absence of ABA but was synthesized when the medium contained ABA. The late-synthesized cotton embryo proteins studied by Dure, Galau, and Greenway (1981) were synthesized in cultured embryos only when the medium contained ABA, and these workers showed that in the absence of ABA the embryos did not contain the mRNA for these proteins but did when the medium contained ABA. It is not known in this case whether ABA initiates transcription or stabilizes the mRNA for these proteins.

Plant embryogeny is a complicated process involving events that are involved with the morphogenesis of the embryo, with its ability to survive desiccation and prolonged inactivity before germination, and with the storage of reserve materials that will be mobilized to support the early development of the post-germination seedling before it develops its own photosynthetic system. It is useful to attempt to separate these many processes into a number of developmental programs as has been done previously by Walbot (1978), and it seems that ABA may function as a regulator of some of these programs.

ACKNOWLEDGMENTS

The research on storage proteins in *Phaseolus vulgaris* described here was supported by grant GM 24775 from NIH and grant 901-0410-8-0174-0 from USDA.

LITERATURE CITED

Crouch, M. L. 1979. *Storage proteins as embryo-specific developmental markers in zygotic, microsporic, and somatic embryos of Brassica napus L.* Ph.D. dissertation, Yale University.

Crouch, M. L., and I. M. Sussex. 1981. *Development and storage protein synthesis in Brassica napus L. embryos in vivo and in vitro.* Planta 153:64–74.

Dure, L. S., G. A. Galau, and S. Greenway. 1981. *Changing protein patterns during cotton cotyledon embryogenesis and germination as shown by in vivo and in vitro synthesis.* Israel J. Botany 29:293–306.

Hsu, F. 1979. *Abscisic acid accumulation in developing seeds of Phaseolus vulgaris L.* Plant Physiol. 63:552–56.

Ihle, J. N., and L. Dure. 1970. *Hormonal regulation of translation inhibition requiring RNA synthesis.* Biochem. Biophys. Res. Comm. 38:995–1001.

King, R. W. 1976. *Abscisic acid in developing wheat grains and its relationship to grain growth and maturation.* Planta 132:43–51.

Quebedeaux, B., P. B. Sweetser, and J. C. Rowell. 1976. *Abscisic acid levels in soybean reproductive structures during development.* Plant Physiol. 58:363–66.

Robichaud, C. S., J. Wong, and I. Sussex. 1980. *Control of in vitro growth of viviparous embryo mutants of maize by abscisic acid.* Devel. Genetics 1:325–30.

Sussex, I. M. 1975. *Growth and metabolism of the embryo and attached seedling of the viviparous mangrove, Rhizophora mangle.* Amer. J. Bot. 62:948–53.

Sussex, I. M., and R. M. K. Dale, 1979. *Hormonal control of storage protein synthesis in Phaseolus vulgaris.* Pages 129–41 in I. Rubenstein, ed., *The Plant Seed: Development, Preservation, and Germination.* Academic Press, New York.

Triplett, B. A., and R. S. Quatrano. 1982. *Timing, localization and control of wheat germ agglutinin synthesis in developing wheat embryos.* Devel. Biol. (in press).

Walbot, V. 1978. *Control mechanisms for plant embryogeny.* Pages 113–66 in M. E. Clutter, ed., *Dormancy and Developmental Arrest.* Academic Press, New York.

13] Hormonal Control of Senescence Allied to Reproduction in Plants

by HAROLD W. WOOLHOUSE*

INTRODUCTION

Some plant species undergo a single cycle of reproduction that is terminated by the senescence and death of the plant; other species flower and fruit repeatedly without the intervention of a senescent phase. These patterns of behavior have long been recognized and were distinguished as monocarpism and polycarpism, respectively (Hildebrand 1881). Hildebrand's categories based on reproduction and lifespan characteristics have the virtues of simplicity of definition and usefulness in an operational sense, in consequence of which they have gained general acceptance. In this review I shall not attempt to overthrow the concepts of mono- and polycarpism or to refute the terminology; I do wish to suggest, however, three aspects of the impact of reproduction on the growth and survival of plants, which suggest that the monocarpic-polycarpic division may be of limited value.

Limitations to the monocarpic-polycarpic concept

1. *Species of variable behavior that do not precisely fit the monocarpic or polycarpic categories.* The common garden weed *Poa annua* provides an interesting example of variable monocarpism. The species includes a continuum of forms from upright, fully monocarpic annuals to prostrate semi-perennials that root at the nodes. The creeping form has been assigned subspecific rank as *P. annua var. reptans* (Hauskins) Timm., but there is a continuum of intermediate types in which the shorter-lived forms have fewer tillers, fewer leaves per tiller, a smaller number of nodes on each stem, and fewer adventitious roots than the perennial types. It is probable that the source of this variation in longevity in *P. annua* may derive from its ancestry as an allotetraploid (n = 14) which arose from a hybrid between the annual *P. infirma* H.B.K. and the creeping perennial *P. supina* Shrad.

*John Innes Institute, Norwich, Norfolk NR4 7UH, UK.

STRATEGIES OF PLANT REPRODUCTION (BARC Symposium number 6—Werner J. Meudt, ed.)
Allanheld, Osmun, Totowa

2. *Species showing partial somatic senescence associated with reproduction.* A further limitation to the monocarpic-polycarpic division is that they fail to take account of the range of types in which there may be senescence of some part of the vegetative soma associated with senescence, even though the whole plant does not die (Leopold 1961). In herbaceous perennials, for example, the whole aerial shoot dies back, leaving a perennating underground system which may be simply a fibrous root system but more often consists of modified roots, stems, or leaves carrying out a storage function. In species of genera such as *Polygonatum,* a well-defined abscission layer is formed at the junction of the senescent shoot and the subterranean parts, while in others (*Aster* sp., *Monarda*) no anatomically distinct abscission zone is developed. In species of *Sambucus* the compound inflorescence dies back to the node immediately below, at which an abscission layer is formed. In *Syringa* species the inflorescence dies back, but no distinct abscission layer is formed. If one views these types of postreproductive senescence as a series, from the monocarpic plants at one extreme through those in which progressively less of the plant senesces, then the other extreme comprises the many species in which only the actual reproductive structures ripen and decay. The series may be viewed as representing different degrees of investment in the reproductive process.

3. *The polyphyletic origin of monocarpism.* Students of photosynthesis are now well aware that the so-called Kranz anatomy and C_4 pathway of photosynthesis is of polyphyletic origin. Within this broad category of photosynthesis there are many variations in both anatomy and details of the biochemical pathways. As we shall see when considering the monocarpic behavior in different groups of plants, there are substantial reasons for supposing that the hormonal controls and biochemical details of this phenomenon may, like C_4 photosynthesis, exhibit considerable variation along the different lines of its polyphyletic ancestry. For this reason it is important that we should consider briefly the ways in which natural selection may operate in giving rise to the monocarpic habit.

NATURAL SELECTION IN RELATION TO REPRODUCTION AND SENESCENCE

The common denominator in the various manifestations of senescence associated with reproduction is the problem of how the senescence process may be contributing to the achievement of fitness in a given environment. Cole (1954) showed how the reproductive capacity and other life-cycle characteristics of a species may serve to maximize the fitness of that species. Much of the subsequent theoretical work on this problem in plants has made the tacit assumption that group selection is not a significant factor. In consequence, the problem has been reduced to one of optimizing the allocation of resources between three major functions (Cohen 1971): (a) vegetative growth in order to acquire photosynthates and nutrients; (b) development of defenses against predation, such as toxins, prickles, and spines; and (c) reproductive attributes, such as timing of sexual development, numbers of seeds produced, and their rates of maturation (Lewontin 1965; Woolhouse 1974; Harper 1977). The difficulty confronting theorists working on this problem, which is particularly pertinent to the subject of potential hormonal controls, is that of how to deal with the fact that in order to maximize fitness over the whole lifespan it is not necessary to maximize fitness at every moment of time. For example, it may be suboptimal in terms of predation by grazing to produce a heavy growth of leaves, but advantageous at a later stage in terms of photosynthetic capacity for seed filling. The solution to these problems has been sought by the application of optimal control theory (Vincent and Pulham 1980). But our present concern is with the underlying physiology; I have alluded to this background, however, because it provides the most clear-cut reasons for a change of emphasis in the physiological work. The shift that appears to be necessary is away from the generalizing of the reproduction and senescence phenomena and toward a more precise description of the changes that occur in each species or group of related species. In order to develop this idea, I have chosen to place the main emphasis in this chapter on case studies of reproduction and senescence in different groups of plants. It is, however, central to both the theoretical work concerning fitness and the earlier physiological work in particular that we confront this question of the allocation of resources.

REPRODUCTION AS AN ALTERNATIVE TO VEGETATIVE GROWTH

The idea of reproductive effort as a completing element for the resources available to an organism is an old one. The origins of the concept may be found in observations such as those of Reichart (1821), who wrote of the wallflower *Cheiranthus cheiri,* "If by always removing the buds we prevent flowering in the second year of a stock which is by nature not inclined to produce lateral shoots, leaf-bearing lateral branches will then arise at the apex. If we permit only the most vigorous of these to remain and remove any blossoms which it may be inclined to produce, then by thus allowing only the most vigorous shoot at the apex to develop we can produce quite a tree. If we wish such a plant to flower in the third year or even later we do not disturb later lateral shoots which give rise to fine flower clusters."

Mattirolo (1899) showed that removal of the flowers of *Vicia faba* produced more growth of leaves, stems, and roots; there was more branching and the plants lived longer than those on which fruits were allowed to develop (Table 13.1). Of particular interest here is the finding that the total weight of material produced in plants from which the flowers were removed was approximately twice that of intact fruiting plants. Doflein (1919) coined the term "Stoffwechseltod" ("metabolic death") to describe the fate of the monocarpic perennial *Agave americana*, which in its natural habitat grows to a considerable size over a period of about ten years before it flowers, fruits, and dies. Molisch (1938) removed the flowers from plants of *Reseda odorata*, thereby inducing a transition from a small annual to a perennating shrub some 2 m in height. Molisch argued that in the short-lived fruiting plants the essential feature of

Table 13.1 The Effect of Removal of Flowers of *Vicia Faba* on Fresh Weight of Various Plant Parts

	Total fresh weight (gr) (27 plants)	
Plant parts	Intact plants	Plants with flowers removed
Stem	1286.0	3513.0
Root	1330.0	2783.0
Nodules	42.8	102.4

Source: Abstracted from Mattirolo (1899).

this kind of death is complete exhaustion of the stored material for the reproductive organs; that is, the developing flowers and fruits draw upon the valuable food materials to such an extent that nothing remains for the other organs and they perish. The corollary to this analysis was that Doflein's terminology was inappropriate, and it was concluded that the term "Erschöpfungstod" ("exhaustion death") was more descriptive of what happened. An abundance of suggestive circumstantial evidence may be marshaled in favor of this hypothesis; in trees, for example, Harper (1977) cites from the work of Rohmeder (1967) the cases of *Fagus sylvatica*, in which the growth of annual rings was inversely correlated with extent of fruiting in that year, and *Fraxinus excelsior*, in which male plants produced greater increments of growth than female plants, presumably because of the expenditure on fruiting in the females.

In cotton there is a pronounced decline in growth rate that is precisely correlated with onset of reproduction and the numbers of flowers and fruits produced (Balls and Holton 1915 a,b; Balls 1918). Mason (1922a,b) found a close relationship between the number of flowers produced in cotton and the slowing down of elongation of the main stem; as the older fruits developed there was a pronounced abcission of younger fruits and flowers, which the author attributed to a shortage of carbohydrates in these organs as assimilates were diverted to the maturing fruits. Ewing (1918) observed that after most of the fruits on a cotton plant had reached maturity, there was often a renewed burst of growth in which more flowers and fruits were developed; plants growing under particularly favorable conditions of water and nutrient supply flowered and fruited continuously, and if the buds were removed as they formed the size of the plants was almost doubled. All of this work lent general support to the view, already widely current among gardeners and fruit growers in particular, that development and maturation of fruits depresses vegetative growth, and to this extent it is in accordance with the hypothesis of competition for allocation of resources. Murneek (1926) examined the effects of fruiting on growth in tomato *(Lycopersicon esculentum)*, including comparisons between plants receiving high and low levels of nitrogen. Figures 13.1 and

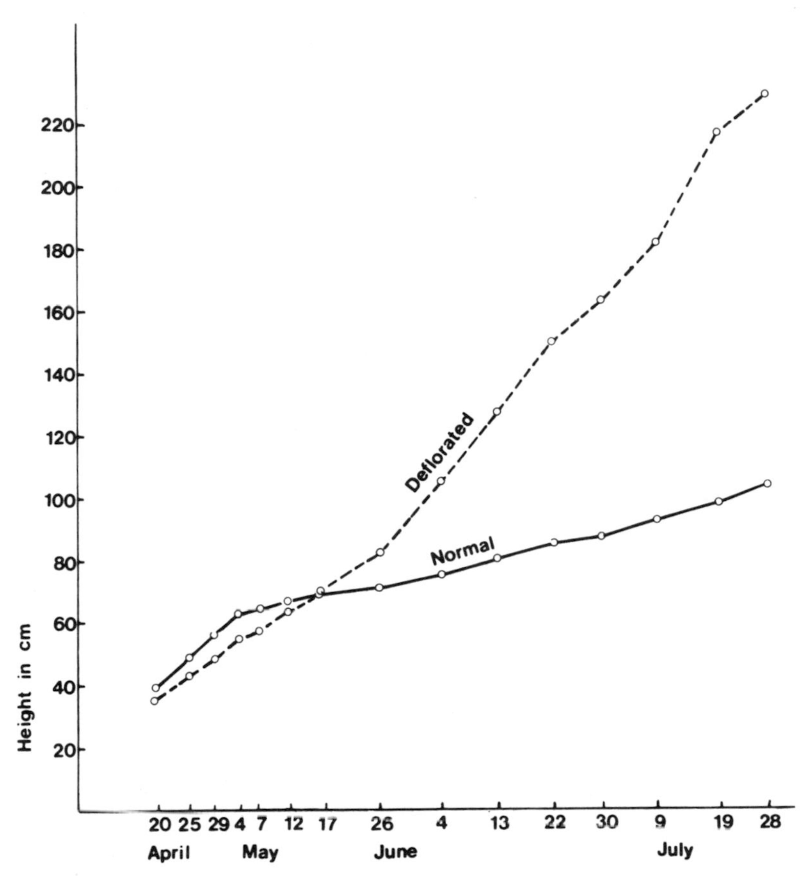

Figure 13.1. Comparison of growth in height of tomato plants bearing flowers and fruits and plants from which the flowers were systematically removed as they opened (after Murneek 1926).

13.2 show the growth in height of plants receiving high and low levels of nitrogen; Murneek analyzed carbohydrate- and nitrogen-containing fractions from the plants and drew the conclusion that the high demand for nitrogen by the fruits was imposing a limitation on the growth of fruiting plants. The author was evidently uneasy with this conclusion, since the effects on fruits were manifest even in plants receiving excess nitrogen. He surmised that the tomato plant might have a capacity for the storage of nitrogen in the vegetative tissues and that growth might be restricted because the plants had reached the limits of their capacity for uptake of nitrogen to meet the demands imposed by the fruits; but there is no experimental evidence that rates of uptake of nitrogen are limiting for growth when the level of supply to the roots is adequate. At the conclusion of his discussion Murneek grapples with the question of why the developing fruit can monopolize the nitrogenous food supply and concludes,

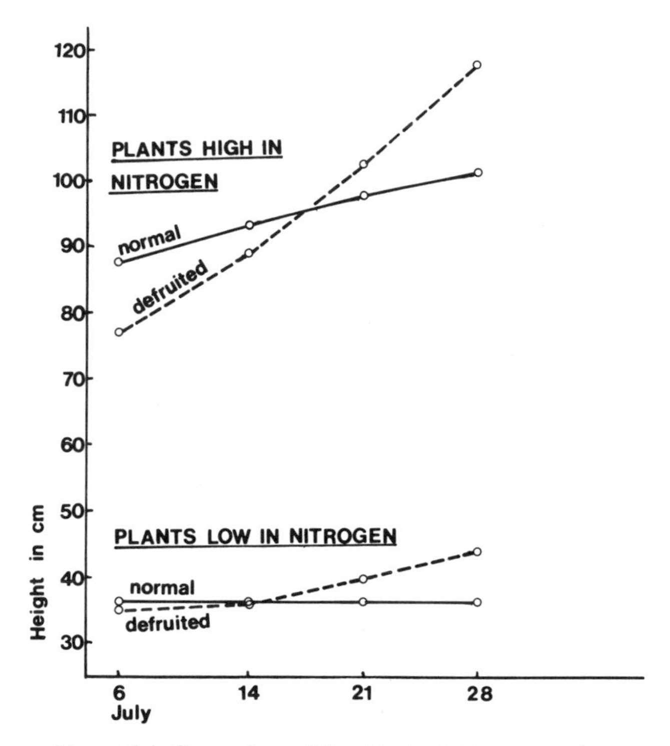

Figure 13.2. Comparison of the effects of nitrogen supply on the growth in height of tomato plants bearing flowers and fruits and plants from which the fruits were systematically removed (after Murneek 1926).

"It is conceivable that plants may have a controlling glandular organization or a system of secretions similar to that existing in animals. . . . If such an organization of secretions or hormones were to come into play at certain stages of development of particular organs, one would more or less be able to account for the control by and diversion to the fruit of certain food constituents. That plants may secrete hormones has been suggested by several botanists, particularly Errera (1905) and Haberlandt (1921). Evidently the basic part of the problem may be accounted for by two possibilities: (1) Either it is a case of localised nutrition brought about by genetic union and the consequent rejuvenescence and initiation of a metabolic gradient or (2) it may be that we have to deal here with a hormonic control of the physiological substrate, the indispensible chemical, possibly nitrogenous, constituents of the plant. In both instances we enter upon almost completely unexplored domains of plant physiology."

We shall return to this question of relative allocations of resources in considering different species. While dealing with general aspects of the phenomenon, however, it is pertinent to examine the overall relationship of sources and sinks involved in the partitioning of materials between reproductive and other structures. It is also relevent to examine the concomitant anatomical changes that are involved in the switch from vegetative to reproductive growth, since these afford significant pointers to the nature of the controls that are operating.

SOURCE AND SINK RELATIONSHIPS ACCOMPANYING SENESCENCE

It is generally agreed among plant physiologists that the movement of assimilates is from sources to sinks along gradients of concentration. The locations of sources and sinks may change in the course of development. As an example we may take the case of the jerusalem artichoke *(Helianthus tuberosus)*, in which carbohydrates pass initially from their source in the leaves to sinks in the actual growing regions and a storage pool in the main stem of the plant. Following the induction of tuberization, which requires short days, the young tubers take over as the most powerful sinks, for which the storage pool in the stems and roots now becomes a source. As filling of the tubers is completed, the aerial stems and roots senesce and die. In the following spring new roots and shoots are developed and function initially as sinks, drawing on the overwintered tubers which now function as sources and in turn become senescent as their storage reserves become exhausted. The behavior of *H. tuberosus* raises a number of points:

1. It provides further support of a general kind for Molisch's exhaustion hypothesis in that as the leaves, stems, and tubers each in their turn change from sinks into sources, so they come to a phase of senescence that is correlated with the exhaustion of their contents.
2. *H. tuberosus* also permits us to generalize the problem somewhat because, in higher latitudes where it is grown for culinary purposes, the requisite combination of short days and high temperatures needed for the induction of flowering is never achieved, but the impact of the tubers as a sink is sufficient to induce the senescence of the shoots. If the tubers are removed as they are formed the shoot will remain alive in a healthy vegetative condition. Evidently the death of the shoot is an event geared to the existence of any competing sink; the development of fruits and seeds is not unique in this respect.
3. The interpretation of the behavior of *H. tuberosus* in terms of changing patterns of sources and sinks shifts the emphasis of physiological enquiry to the question of what determines how a particular organ may serve successively as sink and source and what determines the relative strengths of competing sinks.

ANATOMICAL ASPECTS OF THE SWITCH FROM VEGETATIVE TO REPRODUCTIVE GROWTH

In considering reproduction and senescence from the standpoint of allocation of resources, emphasis has been placed on quantitative aspects of growth and the movement of materials to new sinks in the developing fruits. It is important to recognize, however, that as these changes take place, they may be accompanied by correlated anatomical changes that are an integral part of the switch which is taking place, and they may be an important factor in determining the death of the plant at the end of the reproductive phase.

Changes in the shoots that accompany onset of flowering. It has been shown that in many species anatomical changes occur in the stems as the plant enters the reproductive phase (Roberts and Wilton 1936; Wilton and Roberts 1936; Wilton 1938; Struckmeyer and Roberts 1939; Struckmeyer 1941). Wilton and Roberts (1936) examined the flowering stems of 28 species and found that in most of them the cambial region was devoid of meristematic cells. For example, in *Chrysanthemum morifolium* the zone of active cambium was six cells deep in vegetative plants, but there were no active cambial cells in the flowering stems. (Figure 13.3 shows a comparable situation in the stems of vegetative and flowering plants of *Perilla frutescens.*) Wilton (1938) grew 20 species under various regimes of temperature and daylength in order to vary the extent of reproductive development. Some plants developed flower buds at most of the growing points, and in these vegetative growth ceased because all of the primordia had differentiated as flower buds; plants developing under different conditions produced fewer flowers and continued to develop vegetatively while the flowers and fruits were developing. In species of *Cosmos* and *Amaranthus* which differentiated flower buds at a large proportion of apices, no meristematic tissue remained at the time of full flowering; in species which differentiated only a proportion of the apices as flowers under the conditions in which the plants were grown, active meristematic tissue persisted in the cambial zone. Prior to Wilton's work it had been widely supposed that the decrease of cambial activity in stems was associated with chronological age rather than an effect of flowering. Wilton examined sections from successive internodes of the stems of *Cosmos sulphureus* Cav. var. Klondike and found that in vegetative plants the basal internodes were the first to mature and stop differentiating new tissues, while in plants bearing flowers and fruits the converse was true—the uppermost internodes proximal to the flowers were the first to lose their cambial activity (Wilton 1938). Studies on *Cosmos sulphureus, Xanthium echinatum, Salvia splendens* and *Glycine max* showed that decline of cambial activity was associated with the early stages of development of the flowers (Struckmeyer 1941). Plants of *Glycine* subjected to SD cycles had initiated flower primordia after 12 inductive cycles, by which time the cambium was already less active than in vegetative stems.

The decline of cambial activity in stems bearing flowers leads to a reduction in the amount of secondary tissue that develops, which is particularly pronounced in the phloem. Struckmeyer and Roberts (1939) showed that the phloem of flowering stems contained fewer and smaller sieve tubes and companion cells than vegetative stems in species of *Ageratum, Hedalgo*, and *Delphinium*. In *Ricinus communis* and *Geranium* sp, parenchyma rather than sieve tubes were differentiated in the phloem of flowering stems, the companion cells had thicker walls, and more callose was present in the sieve tubes.

The extent of these various modifications in the phloem of stems bearing flowers was found to vary with the species.

This group of papers on effects of flowering on the anatony of stems has important implications for the senescence of monocarpic plants, but certain aspects of the work are not entirely satisfactory. In many cases the investigations suffered from a lack of adequate controls in the form of strictly vegetative plants; for example, in many cases vegetative and flowering stems were taken from the same plant. In some cases plants that had a persistent tendency to flower were defoliated in order to prevent flowering, but the effects of defoliation per se were not clearly distinguished. Sampling of the internodes appears to have been somewhat haphazard in many of the investigations; in all cases the anatomical observations were qualitative, and no attempt to estimate actual areas of active cambium or phloem appear to have been carried out.

Al-Hadithi (1977) returned to the study of effects of flowering on activity of the cambium and phloem in *Perilla frutescens* L. Britt., a monocarpic species requiring approximately ten SD cycles for the induction of flowering. Attention was concentrated on the anatomy of the 5th internode between the 5th and 6th pairs of true leaves; Figure 13.4 shows the layout of the stem and the relevant areas of internodes that were sampled. Figure 13.3 shows the depth of the cambium in the stems of plants that were induced to flower and vegetative controls which received the same daylength but three short night-breaks to prevent flowering. In flowering plants the depth of the cambium began to decline at 20 days, by which time the young flower buds were just discernible, and was scarcely recognizible at 40 days, by which time the fruits were beginning to develop. The depth of the cambium at the same node in vegetative plants increased gradually throughout the same period.

That the differences in depth of the cambial zones between vegetative and flowering plants are related to differences in meristematic activity is clearly shown by the divergence in the amounts of primary xylem and phloem formed in each case (Figs. 13.5 and 13.6). The figures show that in vegetative plants the amount of secondary xylem increased continuously throughout the period of the experiments; there was only a slight increase in the depth of the secondary xylem in flowering plants over the same period. The depth of the secondary phloem in vegetative plants increased from about 50 to 110 μm during the first 30 days of the experiment and then remained constant, whereas the depth of phloem in the induced plants remained constant at about 50 μm throughout the experiment. The area of active phloem and state of the sieve plates was measured by a fluorescence technique following staining with aniline blue. Figure 13.7 shows the total area of active phloem following the same patterns as the depth of phloem tissue in the vegetative and flowering plants. The fluorescence technique also makes it possible to distinguish the sieve plates clearly and thereby to determine any changes in the numbers of sieve tubes per unit area of phloem and whether or not they are open or covered by a continuous deposit of callose (Figure 13.8). Figures 13.9 and 13.10 show the number of sieve plates and the total area of sieve plates in a cross section at the 5th internode, increasing in proportion with the total area of phloem in each treatment. In the flowering plants there was a small but

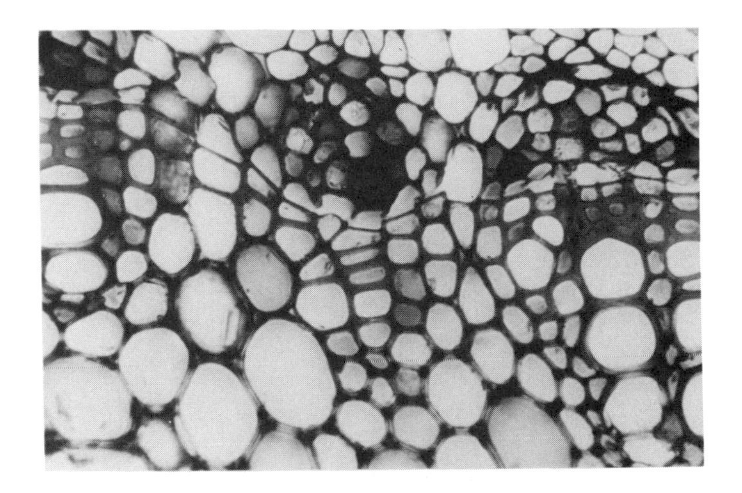

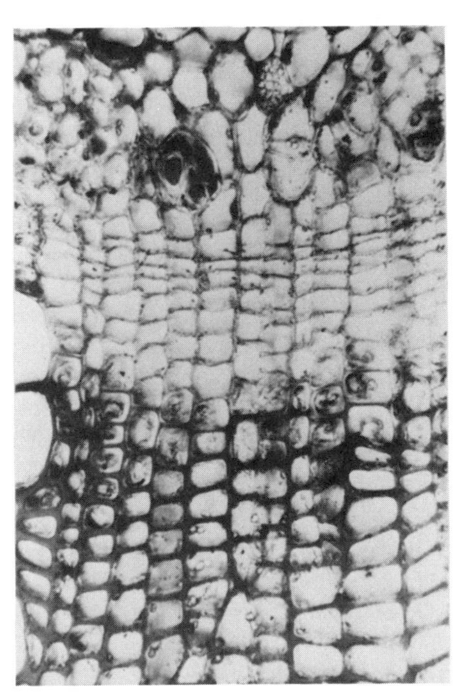

Figure 13.3. Effect of flowering on the activity of the cambium in stems of *Perilla frutescens* (L) Britt. *(a)* T.S. fifth internode of a flowering plant after 40 inductive SD cycles. Note that the cambium has been virtually eliminated as cell division ceased and differentiation of xylem and phloem continued. Magnification X 473. *(b)* T.S. fifth internode of a vegetative plant of *Perilla* cut at the same time as (a). The activity dividing cambium is 5–6 cells deep. Magnification X 351.

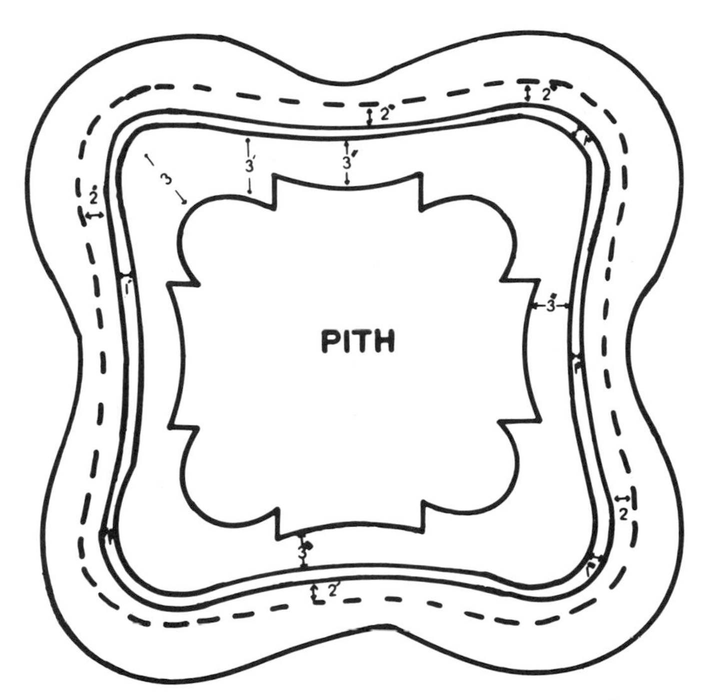

Figure 13.4. Plan of the T.S. of the Stem of Perilla showing positions at which sample measurements were made. 1–1 represents some positions at which the depths of the cambial zone were measured; 2–2 represent some positions at which the depths of the phloem zone were measured; 3–3 represent some positions at which the depth of the xylem zones were measured; and - - - represent phloem fibres.

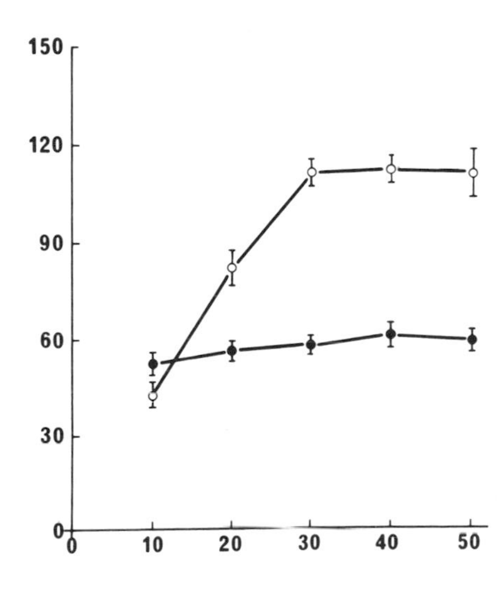

Figure 13.5. The effect of floral induction on the depth of the phloem zone at internode 6 in *Perilla*. Each point represents the mean of 150 measurements—with standard deviation—on 3 replicate sections. Key: induced plants (●——●); non-induced plants (○——○).

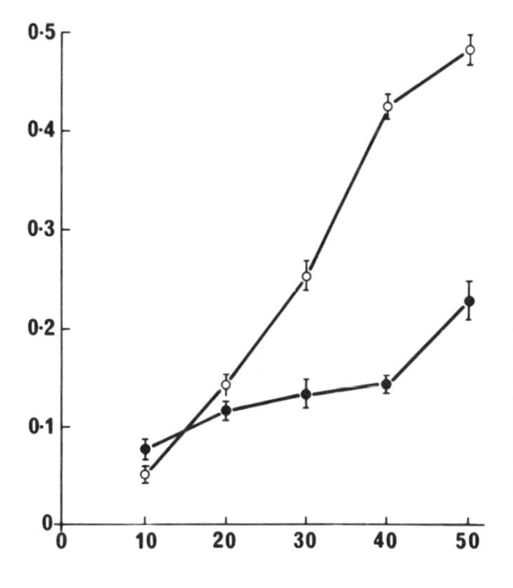

Figure 13.6. The effect of floral induction on the depth of the xylem zone at internode 6 in *Perilla*. Each point represents the mean of 150 measurements—with standard deviation—on 3 replicate sections. Key: induced plants (●——●); non-induced plants (○——○).

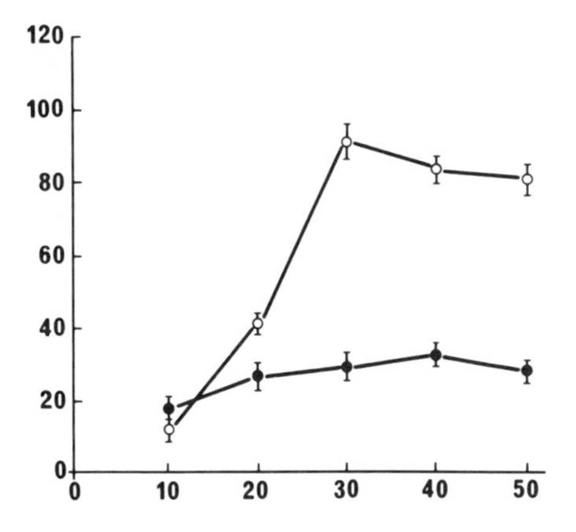

Figure 13.7. The relationship between the inductive conditions and the area of phloem in *Perilla*. Each point represents the mean of 3 sections—with standard deviation. Key: induced plants (●——●); non-induced plants (○——○).

significant increase in the number of sieve tubes blocked by definitive callose in the later stages of flowering and fruiting (Fig. 13.11). The implications of these results with respect to translocation and hormonal changes in the flowering plants will be considered later.

Association of the reproductive phase with a decreased rate of growth of roots. Decreases in the rate of growth of roots in the time of flowering and fruiting occur in *Lycopersicum esculentum* (Smith 1924; Murneek 1926; Strijbosch 1954; Cooper 1955; Leonard and Head 1958; Hudson 1960; Van Dobben 1962); *Gossypium hirsutum* (Eaton 1927, 1931; Andrews and Clouston 1937; Eaton and Ergle 1952); *Zea mays* (Loomis 1935); *Cucumis sativus* (de Stigter 1969);

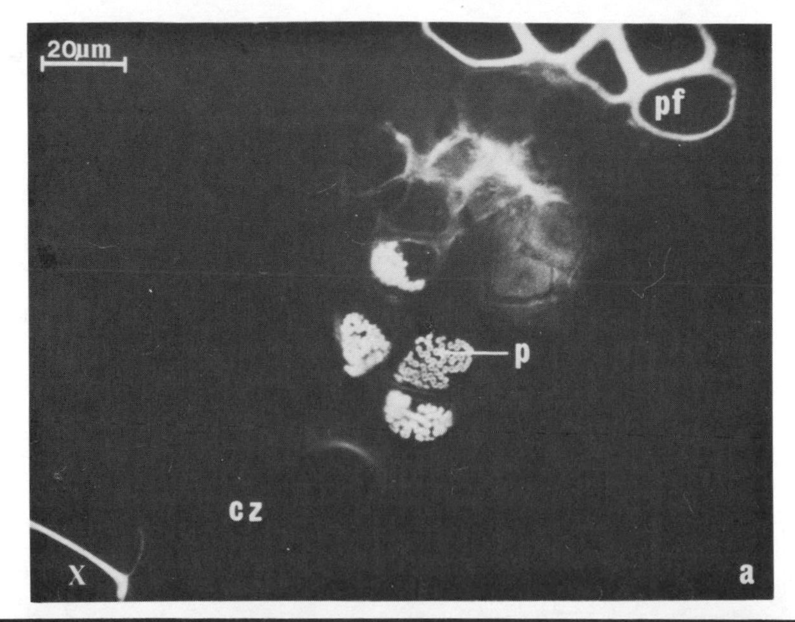

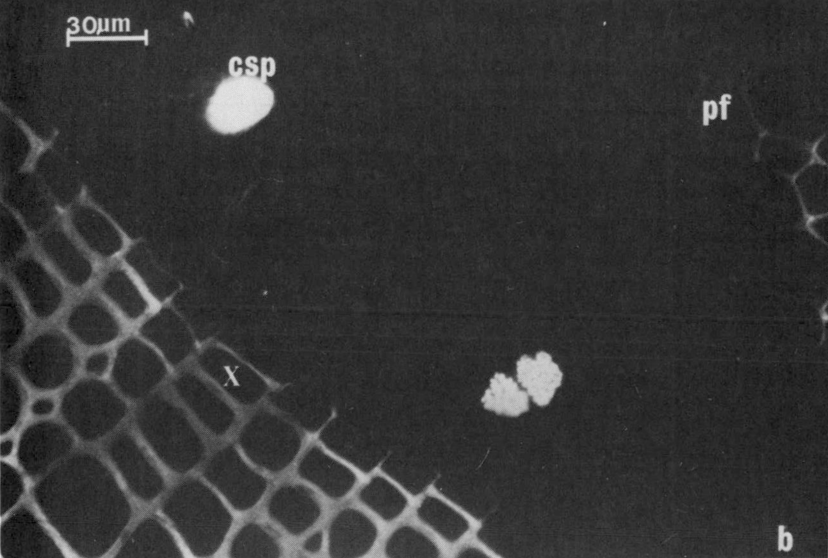

Figure 13.8. Effects of floral induction on the phloem in the central region of the stem at internode 5 in *Perilla*. Fluorescent micrographs of transverse sections, approximately 30 μm thick, showing the extent of the fluorescent zone in the phloem, the fluorescence of callose lining the pores, and the thick deposits of callose between the cellulose walls, covering the entire surface of the sieve plate. *(a)* A fluorescent micrograph of non-induced plant. Taken 40 days after the start of induction (the first 25 days of long night with light-break and the last 15 days of long day in the greenhouse). Note that the fluorescent zone is occupying most of the phloem area. The pores of the sieve plates are opened and in one sieve plate they have a uniform deposit of callose around them. *(b)* A Fluorescent micrograph of an induced plant which received 25 inductive cycles followed by 15 days in the greenhouse. The fluorescent zone occupies a strip of phloem close to the xylem tissue. The rest of the phloem area contains nonfunctional (nonfluorescent) elements. Note the considerable depth of the nonfunctional area that is present between the fluorescent area and the phloem fibers. The sieve plate at the upper side of the micrograph is covered with massive deposits of definitive callose, while the other two at the lower side are open but their pores are somewhat constricted by callose and therefore look closed. (Kodak panatomic-x film. Exciter filter BG3, barrier filters -65, 47.)

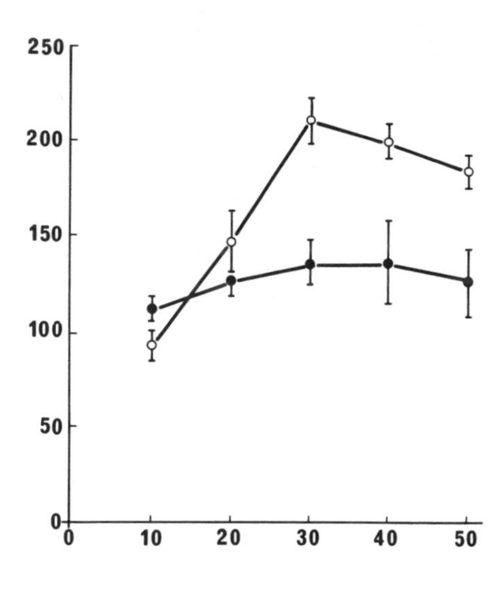

Figure 13.9. The effect of floral induction on the total number of sieve plates (closed + open) present in a transverse section of phloem at stem internode 6 in *Perilla*. Each point represents the mean of 3 measurements, with standard deviation. Key: induced plants (●------●); non-induced plants (○------○).

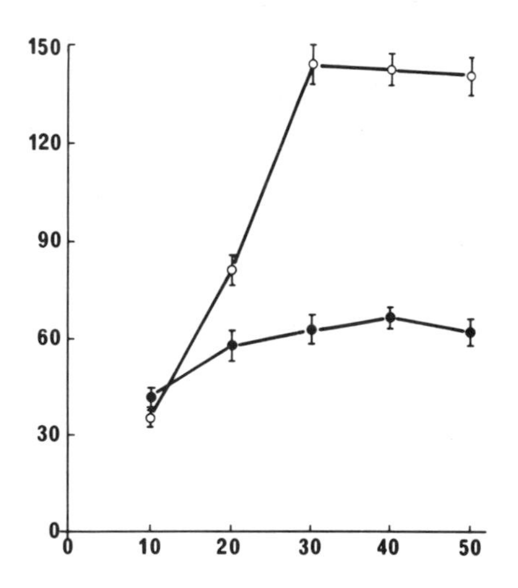

Figure 13.10. The relationship between floral induction and the cross-sectional area of sieve plates (sieve tube lumen) at stem internode 6 in *Perilla*. Each point represents the mean of 150 measurements in μm—with standard deviation on 3 replicates. Key: induced plants (●------●); non-induced plants (○------○).

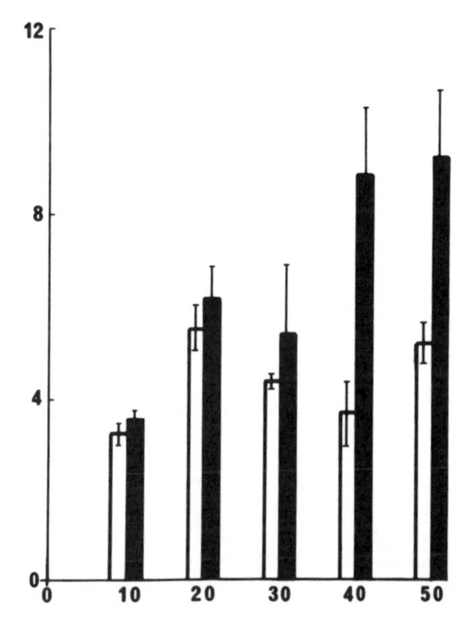

Figure 13.11. The relationship between the inductive conditions and the percentage of closed sieve plates per transverse section of phloem in the main stem of *Perilla*. Each column represents the mean of 3 replicates, with standard deviation. The solid column represents the induced plants; the blank column represents the non-induced plants.

Raphamus sativus (Nightingale 1923); *Pisum sativum* (Brouwer 1962); *Cucumis melo* (van der Post 1968); *Perilla frutescens* (Beever and Woolhouse 1975); cereals and other *Graminae* (Brenchley and Jackson 1921; Stuckey 1941; Petinov and Berko 1961) and various ornamentals (Roberts and Struckmeyer 1946).

Many of the studies referred to above have been concerned only with relatively gross measures of root growth, such as rates of increase in dry matter, and in most cases the species used have been day-neutral species in which flowering cannot be manipulated experimentally, so that adequate controls in the form of vegetative plants were not available. The tomato might be described as a quasi-monocarpic or indeterminate species: it does not perennate nor does it die as soon as the first flowers or fruit are produced, and it generally seems to undergo a more lingering death as flowering and fruiting proceeds. Van Dobben (1962) found that root growth almost ceased in tomato plants at the time of first fruit formation, and growth is resumed subsequently even though the plant continues flowering and fruiting. De Stigter (1969) developed a versatile water culture technique for studies of root growth in which the plants were grown on inclined sheets of black nylon irrigated with a percolating nutrient solution, permitting ready access to the roots which could thus be measured at frequent intervals. Using this technique it was found that in cucumber c. v. Lentse Gele, root growth began to slow down some 3–4 days after pollination and eventually stopped, there was some resumption of slow growth of roots at a later stage, and this was greatly accelerated if the fruits were removed.

De Stigter's technique has been applied to the study of root growth in *Perilla frutescens*. Plants were subjected to SDs in order to induce flowering, and the rates of elongation of individual roots were measured at 24-hr intervals (Figure 13.12); controls consisted of plants receiving the same length of day with short night-breaks to prevent induction of flowering.

The rate of elongation of roots began to slow down after 14 inductive cycles, coincident with the time at which young flower buds first became visible to the naked eye and the cambium ceases to be active. Root growth ceased after 40–50 inductive cycles, by which time the first fruits were beginning to ripen and some of the sieve plates were becoming blocked by definitive callose. As the rate of root elongation declined in plants induced to flower, there occurred a strongly correlated decline in the rate of respiration of the roots (Figure 13.13). While these findings do not provide proof of a definitive role of the root system in the mechanism of monocarpic senescence, they do indicate strongly correlated changes in the root system that cannot be ignored.

HORMONAL CONTROLS AFFECTING MONOCARPIC SENESCENCE

Summarizing the evidence presented so far, it may be concluded that senescence associated with reproduction has the following attributes:

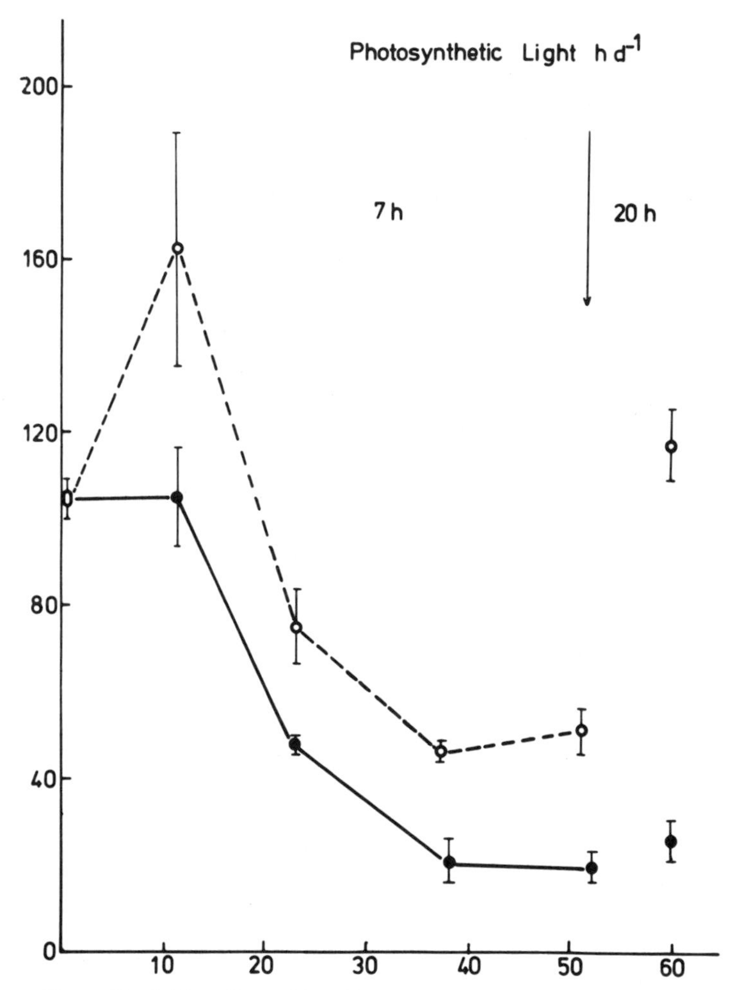

Figure 13.12. The effect of floral induction on the rate of respiration of the roots of *Perilla*. Each point is the mean ± s.e. of 8–10 replicates on two plants. The S D treatment was terminated on day 52 after the start of induction and all the plants were returned to 20h photosynthetic light. Key: inducing conditions (●------●); non-inducing conditions (○------○).

1. Senescence is usually associated with the development of a sink of progressively increasing strength in the developing fruits.
2. Senescence is frequently associated with a conversion of vegetative to floral apices.
3. Senescence is generally correlated with a decline in the rate of growth and respiration of the roots.

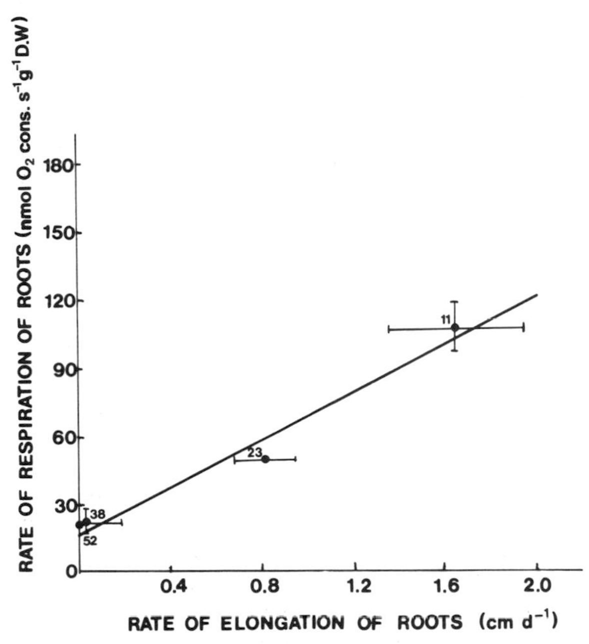

Figure 13.13. Relationship of respiration rate to rate of root elongation in *Perilla*. The numbers against the points indicate the number of inductive cycles which the plants had received at the time the measurements were made.

4. Removal of the flowers as they arise arrests the senescence process and permits a continuation of vegetative growth.
5. Onset of senescence is strongly correlated with a loss of cambial activity in the stems.

Each of these general facts about reproduction and senescence may be seen to contain implications suggesting the involvement of hormones in the regulation of the processes involved, which may be summarized as follows:

1. Diversion of assimilates. It has been found that in the fruits of many species, cytokinin levels are high in the early stages of development when cell division is at a maximum. In the later stages of development when cell division is less evident and cell enlargement is the principle mode of growth, cytokinin levels decline while IAA and gibberellins increase. These observations are closely commensurate with the role of the fruit as a powerful sink, since it is known that these hormones, when applied exogenously to the sites of excised ovules or vegetative apices, may replace, in part at least, the ovules and apices, in the sense that they are able to sustain the transport of nutrients and assimilates to these sites (Booth et al 1962; Wareing and Seth 1967; Jeffcoat and Harris 1972).

There has been a continuing controversy over whether the presence of high levels of hormones in developing fruits promotes the metabolic activity, which generates a sink demand, or whether there is a distinct phenomenon of hormone-directed transport operating (Seth and Wareing 1967). Reviewing this problem, Noodén (1980) concedes that metabolic sink activity may be involved but stresses the importance of hormone-directed transport on three grounds. First, it is noted that kinetin promotes the movement of γ-amino butyric acid toward the site of application in a detached leaf, even though this amino acid is not incorporated into protein (Mothes, Engelbrecht, and Schütte 1961). It must be recognized, however, that in these experiments the region of the leaf that is treated with kinetin remains green and metabolically active, while the surrounding tissues become yellow and senescent. As this happens there is an active transfer of sugars and other metabolites from the senescent to the viable regions, and it may well be that the γ-amino butyrate is simply carried along in this mass transfer of materials to the site of higher metabolic activity, and no specific hormone-directed transport need be involved. Second, it is argued that "auxin appears to stimulate transport to the site where it is applied more than it promotes sink activity." In this proposition the word *appears* is crucial, since there is in practice no adequate method for obtaining satisfactory measurements of transport or of "sink activity." Noodén's third argument in favor of a primary role for hormone-directed transport in the sink activity of fruits rests on the observation that auxin may induce the differentiation of channels of transporting tissue (Sachs 1975). The evidence for this effect of auxin has gained a general measure of acceptance, but it is not remotely clear how this observation on wound-induced differentiation in stems of *Phaseolus vulgaris* bears upon the movement of materials into the developing fruits. Thus it seems reasonable to rest with the long-established view that the hormonal status of developing fruits is an important factor in determining their capacity to function as sinks for assimilates and other nutrients (which may in some species constitute a drain upon the rest of the plant which contributes to the process of senescence), but there is as yet no evidence concerning the mechanisms underlying this effect.

2. Conversion of vegetative shoot apices to inflorescences. If comparisons are made of a range of monocarpic plants it is found that one may distinguish what may be viewed as degrees of monocarpism, which differ according to the extent to which the generative capacity is lost. Thus, in many annual weeds and desert ephemerals and the common cereals, all the vegetative apices are converted to inflorescences and there are no remaining meristems to continue the vegetative growth.

Epilobium montanum is typically a perennial herb, flowering in June to July, in which the flowering stems die as the seeds mature. As the aerial stems die down, short, overwintering stolons are produced at ground level in response to the shortening daylength of autumn. If, however, the daylength is artificially maintained as the fruiting stage is reached, the lateral buds at the base of the stems do not develop and the plant behaves as a monocarpic annual.

Other species in which the monocarpic habit may be less clearly defined are those in which the terminal meristems of the shoot are not converted to inflorescences, for example, in soya bean, cotton, and peas. Varieties are known, in each of these species, in which there is a tendency to develop lateral shoots from the base of the stem as the original fruiting shoots become senescent. Evidently there is genetic variation in the residual capacity to reactivate these lateral buds at the base of the stem, but it is not clear whether these genotypes differ in retention of the meristematic potential in the basal buds or in the capacity to produce the hormones necessary for their activation.

It has been suggested that a major contributory factor to senescence in species in which all of the vegetative apices are converted to floral structures is the elimination of supplies of auxin necessary for continued growth (Beever and Woolhouse 1974, 1975). Evidence for this hypothesis comes from studies on *Coleus* sp. and the related genus *Perilla*. When *Perilla* plants are exposed to short days, which induce flowering, there is a decrease in the auxin content of the shoots (Chailakyan and Zhadanova 1938; Zhadanova 1945; Harada 1962). The decreased rate of stem elongation when *Perilla* is induced to flower is almost certainly due to a decreased auxin supply from the shoot apex and younger leaves, which are known to be the major source of diffusible auxin entering the young stems (Jacobs 1952; Jacobs and Bullwinkel 1953). This effect becomes accentuated as no further leaves are differentiated and the existing ones grow older. There is a concensus of opinion to suggest that although excised roots in culture may become autonomous for auxin production (van Overbeek 1939; Thurman and Street 1960), but attached roots receive the auxin required for growth (Thimann 1936; Torrey 1963; McDavid, Sagar, and Marshall 1972) by acropetal polar transport from the shoots (Morris, Briant, and Thomson 1969; Iversen, Aasheim, and Pedersen 1971; Scott 1972), which probably takes place in the phloem (Torrey 1976). On the basis of this evidence we may construct the first element of a feed-back loop (Figure 13.14) in which auxin supply from the shoot has a promotive role in growth of the roots. Recent studies of geotropism provide strong evidence that growth of roots is modulated by a growth inhibitor, possibly ABA, originating in the root cap. Thus we may envisage root growth as being controlled by the balance of auxin supplied from the shoot and inhibitor from the root cap; when flowering diminishes the auxin supply, the balance favors the inhibitor and root growth is suppressed. The feed-back loop is completed by reciprocal effects of hormones produced in the roots, such as cytokinins, stimulating shoot apical growth, which in turn affects the supply of auxin to the roots (Sachs 1972; Figure 13.14). Support for both sides of this suggested feed-back loop is provided by grafting experiments. Lindemuth (1901) grafted a shoot of the indeterminate perennial *Abutilon thompsonii* onto a stock of the normally determinate monocarpic plant *Modiola caroliniana*. The *Abutilon* scion kept the *Modiola* growing for a further two and a half years; this result could be explained in terms of the perennial scion providing a continued supply of auxin for the roots of *Modiola*, which is normally lost when the shoot of *Modiola* is switched to flowering.

In *Phaseolus vulgaris* and *Glycine max* leaf senescence and abscission are subject to genetic variation (Probst 1950; Honma, Bouwkamp, and Stojianov

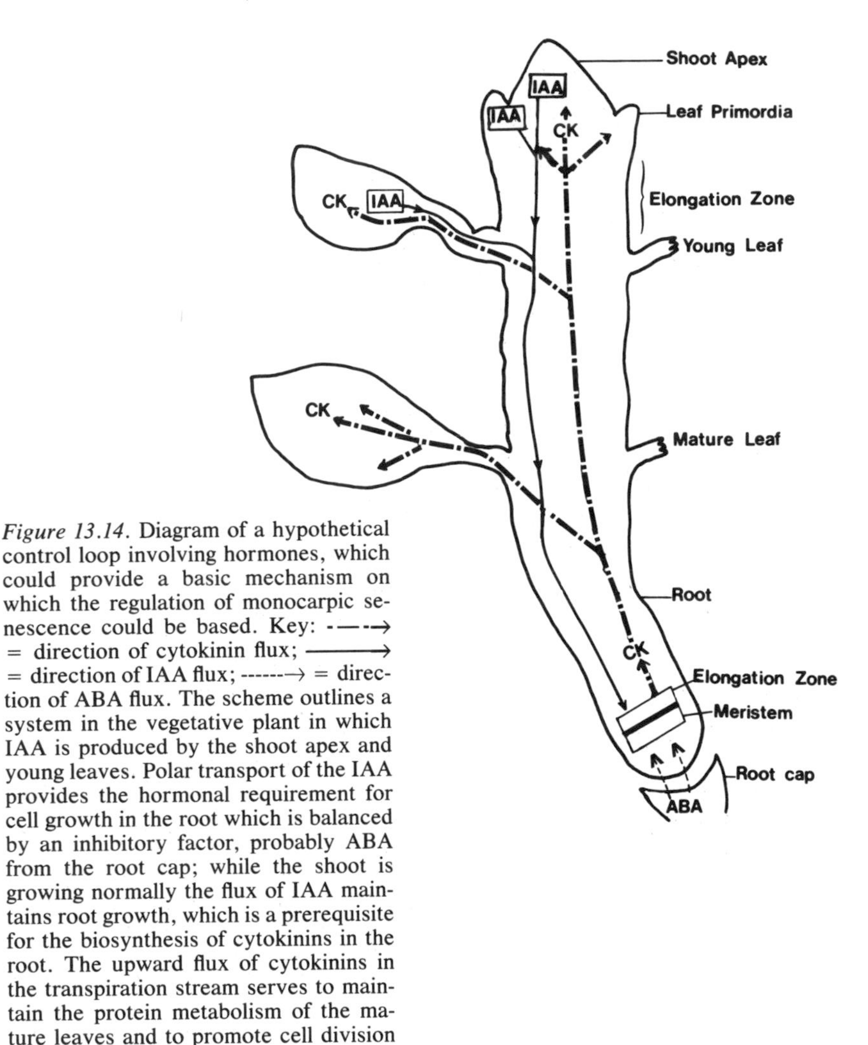

Figure 13.14. Diagram of a hypothetical control loop involving hormones, which could provide a basic mechanism on which the regulation of monocarpic senescence could be based. Key: -·—·→ = direction of cytokinin flux; ———→ = direction of IAA flux; ------→ = direction of ABA flux. The scheme outlines a system in the vegetative plant in which IAA is produced by the shoot apex and young leaves. Polar transport of the IAA provides the hormonal requirement for cell growth in the root which is balanced by an inhibitory factor, probably ABA from the root cap; while the shoot is growing normally the flux of IAA maintains root growth, which is a prerequisite for the biosynthesis of cytokinins in the root. The upward flux of cytokinins in the transpiration stream serves to maintain the protein metabolism of the mature leaves and to promote cell division in the shoot apex and young leaves.

1968). Hardwick (1979) carried out reciprocal grafting of genotypes showing early and delayed senescence. In *Phaseolus* the grafting of varieties showing delayed senescence and abscission "retainers" onto rootstocks of varieties showing early senescence and abscission "droppers" led to some acceleration of senescence. Likewise in the reciprocal cross, the rootstocks of "retainer" varieties exerted some influence in delaying senescence of "dropper" scions but not to the level of ungrafted "retainer" controls. Of particular interest in these experiments was the remarkable correlation between capacity for root initiaiton on the petioles of excised leaves, as measured by a modification of

Luckwill's (1956) bioassay for auxin, and the extent of leaf retention. It would clearly be desirable to have measurements of diffusible auxin from the leaves and of cytokinin fluxes from the roots of these different genotypes, since the results of the grafting certainly invite interpretation in terms of the feed-back loop depicted in Figure 13.14. Similar results were obtained in grafting experiments using six different genotypes of *Glycine,* with the scion effect predominating.

3. The role of the root system in monocarpic senescence. The first indications that roots were the source of cytokinin production came from experiments in which the senescence of excised tobacco leaves was delayed by either kinetin treatment, the formation of roots on the petiole (Mothes 1960), or treatment with xylem sap exuding from the stems. Many subsequent studies have provided further evidence for the roots as a major source of cytokinin production (Kende 1971), indeed possibly the sole source (van Staden and Davey 1979), so that we may confidently subscribe to the upward flux of cytokinin shown in the feed-back loop of Figure 13.14. The problem that now arises, however, is to determine the extent to which the flux of cytokinin from the roots contributes to the regulation of senescence; it seems probable that the detailed pattern of events may vary according to the species.

In *Xanthium strumarium* L., the cytokinins of leaves and buds appear to be in a state of rapid turnover, requiring continuous renewal by the roots. When roots are removed, cytokinin levels in the leaves and buds promptly decline (van Staden and Wareing 1972) and exogenously applied cytokinins are rapidly metabolized. A single photoinductive cycle led to a drop in the cytokinin output from the roots of *Xanthium* and a corresponding decrease in the level in the shoots (Henson and Wareing 1974, 1977a, b). In *Perilla frutescens,* which requires ten inductive cycles, the position is somewhat different in that induction of flowering leads to a transient rise in cytokinin output from the roots and a corresponding stimulation of leaf metabolism (Beever and Woolhouse 1973, 1974, 1975), although the cytokinin flux from the roots finally falls to a very low level as the fruits mature.

The question now arises whether the high level of cytokinins found in developing fruits originates in these organs or in the roots, in which case the fruits may be affecting the rest of the shoot by competing for cytokinin (Figure 13.15). Van Staden and Davey (1979) concluded that the evidence for cytokinin synthesis within the fruits was not convincing, in which case the possibility of the competitive effect depicted in Figure 13.15 may arise. When developing fruits were removed from young vines the cytokinin content of the leaves increased (Hoad, Loveys, and Skene 1977), which has been interpreted in terms of diversion from the fruits (van Staden and Davey 1979). It has been found that in other species, however, removal of the terminal and axillary buds or of the flowers increased the flux of cytokinins from the roots (Beever and Woolhouse 1974; Colbert and Beever 1981), suggesting that it may not be correct to interpret effects of fruit removal simply in terms of removal of a source of competition for cytokinins.

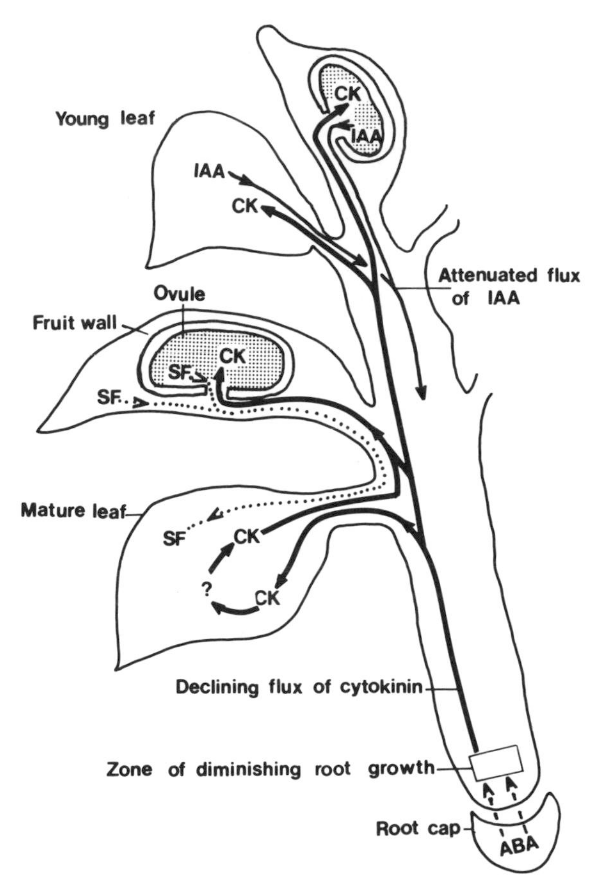

Figure 13.15. Diagram of a hypothetical scheme for the hormonal regulation of monocarpic senescence. The scheme envisages the control loop depicted for the vegetative plant (Figure 13.14) broken in two ways, (a) by the elimination of the sources of IAA as the vegetative shoot apices switch to differentiating as flowers and the young leaves mature, and (b) a tentative senescence factor which may be produced by the developing fruits in some species and promote senescence in the immediately adjacent tissues. For discussion of further constraints upon this model, see text. Key: ····→ = direction of senescence factor flux.

4. The role of the flowers and fruits in monocarpic senescence. In attempting to analyze further the involvement of flowers and fruits in the induction of senescence, we shall refer particularly to the large-seeded legumes *Phaseolus vulgaris, Glycine max,* and *Pisum sativum;* they should probably be treated as separate case studies, for there are strong indications that the regulation of senescence may be different in each.

In *Glycine,* stem growth continues after the onset of flowering of indeterminate varieties, but in determinate genotypes stem growth is terminated by the formation of an inflorescence. Determinacy is controlled by a two-gene system, but does not greatly affect the timing of monocarpy or yield (Bernard 1972). Sinclair and de Wit (1975, 1976) applied a simulation modeling technique to the movement of nitrogen to the developing seeds of *Glycine* and concluded that the plants must be self-destructive "since they need to translocate large amounts of nitrogen from vegetative tissues during seed fill." Notwithstanding the elegance of this model, its underlying assumptions appear unduly naive. There is no evidence to suggest that the plants *need* to translocate any given amount of nitrogen; the demonstration by Hardwick (1979) of increased yield in *Glycine* as a function of leaf retention suggests rather that it is the retention of the leaves, permitting seed filling to continue longer, that is the crucial step. In cereals such as wheat there is also a strong correlation of grain yield with leaf area duration after ear emergence (Evans, Wardlaw, and Fischer 1975), so that there is again no need to invoke a direct exhaustion mechanism. Indeed, the limiting factor to ultimate yield in this case is often the relatively early maturation of the seed, a process which may be regulated from within the seed itself; high temperatures in the grain particularly favor early maturation. Reference to the timing of maturation of the fruit as a factor in monocarpic senescence refocuses attention on the time course of the impact of the developing fruit in the proposed hormonal feed-back loop (Figure 13.15). First, it should be noted that if a hormonal competition model, as for cytokinins, is considered, such a proposition raises as many questions as it answers: How, for example, should the fruit be successful as a competitor for cytokinin carried in the xylem, since the leaves will continue to receive the larger proportion of the transpiration stream? Do the older leaves reexport their cytokinins and pass them along with photosynthates to the fruits?

A further basic question that arises is whether there is any additional positive signal emanating from the developing fruits to shut down metabolism and induce senescence in the rest of the plant. The idea of a positive senescence signal appears to have its origin in the finding that in dioecious species, such as *Cannabis sativa* L. (Mothes and Engelbrecht 1952) and spinach (Leopold, Niedergang-Kamien, and Janick, 1959), the male as well as female plants die after flowering. In the male plants of these species there is little likelihood of exhaustion of materials being a cause of senescence. It has been suggested that such male plants probably senesce in response to a signal from the flowers (Noodén 1980), but there is no positive evidence of this. In *Glycine max* c.v. 'Anoka,' senescence may begin when most of the fruits on the plant are filling rapidly (Lindoo and Noodén 1976), but the senescence can be almost totally suppressed by depodding as late as 60 days after induction of flowering, when the weight of the seeds is already 90% of the final dry weight (Nooden, Rupp, and Derman 1978). When Y-shaped plants were produced by removing the apex above the unifoliate pair of leaves and then depodded on one of the branches, the leaves turned yellow and senesced on the fruiting branch but remained green upon the other (Lindoo and Noodén 1977; Figure 13.16); the authors concluded that the yellowing of leaves on the fruiting branch was due to a senescence signal of limited mobility passing outward from the fruits. How very local this signal must be was indicated by the fact

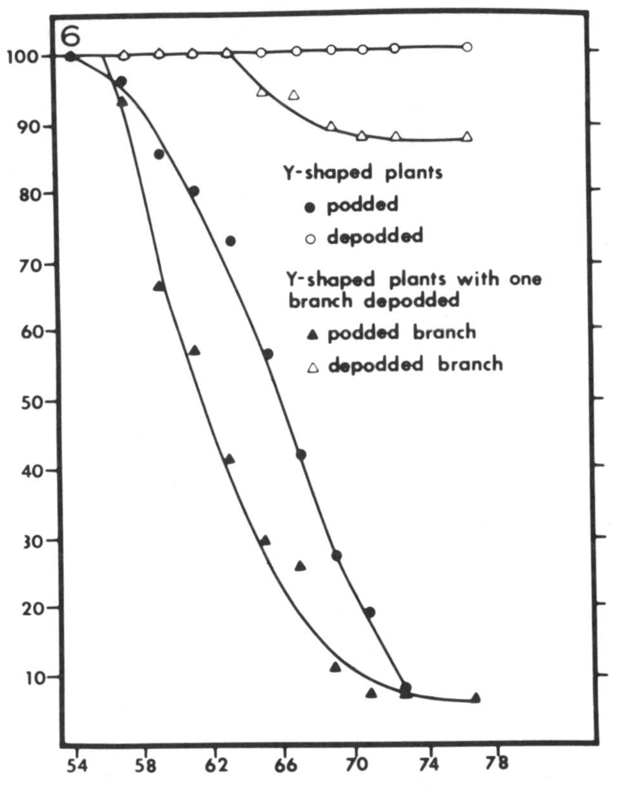

Figure 13.16. Y-shaped plant produced by removing the shoot apex above the unifoliate pair of leaves. The graph shows the leaves senescing on the fruiting branches but remaining green upon the branches from which the fruits were removed.

that when fruits were removed from the axils of alternate leaves only, those leaves which remained subtending fruits became senescent. Application of cytokinins to the leaves of fruiting soybeans delayed, but did not prevent, monocarpic senescence (Lindoo and Noodén 1978) as in *Phaseolus vulgaris* (Fletcher 1969; Adedipe, Hunt, and Fletcher 1971). Abcisic acid applied to the leaves of fruiting plants accelerated senescence but was without effect on depodded plants; the levels of endogenous ABA rose in senescing leaves and the level of cytokinins decreased. The results from Noodén's group are interesting and warrant serious consideration, but caution is required in their interpretation. The fruit of soya bean is a relatively massive structure; as it develops the ABA content rises to high levels, and the possibility of a back diffusion of some ABA or allied promoter of senescence to the adjacent leaf is therefore not unlikely. It must also be noted, however, that there is no information concerning the effects of the various surgical treatments on the transpirational flux through the adjacent leaves. If, as suggested by Thimann (1980), stomatal closure is a causal factor in leaf senescence, then it may be that the leaves subtending fruits receive a smaller proportion of the total transpirational flux and accordingly a reduced supply of cytokinins. In short, the experiments to date with *Glycine* are insufficient to prove any massive contribution from a positive senescence signal and may be largely accountable in terms of the hormonal feed-back loops in Figures 13.14 and 13.15, variously modified by the surgery applied. Further support for such a view is the finding that foliar applications of α-naphthyl acetic acid (NAA), together with the cytokinin benzyl adenine, prevent monocarpic senescence in *Glycine*. Most notable is the fact that loss of starch and nitrogen from the leaves is prevented by this treatment and yet the yield is not depressed (Noodén, Kahanak, and Oktan 1979); the results provide further evidence against the exhaustion concept of Sinclair and de Witt and afford indirect support for the feed-back concept, since the auxin may well be acting to sustain root growth and hence the assimilation of nitrogen, which apparently continues following NAA and cytokinin treatment.

Defruiting and grafting experiments have also been used to study monocarpic senescence in *Pisum sativum* (Malik and Berrie 1975), with broadly similar results to those in *Glycine*. Fruit removal delays senescence, and in grafting experiments depodded scions were able to form successful unions but scions bearing fruit were not; Malik and Berrie interpreted their results in terms of a senescence signal originating in the fruits. In peas, however, the pattern of events is complicated by the initial senescence of the stem apex, which appears to be a complex matter involving a range of genetic controls; some of the salient facts are as follows:

In varieties such as Alaska, apical senescence proceeds irrespective of whether the plants are fruiting and may be delayed by applications of gibberellic acid (GA_3) (Lockhard and Gottschall 1961) and accelerated by AMO-1618, and inhibitor of GA biosynthesis.

In other genotypes of pea, as for example the early line G2, apical senescence occurs only in long photoperiods and is dependent on the presence of fruits; fruits are also produced in short photoperiods in this line, but the shoot apex continues to grow normally under these conditions.

The photoperiodic sensitivity of apical sensescence in the line G2 is determined by the presence of dominant alleles at two genetic loci, Sn and Hr (Murfet and Marx 1976).

Presence of the allele Sn caused the production of a graft-transmissible stimulus which inhibits flower initiation and development and delays apical senescence (Murfet 1971; Murfet and Reid 1973).

In line G2 the fruits produce a factor that promotes apical senescence in LDs (Davis, Van Bavel, and McCree 1977); in SDs this senescence-inducing factor from the fruits is either lacking or its effect is counteracted; in SDs leaves of G2 produce a graft-transmissible substance that delays apical senescence (Proebsting, Davies, and Marx 1977), which could be acting directly or by counteracting the senescence-stimulus from the fruits.

Photoperiod is well known to alter the levels of gibberellins in peas as in

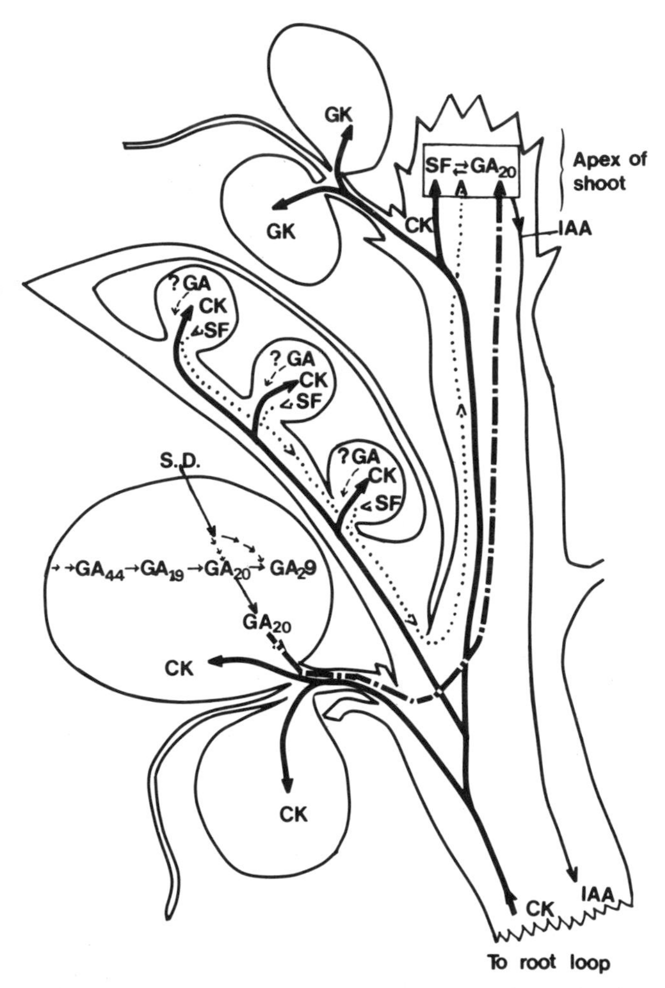

Figure 13.17. Diagram of a hypothetical scheme for the regulation of apical senescence in *Pisum sativum* genotype G2. On to a basic cytokinin–IAA loop, as depicted in Figure 13.14, is superimposed a daylength-dependent GA-mediated promotive signal from the leaves and an inhibitory signal originating in the fruits. The symbol SF GA$_{20}$ serves to indicate the presence of opposing factors in the apical zone; no interconversion or direct interaction of the factors is implied. Key: ▪▪▪▪→ = director of senescence factor flux; -—-→ = direction of promotive gibberellin flux; ——→ = direction of cytokinin flux; ——→ = direction of IAA flux.

other species (Jones 1973), and in line G2 application of gibberellic acid (GA_3) to the apical buds of plants in LDs delayed senescence indefinitely. Of the gibberellins known to be present normally in peas, GA_{20} was most effective in delaying apical senescence, other gibberellins were less effective, and auxins and cytokinins had no effect (Proebsting, Davies, and Marx et al 1978).

Short-day treatment of G2 plants alters the pattern of endogenous gibberellins in the leaves (Proebsting, Davies, and Marx, 1978) and the cotyledons and pods (Ingram and Browning 1979). It must be emphasized that both the pattern and the underlying mechanism of changes involved are still not fully understood, but it is generally agreed that there are increases in levels of GA_{20}. It is an attractive hypothesis to suggest that in a reaction sequence

$$\text{ent-Kaurene} \rightarrow GA_{12} \rightarrow \text{13-hydroxy } GA_{12} \rightarrow GA_{44} \rightarrow GA_{19} \rightarrow GA_{20} \rightarrow GA_{29}$$

(Frydman and Macmillan 1975; Hedden, Macmillan, and Phinney 1978; Ropers et al 1978), the SD effect is operating to alter the relative rates of the reactions at the end of this pathway, leading to an increase in the level of GA_{20}, which passes to the apex and delays senescence. An overall picture of the possible pattern of events in line G2 is given in Figure 13.17.

It cannot be too heavily emphasized that this account is greatly simplified; in the interests of clarity the modifying effects of other loci, E and L have been omitted and the probable involvement of phytochrome and uncertainties concerning the chemistry have been set aside. There are, however, several important general facts that can be drawn from even this simplified account. We see here for the first time a really powerful union of genetic analysis with biochemical work, which suggests that this is probably the only way forward if real progress is to be made in unraveling the controls upon senescence. It is evident also that there exists a combination of signals, both inhibitors from the fruits and promoters of apical senescence from the leaves.

OTHER FACTORS AFFECTING MONOCARPIC SENESCENCE

In the course of this chapter reference has been made at various points to the involvement of cytokinins, auxins, gibberellins, and ABA in the regulation of monocarpic senescence. It is perhaps surprising, therefore, that the other known plant hormone, ethylene, was not featured in this discussion, particularly in view of the known effect of ethylene in promoting such manifestations of senescence as the ripening of fruits. The leaves of herbaceous monocarpic species such as *Xanthium pennsylvanicum* and *Phaseolus vulgaris* do not appear to produce increased levels of ethylene in the course of senescence (Osborne 1978). Production of ethylene by the fruits of the monocarpic species discussed in this article does not appear to have been measured, and the limited mobility of the senescence signal from fruits of *Glycine* inferred by Noodén et al (1979), if correct, is not compatible with a gaseous substance such as ethylene.

CONCLUDING REMARKS

From the foregoing discussion I would seek to emphasize the following points. The monocarpic habit is certainly of polyphyletic origin, and when studied in detail it is seen not to stand in such sharp contradistinction to polycarpism as some authors have suggested. The powerful mathematically based theoretical work on natural selection for polycarpism offers attractive guides to general considerations concerning the phenomenon, but the initial premises concerning competing allocations for resources should be clearly understood to be assumptions that may not be good starting points for the physiologist.

Some progress has been made in the physiological work, and the central role of hormones in regulating monocarpic senescence is beyond dispute. Consideration of a range of species does not offer clear evidence of common regulatory processes; this may be a reflection of the incomplete state of our knowledge of the full spectrum of hormone changes in any one species, but it may go much deeper. There is no doubt that the monocarpic habit has arisen many times, in response to different selection pressures, and is of different antiquity in different groups. In *Zea mays,* for example, it seems probable that the immediate ancestors were perennials (Doebly and Iltis 1980), the particular agent of selection was man, and the emergent monocarpic species has existed for but a few thousand years, less time even than the oldest individuals of the bristle cone pine. Given these varied circumstances in the origins of the monocarpic habit and the varying extent of its development in different groups, it is scarcely surprising that one encounters the multiplicity of controls that have been referred to here. This intrinsic variation in the events concerned in senescence becomes even more apparent when we probe the underlying physiological and biochemical changes (Thomas and Stoddart 1980; Woolhouse 1981a, b). Therefore, it would appear wise at this stage, to investigate monocarpic senescence systematically in each species and to eschew the temptation to facile generalizations.

LITERATURE CITED

Adedipe, N. O., L. A. Hunt, and R. A. Fletcher. 1971. *Effects of benzyladenine on photosynthesis, growth and senescence of the bean plant.* Physiol. Plant. 25:151–53.

Al-Hadithi, T. R. A. M. 1977. *The interaction between flowering and root-growth in Perilla frutescens with particular reference to the functioning of the phloem transport system.* PhD dissertation, University of Leeds.

Andrews, F. W., and T. W. Clouston. 1937. *Section of Botany and Plant Pathology.* Rep. Agric. Serv. Sudan 1936:40–41.

Balls, W. L. 1918. *Analysis of agricultural yield. III. The influence of natural environmental factors upon the yield of Egyptian cotton.* Phil. Trans. B. 208:157–223.

Balls, W. L., and F. S. Holton. 1915a. *Analysis of agricultural yield. II. The sowing-date experiment with Egyptian cotton, 1912.* Phil. Trans B. 206:103–80.

———. 1915b. *Analysis of agricultural yield. II. The sowing-date experiment with Egyptian cotton, 1913.* Phil. Trans. B. 206: 403–80.

Beever, J. E., and H. W. Woolhouse. 1973. *Increased cytokinin from roots of Perilla frutescens during flower and fruit development*. Nat. New Biol. 246:149–50.

———. 1974. *Inter relationships of flowering, root growth and senescence in Perilla frutescens*. Pages 681–86 in R. L. Bieleski, A. R. Ferguson, and N. M. Cresswell, eds. *Mechanisms of Regulation of Plant Growth Bulletin, 12*. Roy. Soc. New Zealand, Wellington.

———. 1975. *Changes in the growth of roots and shoots when Perilla frutescens L. Britt is induced to flower*. J. Exp. Bot 26:451–63.

Bernard, R. L. 1972. *Two genes affecting stem termination in soybeans*. Crop Sci. 12:235–39.

Booth, A., J. Moorby, C. R. Davies, H. Jones, and P. F. Wareing. 1962. *Effects of indolyl-3-acetic acid on the movements of nutrients within plants*. Nature (London) 194:204–5.

Brenchley, W. E., and V. G. Jackson, 1921. *Root development in barley and wheat under different conditions of growth*. Ann. Bot. (London) 35:533–56.

Brouwer, R. 1962. *Nutritive influences on the distribution of dry matter in the plant*. Neth. J. Agric. Sci. 10:399–408.

Chailakyan, M. Kh., and L. P. Zhadanova. 1938. *Hormones of growth in formation processes. 1. Photoperiodism and creation of growth hormones*. Dokl. (Proc.) Acad. Sci. USSR: 19:107–11.

Cohen, D. 1971. *Maximising final yield when growth is limited by time or by limiting resources*. J. Theor. Biol. 33:299–307.

Colbert, K. A., and J. E. Beever, 1981. *Effect of disbudding on root cytokinin export and leaf senescence in tomato and tobacco*. J. Exp. Bot. (in press).

Cole, L. C. 1954. *The population consequences of life history phenomena*. Q. Rev. Biol. 29:103–37.

Cooper, A. J. 1955. *Further observation on the growth of the root and the shoot of the tomato plant*. Proc. 14th Int. Hort. Congr.:589–95.

Davis, S. D., C. H. M. van Bavel, and K. J. McCree. 1977. *Effect of leaf ageing on stomatal resistance in bean plants*. Crop Sci. 17:640–45.

Doebley, J. F., and H. H. Iltis. 1980. *Taxonomy of Zea (Gramineae). 1. A subgeneric classification with key to taxa*. Amer. J. Bot. 67:982–93.

Doflein, F. 1919. *Das Problem des Todes und der Unsterblichkeit bei den Pflanzen und Tieren*. 1–119.

Eaton, F. M. 1927. *Defruiting as an aid to cotton breeding*. J. Hered. 18:456–60.

———. 1931. *Root development as related to character of growth and fruitfulness of the cotton plant*. J. Agric. Res. 43:875–83.

Eaton, F. M., and D. R. Ergle. 1952. *Fiber properties and carbohydrate and nitrogen levels of cotton plants as influenced by moisture supply and fruitfulness*. Plant Physiol. 27:541–62.

Errera, L. 1905. *Conflits de préséance et excitation inhibitories chez les végétaux*. Bull. Soc. Roy. Bot. Belgium 42:27–43.

Evans, L. T., I. F. Wardlaw, and R. A. Fischer. 1975. *Wheat*. Pages 101–49 in L. T. Evans, ed. *Crop Physiology: Some Case Histories*. Cambridge Univ. Press, London.

Ewing, E. C. 1918. *A study of certain environmental factors and varietal differences influencing the fruiting of cotton*. Tech. Bull. Miss. Agric. Exp. Stn. 8:1–95.

Fletcher, R. A. 1969. *Retardation of leaf senescence by benzyladenine in intact bean plants*. Planta 89:1–8.

Frydman, V. M., and J. Macmillan. 1975. *The metabolism of gibberellins A_9, A_{20}, A_{29} in immature seeds of Pisum sativum cv*. Progress No. 9, Planta 125:181–95.

Haberlandt, G. 1921. *Wundhormone als Erreger von Zellteilungen*. Beitr. Allg. Bot. 2:1–53.

Harada, H. 1962. *Etude des substances naturelles de croisance en relation avec la floraison. Isolement d'une substance de montaison*. Revua Gen. Bot. 69:201–97.

Hardwick, R. C. 1979. *Leaf abscision in varieties of Phaseolus vulgaris L. and Glycine max (L) Merrill—a correlation with propensity to produce adventitious roots*. J. Exp. Bot. 30:795–804.
Harper, J. L. 1977. *Population Biology of Plants*. Academic Press, London/New York.
Hedden, P., J. Macmillan, and B. O. Phinney. 1978. *The metabolism of the gibberellins*. Ann. Rev. Plant Physiol. 29:149–92.
Henson, I. E., and P. F. Wareing. 1974. *Cytokinin in Xanthium strumarium: a rapid response to short day treatment*. Physiol. Plant. 32:185–87.
———. 1977a. *An effect of defoliation on the cytokinin content of buds of Xanthium strumarium*. Plant Sci. Lett. 9:27–31.
———. 1977b. *Cytokinins in Xanthium strumarium L. the metabolism of cytokinins in detached leaves and buds in relation to photoperiod*. New Phytol. 78:27–33.
Hildebrand, F. 1881. *Die Lebensdauer und Vegetationsweise der Pflanzen, ihre Ursache und ihre Entwicklung*. Bot. Jahrb 2:51–135.
Hoad, G. V., B. R. Loveys, and K. G. M. Skene. 1977. *The effect of fruit removal on cytokinins and gibberellin-like substances in grape leaves*. Planta. 136:25–30.
Honma, S., J. C. Bouwkamp, and M. J. Stojianov. 1968. *Inheritance of dry pod-color in snap beans*. J. Hered. 59:243-44.
Hudson, J. P. 1960. *Relations between root and shoot growth in tomatoes*. Sci. Hort. 14:49–54.
Ingram, T. J., and G. Browning. 1979. *Influence of photoperiod on seed development in the genetic line of peas G_2 and its relation to changes in endogenous gibberellins*. Planta 146:423–32.
Iversen, T-H., T. Aasheim, and K. Pedersen. 1971. *Transport and degradation of auxin in relation to geotropism in roots of Phaeseolus vulgaris*. Physiol. Pl. 18:941–44.
Jacobs, W. P. 1952. *The role of auxin in differentiation of xylem around a wound*. Amer. J. Bot. 39:301–9.
Jacobs, W. P., and B. Bullwinkel. 1953. *Compensatory growth in Coleus shoots*. Amer. J. Bot. 40:385–92.
Jeffcoat, B., and G. P. Harris. 1972. *Hormonal regulation of ^{14}C-labelled assimilates in the flowering shoot of carnation*. Ann. Bot. (London) 36:353–61.
Jones, R. L. 1973. *Gibberellins: their physiological role*. Ann. Rev. Pl. Physiol. 24:571–98.
Kende, H. 1971. *The cytokinins*. Int. Rev. Cytol. 31:301–38.
Leonard, E. R., and G. C. Head. 1958. *Technique and preliminary observations on growth of the roots of glasshouse tomatoes in relation to that of the tops*. J. Hort. Sci. 33:171–85.
Leopold, A. C. 1961. *Senescence in plant development*. Science 134:1727–32.
Leopold, A. C., E. Niedergang-Kamien, and J. Janick. 1959. *Experimental modification of plant senescence*. Plant Physiol. 34:570–73.
Lewontin, R. C. 1965. *Selection for colonizing ability*. In H. G. Baker and G. L. Stebbins, eds., *The Genetics of Colonising Species*. Academic Press, New York.
Lindemuth, H. 1901. *Das Verhalten durch Copulation verbundener Pflanzenarten*. Ber. Deut. Bot. Ges 19:515–29.
Lindoo, S. J., and L. D. Noodén. 1976. *The interrelation of fruit development and leaf senescence in "Anoka" soybeans*. Bot. Gaz. 137:218–23.
———. 1977. *Behaviour of the soybean senescence signal*. Plant Physiol. 59:1136–40.
———. 1978. *Correlations of cytokinins and abscisic acid with monocarpic senescence in soybean*. Plant Cell Physiol. 19:997–1006.
Lockhard, J. A., and V. Gottschall. 1961. *Fruit-induced and apical senescence in Pisum sativum*. Plant Physiol. 32:204–7.
Loomis, W. E. 1935. *Translocation and growth balance in woody plant*. Ann. Bot. 49:247–72.

Luckwill, L. C. 1956. *Two methods for the bioassay of auxins in the presence of growth inhibitors*. J. Hort. Sci. 31:89–98.
McDavid, C. R., G. R. Sagar, and C. Marshall. 1972. *The effect of auxin from the shoot on root development in Pisum sativum L.* New Phytol. 71:1027–32.
Malik, N. S. A., and A. M. M. Berrie. 1975. *Correlative effects of fruits and leaves in senescence of pea plants*. Planta 124:169–75.
Mason, T. G. 1922a. *Growth and correlation in Sea Island cotton*. West Indian Bull. 19:214–38.
———. 1922b. *Growth and abscission in Sea Island cotton*. Ann. Bot. 36:457–84.
Mattirolo, O. 1899. *Sulla influenza che la estirpazione die fiori esercita sui tubercoli radicali delle piante leguminose*. Malpighia 13:382–421.
Molisch, H. 1938. *Die lebensdauer der pflanzen*. The longevity of plants trans. by F. H. Fulling. Published by translator, New York Botanic Gardens.
Morris, D. A., R. E. Briant, and P. G. Thomson. 1969. *The transport and metabolism of ^{14}C-labelled indoleacetic acid in intact pea seedlings*. Planta 89:178–97.
Mothes, K. 1960. *Uber das Altern der Blatter und die Moglichkeit ihrer Wiederverjüngung*. Naturwissenschaften 47:337–51.
Mothes, K., and L. Engelbrecht. 1952. *Über gesschlechtsverschieden Stoffwechsel zweihaüsiger einjähriger Pflanzen. I. Untersuchungen über den Stickstoffumsatz beim Hauf (Cannabis sativa L.)*. Flora (Jena) 139:1–27.
Mothes, K., L. Engelbrecht, and H. R. Schütte. 1961. *Über die Akkumulation von α-* Aminoisobuttersäure in Blattgewebe unter dem Einfluss von Kinetin. Physiol. Pl. 14:72–75.
Murfet, I. C. 1971. *Flowering in Pisum: Reciprocal grafts between known genotypes*. Austral. J. Biol. Sci. 24:1089–1101.
Murfet, I. C., and G. A. Marx. 1976. *Flowering in Pisum: Comparison of the Geneva and Hobart systems of phenotypic classification*. Pisum Newslett. 8:46–47.
Murfet, I. C., and J. B. Reid. 1973. *Flowering in Pisum: Evidence that gene Sn controls a graft transmissible inhibitor*. Austral. J. Biol. Sci. 26:675–77.
Murneek, A. C. 1926. *Effect of correlation between vegetative and reproductive functions in the tomato (Lycopersicon esculentum Mill.)*. Pl. Physiol. 1:3–56.
Nightingale, G. T. 1923. *Light in relation to the growth and chemical composition of some horticultural plants*. Proc. Amer. Soc. Hort. Sci. 19:18–29.
Noodén, L. D. 1980. *Senescence in the whole plant*. In K. V. Thimann, ed. *Senescence in Plants*. C.R.C. Series in Aging. C.R.C. Press, Boca Raton, Florida.
Noodén, L. D., D. C. Rupp, and B. D. Derman. 1978. *Separation of seed development from monocarpic senescence in soybeans*. Nature (London) 271:354–57.
Noodén, L. D., G. M. Kahanak, and Y. Oktan. 1979. *Prevention of monocarpic senescence in soybeans with auxin and cytokinin: An antidote for self-destruction*. Science 206:841–43.
Osborne, D. J. 1978. *Ethylene*. in *Phytohormones and Related Compounds: A Comprehensive Treatise,* vol. 1. D. S. Letham, P. B. Goodwin, and T. J. V. Higgins, eds. Elsevier/North Holland Biomedical Press.
Petinov, N. S., and N. F. Berko. 1961. *Effect of water supply on the absorbing activity and respiration rate of the root systems of corn*. Fiziol. Rast. 8:51–57 (in Russian). (Plant Physiol 8:34–38, English translation.)
Probst. A. H. 1950. *The inheritance of leaf abscission and other characteristics in soybeans*. Agron. J. 42:35–45.
Proebsting, W. M., P. J. Davies, and G. A. Marx. 1977. *Evidence for a graft-transmissible substance which delays apical senescence in Pisum sativum L.* Planta 135:93–94.
———. 1978. *Photoperiod induced changes in gibberellin metabolism in relation to apical growth and senescence in genetic lines of peas (Pisum sativum L.)*. Planta 141:231–38.

Reichart, C. 1821. *Land und Gartenschatz*. Praktisches Handbuch fur dem Blumen-und Zierpflanzen-Gartenbau. 6, Aufl. 5. Tiel. P.114.
Roberts, R. H., and B. E. Struckmeyer. 1946. *The effect of top environment and flowering upon top-root ratios*. Pl. Physiol. 21:332–44.
Roberts, R. H., and O. C. Wilton. 1936. *Development and blossoming*. Science 84:391–92.
Rohmeder, E. 1967. *Beziehungen zwischen Frucht-bzw. samenerzeugung und Holzerzeugung der Waldbäume*. Allg. Furstzeitschr. 22:33–39.
Ropers, H-J., J. E. Graebe., R. Gaskin, and J. Macmillan. 1978. *Gibberellin biosynthesis in a cell-free system from immature seeds of Pisum sativum*. Biochem. Biophys. Res. Commun. 80:690–97.
Sachs, T. 1972. *A possible basis for apical organization in plants*. J. Theor. Biol. 37:353–61.
———. 1975. *The induction of transport channels by auxin*. Planta 127:201–6.
Scott, T. K. 1972. *Auxins and roots*. Ann. Rev. Pl. Physiol. 23:235–58.
Seth, A. K., and P. F. Wareing. 1967. *Hormone-directed transport of metabolites and its possible role in plant senescence*. J. Exp. Bot. 18:65–77.
Sinclair, T. R., and C. T. De Wit. 1975. *Photosynthate and nitrogen requirements for seed production by various crops*. Science 189:565–67.
———. 1976. *Analysis of the carbon and nitrogen limitations to soybean yield*. Agron. J. 68:319–24.
Smith, O. 1924. Steam sterilization of greenhouse soil and its effect upon the root system of the tomato. Ph.D. dissertation, Iowa State College.
Stigter, H.C.M. de. 1969. *A versatile irrigation-type water culture for root-growth studies*. Z. Pflphysiol. 60: 289–95.
Strijbosch, T. 1954. *Wortelontwikkeling bij tomaten* [Root development of tomatoes]. Jverzl. Proeft. Zwid-Holland. Glasdist. Naaldwijk 1954:33–34.
Struckmeyer, B. E. 1941. *Structure of stems in relation to differentiation and abortion of blossom buds*. Bot. Gaz. 103:182–91.
Struckmeyer, B. E., and R. H. Roberts. 1939. *Phloem development and flowering*. Bot. Gaz. 100: 600–606.
Stuckey, I. H. 1941. *Seasonal growth of grass roots*. Amer. J. Bot. 28:486–91.
Thimann, K. V. 1936. *Auxins and the growth of roots*. Amer. J. Bot. 23:561–69.
———. ed. 1980. *Senescence in Plants*. C.R.C. Press, Boca Raton, Florida.
Thomas, H., and J. L. Stoddart. 1980. *Tissues, organs and whole plants: leaf senescence*. Ann. Rev. Pl. Physiol. 31:83–130.
Thurman, D. A., and H. E. Street. 1960. *The auxin activity extractable from excised tomato roots by cold 80 per cent methanol*. J. Exp. Bot. 11:188–97.
Torrey, J. G. 1963. *Cellular patterns in developing roots*. Symp. Soc. Exp. Biol. 17:285–314.
———. 1976. *Root hormones and plant growth*. Ann. Rev. Pl. Physiol. 27:435–59.
Van der Post, C. J. 1968. *Simultaneous observations on root and top growth*. Acta Hort. 7:138–43.
Van Dobben. W. H. 1962. *influence of temperature and light condition on dry matter distribution*. Neth. J. Agric. Sci. 10:377–89.
Van Overbeek, J. 1939. *Evidence for auxin production in isolated roots growing in vitro*. Bot. Gaz. 101:450–56.
Van Staden, J., and J. E. Davey. 1979. *The synthesis, transport and metabolism of endogenous cytokinins*. Plant, Cell and Environment 2:93–106.
Van Staden, J., and P. F. Wareing. 1972. Results presented by P. F. Wareing at a review lecture, Hormonal factors in the environmental control of plant growth and development. The Royal Society, April 1972.
Vincent, T. L., and H. R. Pulham. 1980. *Evolution of life history strategies for an asexual annual plant model*. Theor. Pop. Biol. 17:215–31.

Wareing, P. F., and A. K. Seth. 1967. *Ageing and senescence in the whole plant*. Symp. Soc. Exp. Biol. 21:543–58.

Wilton, O. C. 1938. *Correlation of cambial activity with flowering and regeneration*. Bot. Gaz. 99:854–64.

Wilton, O. C., and R. H. Roberts. 1936. *Anatomical structure of stems in relation to the production of flowers*. Bot. Gaz. 98:45–64.

Woolhouse, H. W. 1974. *Longevity and senescence in plants*. Sci. Prog. Oxf. 61:123–47.

———. 1981a. *Regulatory aspects of nucleic acid and protein turnover in relation to plant senescence*. In preparation.

———. 1981b. *The molecular biology of foliar senescence*. In preparation.

Zhadanova, L. P. 1945. *Geotropic reaction of leaves and content of growth hormone in plants*. Dokl (Proc.) Acad. Sci. USSR 49:62–65.

five

MANAGEMENT AND CONTROL OF PLANT REPRODUCTION

14] Opportunities and Needs to Control Plant Reproduction

by M. N. CHRISTIANSEN* and
GEORGE L. STEFFENS*

ABSTRACT

Benefits to agriculture would accrue if plant functions that contribute to reproduction could be controlled. A number of aspects of plant growth and development that relate to reproduction may be amendable to modification by various means. These include germination, morphology, flowering and fruiting, carbon fixation, mineral nutrition and nitrogen fixation, stress resistance, and sexual compatibility. Opportunities to control plant reproduction lie in two areas, viz. modification of the environment and modification of the plant. The potential for changing the environment to improve plant reproductive functions appears to be less than for changing the plant. We must understand how natural growth control systems operate before such systems can be used to manipulate plant reproduction for man's benefit.

INTRODUCTION

We are often reminded of the Malthusian prediction of dire calamity when human population exceeds the earth's food-producing capacity. Perhaps if our number one priority were control of reproduction—in both plant and man—we could relax. Human population would be controlled and crop productivity greatly enhanced.

Agriculture would benefit in many ways if we could control by chemical, or cultural manipulation, or by genetics the plant functions that contribute to reproduction. Some of the needs to control plant reproduction will be considered before discussing some likely opportunities.

*U.S. Department of Agriculture, Agriculture Research Service, Beltsville Agricultural Research Center, Beltsville, Maryland 20705.

STRATEGIES OF PLANT REPRODUCTION (BARC Symposium number 6—Werner J. Meudt, ed.)
Allanheld, Osmun, Totowa

NEEDS FOR CONTROL

Most advances in crop productivity to the present time are a result of the empirical approach. That approach is currently providing fewer and fewer dividends, and crop yields seem to have reached a plateau. Perhaps it is time to turn to more basic approaches. Perhaps it is time for the biochemist, the physiologist, the molecular biologist, and the genetic engineer to vindicate past support (however sporadic) by providing new approaches. Hardy (1979) has asked the question in prior discussions, "What's wrong with the natural processes that limit plant productivity?" How can we alter photosynthesis, or nitrogen fixation, or other basic functions to remove some of the limiting intermediate steps? Once we have determined what's wrong, can we then address the development of steps to right the deficiencies?

Germination. Most certainly we should learn to control the beginning of the life cycle, namely, germination. Induction or prevention of germination would permit control of natural dormancy in crop plants, or in weed seeds, thereby permitting timing of seed germination for crops to maximize productivity in the best climate available or, conversely, to trigger weed seed germination out of phase with the environment.

Morphology. Need exists for alteration of plant size or shape to improve disease or insect control, and to alter light penetration, thereby increasing light harvest by leaves or to facilitate harvest efficiency. This might well be accomplished with chemicals, but opportunities to select desirable genetic attributes also exist in many crops.

The direct effect of genetic variability in flower morphology offers opportunity for use of obligatory out-crossing in crop plants to produce hybrids.

Flowering and fruiting. If cotton is taken as an example, it is an extremely puzzling plant that produces many flowers but sets few fruits. Many cotton producers have dreamed of controlling time of flowering and fruit set to maximize yield, minimize fruiting time (and hence minimize need for protection from pests), and optimize time of harvest. Such dreams are not restricted to cotton growers; most people in plant science have such dreams.

Multiple cropping practices also emphasize the need for close control of the reproductive process—not only the turning-on of flowering and fruit set, but termination of the fruiting cycle and induction of plant senescence or maturity. Here again we might do much for control of seed production in weed pests if flowering could be controlled.

Carbon fixation. The partitioning of photosynthate to preferential sites in the plant can determine level of productivity via controlling fruitfulness. The general level of photosynthate in the plant can often determine flowering, fruit set, and maturity. Some evidence relates carbohydrate level effect to these processes via ethylene (Guinn 1974, 1976). We need to better control where photosynthate is directed. Some of the answers lie in manipulation of light; some perhaps can be found in chemical regulation of hormone pathways.

Mineral nutrition and nitrogen fixation. Little attention is paid to mineral nutrition these days. But the remaining land areas that will be needed to feed a rapidly growing world population have serious mineral toxicities and deficiencies that adversely affect all plant functions. Chemical amendment of soil is an energy-costly process. We need less costly approaches. Perhaps genetics will provide plants adapted to mineral stress conditions—and plants with greater fertilizer-use efficiency. Certainly we need increased nitrogen fixation in legumes, and also realization of the dream of developing nitrogen fixation capability in nonlegume species. Improved knowledge of nitrogen effect on vegetative-fruiting balance in plants can also be useful to intensive agriculture. Maximizing yields with added nitrogen has created nutritional imbalances that reflect on reproductive patterns.

Stress resistance. After the experience of the 1980 growing season in the United States, words need not be wasted on what drought can do to crop productivity. Dr. W. L. Decker's chapter will show that water availability accounts for as much as 80% of the long-term variability in crop yields. Some stresses impact directly on the reproductive process. Low humidity or heat can cause pollen sterility. Rice, which feeds 60% of the world's population, is extremely sensitive to chilling at anthesis. Other stresses, such as air pollution, salinity, and mineral stress, exert an overall or a general effect on reproduction and crop productivity. We need crop plants with inherent environmental resiliency and we need chemical-cultural techniques to combat the effect of environmental extremes.

Sexual compatibility. The modification of self-fertility, either genetically or chemically, offers a tool for much progress in hybrid crop seed production. In those species where self-sterility has been developed, it has been a useful tool to simplify production of hybrids. Several years ago an experimental compound was evaluated that prevented pollen formation. It was not effective under varying environmental conditions, but it demonstrates a useful approach to hybrid seed production. We need such a tool.

OPPORTUNITIES TO CONTROL PLANT REPRODUCTION

Earlier it was suggested that an empirical approach to agricultural improvement, including conventional methods of plant breeding, was passé. Nevertheless, the possibilities for genetic alteration of plant reproductive processes have increased immeasurably with the development of genetic engineering techniques. New gene combinations with introgression of wild species or altered gene complements are now quite possible. Asexual reproduction via cell culture offers opportunity to drastically reduce cost of ornamental and fruit crop plants. DNA transfer methods may provide opportunity to alter self-fertility and may change many of the processes associated with plant reproduction.

The opportunities that exist in manipulation of plant function by chemical and cultural techniques lie in two areas: modification of the environment, and modification of the plant.

Environmental alteration. There are opportunities to effect change of mineral nutrition; light quality, timing, and quantity; water relations, particularly of plant species originating in Mediterranean climates; temperature; and gaseous components.

Some examples include the management of bud set in tree crops by nitrogen and phosphorus availability. The inverse relation between nitrogen and phosphorous effects on crop maturity has been known for many years. Minor element (copper, zinc, calcium) effects on flower formation have been established for many plant species (Clark 1981). The control of germination, flowering, and plant morphology by light has its origins at Beltsville, and knowledge of the subject was greatly expanded by Sterling Hendricks, to whose memory BARC Symposium VI was dedicated.

Gaseous components, including O_2, and CO_2, and ethylene, are known to effect flowering, tuberization, rooting, and senescence. Controlled use of these substances, particularly in enclosed plant growth systems, has additional possibilities for manipulation of plant function.

Temperature modification of plant response is well documented and includes germination, growth habit, flowering, fruit retention, and senescence. Low-temperature requirements for bud break are well known in stone fruits as well as in seed germination. Germination environment can have long-term effect (Christiansen 1969). High temperature oftimes coupled with low humidity can have drastic effects on pollen viability in grain crops as well as in fruit and fiber plants.

Low temperature, including frost and chilling, has extreme influence on flower viability in citrus and nontropical fruits such as apple and peach. The monetary costs are tremendous, both in terms of damage and in terms of expenditures, for measures to reduce injury, i.e., fans, smudge pots, or sprinkler irrigation. There appears to be little opportunity for development of new techniques for alteration of aerial temperature. Measures such as use of dark, radiant-energy-absorptive surfaces, including black plastic or carbon black, have some potential for increasing soil temperatures.

Water status of the plant and aerial water content can alter pollination by affecting pollen viability and by affecting flower and fruit retention. In some species wetting of pollen can be destructive or high humidity can reduce pollen dispersal from anthers. Thus, in sprinkle-irrigated culture the timing of water application can be critical.

The potential for changing the environment to foster improved reproduction is perhaps less than the possibilities for changing the plant with less energy expenditure likely in changing plants.

Alteration of plant function and reproduction. Earlier chapters have described how hormones act to control flowering and reproduction. Much of the flowering reaction of plants to external stimuli is via hormone systems. Plant growth regulators (synthetic) generally operate as promoters or antagonists of the natural system. A number of theories have been advanced as to how the various known hormones function. Present evidence points to gibberellin induction via enzymes; to cytokinins, probably acting at the gene level; to auxins, which perhaps increase cell wall plasticity; and to abscisic acid, which is induced by stress and acts to close stomates and may induce ethylene, which can in turn induce senescence and leaf abscission. Unfortunately most of these statements are conjectural. Until we can understand the mode of action of natural growth regulator systems the utilization or manipulation of these systems to improve crop production will develop slowly.

The use of growth regulants has slowly increased since the 1930s. Some of these chemicals affect the reproductive system of crop plants and have proven to be useful to agriculture. Weaver (1972) discusses the effect of exogenously applied plant regulators on reproduction and reproductive organs under a number of main areas, including: (a) flower induction and promotion of flowering, (b) prevention or delay of flowering, (c) regulation of sex expression, (d) control of fruit set and development, (e) control of senescence of fruits, and (f) control of flower and fruit abscission. In addition he considers areas involving seed, tuber, and bud dormancy or rest, tuber sprout inhibition, and seed germination. Each area is involved, but when all of these processes are considered, one can see how very complicated and involved the regulation or control of plant reproductive processes really are. Almost any chemical growth regulator used in agriculture somehow effects the reproductive process.

A recent report by Looney (1980) and Chapter 17 by M. W. Williams show how dependent the production of apples is on the use of chemicals for promotion of flower initation and fruit set, chemical blossom and fruit thinning, prevention of fruit drop, improving fruit quality, and regulating ripening. Wittwer (1978) presents a rather long list of reproduction-associated phenomonon that can be modified or controlled by plant growth regulators. Some of these include acceleration of flower bud and seed stalk development to aid in plant breeding, extending the effective harvest period of a number of fruits and horticultural crops, facilitating the mechanical harvesting of fruit crops, and inhibiting the sprouting in potatoes and onions. Several chemicals are used for inhibiting the growth of axillary buds in tobacco (Steffens 1980) or to "pinch" buds on ornamental and woody species which in turn allows growth of axillary buds that produce blossoms. Growth regulators have also been used to inhibit flowering in sugarcane to provide yield increases of sugar (Nickell 1978). The control of flower sex expression in certain types of cucumbers has also allowed for the production of new hybrids adapted to mechanical harvesting (Wittwer 1978).

Plant hormones as well as some synthetic chemicals are used in plant tissue culture systems to mass-produce shoot either adventitiously from a variety of explants or by enhanced axillary branching of excised shoot tips (Murashige

1980). Culture systems with added growth substances are used in the production of specific pathogen-free plants. New plant varieties have been produced via embryo culture and culture of ovules and ovularies before or after pollination. Seeds of several vegetable plants are being produced from specially chosen parents and cloned in large numbers via tissue culture.

A number of other specific examples of the use of chemicals to modify the reproductive process or reproductive organs of plants could be cited. We know little of how these chemicals exert their effects, however. The ability consistently to regulate plant function, particularly reproduction, can be gained only after much further research. When we learn how hormones and other natural and synthetic substances control plant activities, more of the keys necessary to manipulate plant reproduction for man's benefit will be available.

LITERATURE CITED

Christiansen, M. N. 1969. *Season-long effects of chilling treatments applied to germinating cottonseed.* Crop Sci. 9:672–73.

Clark, R. B. 1981. *Plant response to mineral element toxicity and deficiency.* In M. N. Christiansen and C. F. Lewis, eds. *Breeding Plants for Marginal Environments.* Wiley Interscience, New York.

Guinn, G. 1974. *Abscission of cotton floral buds and bolls as influenced by factors affecting photosynthesis and respiration.* Crop Sci. 14:291–93.

———. 1976. *Nutritional stress and ethylene evolution by young cotton bolls.* Crop Sci. 16:89–91.

Hardy, R. W. F. 1979. *Chemical plant growth regulators in world agriculture.* Pages 165–206 in T. K. Scott, ed. *Plant Regulation and World Agriculture.* Plenum Press, New York.

Looney, N. E. 1980. *Growth regulator use in commercial apple production.* Pages 409–18 in F. Skoog, ed. *Plant Growth Substances.* 1979. Springer-Verlag, Berlin-Heidelberg-New York.

Murashige, T. 1980. *Plant growth substances in commercial uses of tissue culture.* Pages 426–34 in F. Skoog, ed. *Plant Growth Substances.* 1979. Springer-Verlag, Berlin-Heidelberg-New York.

Nickell, L. G. 1978. *Plant growth regulators.* Chemical and Engineering News 56:18–34.

Steffens, G. L. 1980. *Applied uses of plant growth substances—Growth inhibitors.* Pages 397–408 in F. Skoog, ed. *Plant Growth Substances.* 1979. Springer-Verlag, Berlin-Heidelberg-New York.

Weaver, R. J. 1972. *Plant Growth Substances in Agriculture.* W. H. Freeman, San Francisco, CA.

Wittwer, S. H. 1978. *Phytohormones and chemical regulators in agriculture.* Pages 599–615 in D. S. Lethan, P. B. Goodwin, and T. J. V. Higgins, eds. *Phytohormones and Related Compounds: A Comprehensive Treatise.* Elsevier/North-Holland Biomedical Press, Amsterdam, Oxford, New York.

15] Genetic Control of Nitrogen Metabolism in Plant Reproduction

by P. B. CREGAN*

ABSTRACT

With continually increasing demand for greater plant protein quantity and quality, genetic enhancement of the productivity and efficiency of plant nitrogen metabolism will remain an important objective of plant research. A comparison of field-grown wheat, maize, and soybean suggests interspecific genetic differences in the nitrogen economy of these grain crops. The seasonal nitrogen accumulation curve of wheat and soybean has a longer lag phase than does maize, while the slope of the linear phase of the seasonal nitrogen accumulation curve is greater in wheat than in maize or soybean. Anthesis occurs in wheat after more than 85% of the total above-ground plant nitrogen is accumulated. In maize and indeterminate soybeans, about 70% and 25%, respectively, of the total nitrogen has accumulated by anthesis. All three crops, however, have similar Nitrogen Harvest Indices ranging from 63% to 72%. These data indicate that wheat has an exceptional capacity to remobilize nitrogen from vegetative tissue to the developing seed. Interspecific comparisons may suggest processes for which genetic improvement may be possible in a crop, but only genetic differences within a species can currently be useful in facilitating actual genetic improvement. Nevertheless, large intraspecific genetic differences are apparent from selection studies such as that begun by the Illinois Agricultural Experiment Station in 1896. From one maize variety, divergent selection for high and low seed protein has produced genotypes with seed protein from a low of 5% to more than 20%. This striking response to selection occurred gradually over more than 60 cycles of selection and was probably the result of the genetic alteration of many steps in the complex series of physiological events beginning with NO_3^- uptake and terminating with the deposition of storage protein in the seed. Among these numerous steps involved in nitrogen metabolism only the genetic control of nitrate

*Cell Culture and Nitrogen Fixation Laboratory, Beltsville Agricultural Research Center, Beltsville, Maryland 20705.

STRATEGIES OF PLANT REPRODUCTION (BARC Symposium number 6—Werner J. Meudt, ed.)
Allanheld, Osmun, Totowa

reduction has been studied in detail. Nitrate reductase activity has been used with limited success as a criterion in the selection of plant genotypes with higher seed nitrogen, protein percentage, or yield. The use of only one biochemical or physiological selection criterion reduces the probability of attaining plant genotypes with significantly enhanced nitrogen metabolism or productivity unless the selection factor is the major limiting factor in most or all genotypes over a range of environmental conditions. Because of this unlikely circumstance, most plant improvement programs have relied upon an end produce, such as seed yield, total nitrogen, or seed nitrogen percentage, which intergrates the effects of genotype, environment, and their interaction over the growing season. This approach offers no knowledge of the physiological or biochemical basis of genetic improvement, however. A properly designed selection system that includes the measurement of a series of critical physiological processes should theoretically allow the identification of limiting physiological factors and also be a good predictor of the end product of nitrogen metabolism. The scientific literature suggests the presence of genetic variability in grain crops for the efficiency of NO_3-uptake or N_2-fixation, NO_3-reduction, protease activity and remobilization of amino acids, and the rate of synthesis of seed storage protein. These and other aspects of nitrogen metabolism for which no genetic information is currently available might be included in a physiological selection system directed toward the development of plant genotypes with more efficient or productive nitrogen metabolism.

INTRODUCTION

BARC Symposium VI addressed the important and biologically complex issue of plant reproduction. We as plant scientists consider plant reproduction from the academic vantage point, but many members of our world community look upon plant reproduction as their next meal or pay check. The consequences of forces that curtail the reproductive processes of a wheat, rice, corn, or bean crop may leave our fellows with empty plates or economic hardship. Plant scientists have the opportunity to accumulate and, in some cases, apply the knowledge necessary to lessen the likelihood of such deprivation.

The important role of nitrogen in plant productivity is obvious. Equally clear is the role of nitrogen in human nutrition. Altschul (1976) described the current debate among human nutritionists over the presence and degree of protein deficiency in the diets of peoples in some parts of the less-developed world. Can human nutritional problems be resolved simply with more food, or are dietary shifts toward foods with improved protein quality required? Whatever the answer to the question, plant nitrogen metabolism and protein production remain extremely important. Increases in world population will require increasingly greater amounts of plant protein in the near future.

The dependence of many aspects of plant N metabolism upon energy is clearly established (Hageman and Flesher 1960; McKee 1962; Hardy, Havelka, and Quebedeaux 1976), and while it may be unrealistic to consider the processes of N metabolism without simultaneous examination of energy supply and assimilation, the complexity of the relationship between N metabolism and energy supply dictates such a separation. Thus my approach will be to consider the process of N metabolism in plants with the aim of reviewing and suggesting approaches that might allow greater biological efficiency or productivity in plant reproduction.

Efforts directed toward the management or control of plant productivity require modification of the reproductive strategy of the plant. Two general approaches may be employed to modify the reproductive processes in crop plants: the alteration of the environment, or the improvement of plant genotype. As a plant breeder, the alteration of the plant genotype and consequently its response to the environment represents a very desirable alternative.

COMPARATIVE NITROGEN FLUX IN SOME MAJOR GRAIN CROPS

Can the role of N metabolism in plant reproduction be genetically managed or controlled in such a way as to improve the biological efficiency or productivity of crop plants? Stated differently, this question asks whether useful genetic variability exists for the process of nitrogen metabolism as a whole or for specific critical aspects of this process. Some insight may be gained by comparing the seasonal N flux in three major U.S. seed-producing crops, *Triticum aestiuum* L. em. Thell. (wheat), *Zea mays* L. (maize), and *Glycine max* (L.) Merrill (soybean). Data gathered from the few available field studies that give sufficient information to determine seasonal N accumulation are plotted in Figures 15.1–15.3. Although the data summarized in these figures were collected over a broad range of geographical areas, soil types, fertility and moisture conditions, and crop cultivars, the sets of nitrogen accumulation curves for each crop were generally quite similar.

Maize appeared to have a relatively short lag phase (Figure 15.4). A long lag phase in wheat would be expected, as this crop is planted early in the spring, or, in the case of winter wheat, initiates growth in the spring under cool conditions. Soybean, however, usually encounters similar or warmer early season temperatures than maize, and thus the longer lag phase for soybean was probably not environmentally induced but represented a physiological difference in the two species. The C_4 photosynthetic system may allow enhanced early season carbon assimilation in maize, resulting in greater energy availability and N accumulation. Alternatively, the soybean accumulates N via NO_3-uptake and reduction early in the season, and these systems may function less efficiently in soybean than in maize.

The slope of the N accumulation linear phase was greater in wheat than in either maize or soybean and suggested that a large proportion of total seasonal nitrogen assimilation occurred in a relatively shorter time in wheat. The linear phase of the sigmoid curves for maize and soybean had approximately equal slopes, indicating that an equal proportion of total seasonal nitrogen was assimilated during the linear phase in these two crops.

The relative times of anthesis demonstrated a major difference between the

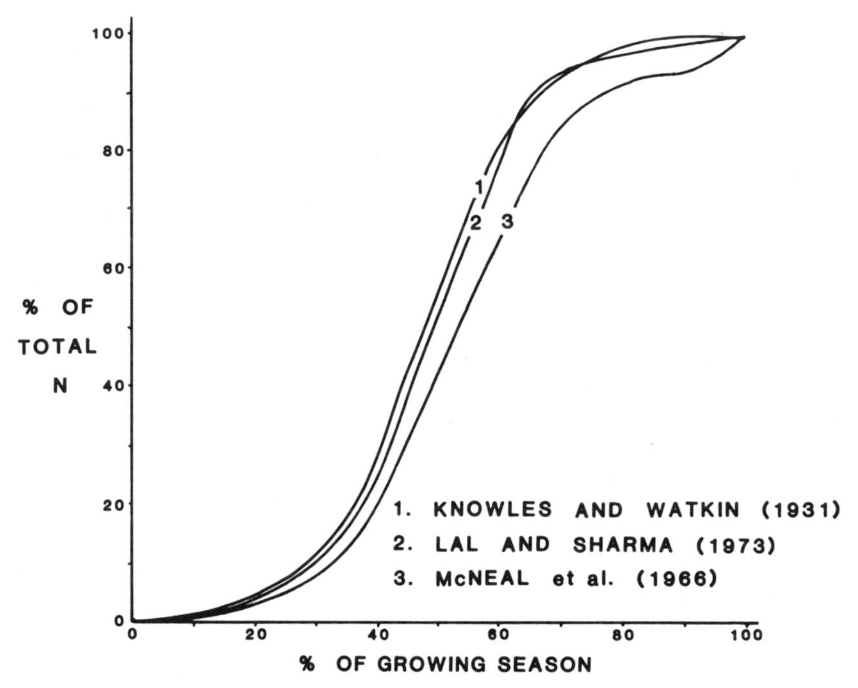

Figure 15.1. Estimated seasonal nitrogen accumulation in field-grown wheat. (Data from Knowles and Watkin, 1931, include spring growth only, estimated to have begun April 1; data from Lal and Sharma, 1973 are based upon the two-year mean of the variety 'Kalyan Sona.')

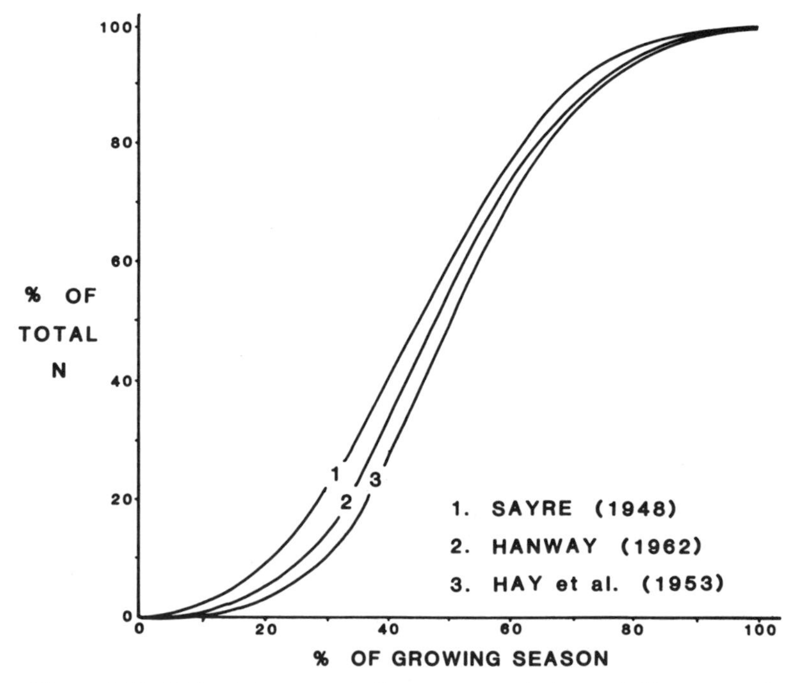

Figure 15.2. Estimated seasonal nitrogen accumulation in field-grown maize. (Data from Hanway, 1962a,b are based upon treatment [Plot 1004] in which no apparent nutrient deficiencies existed.)

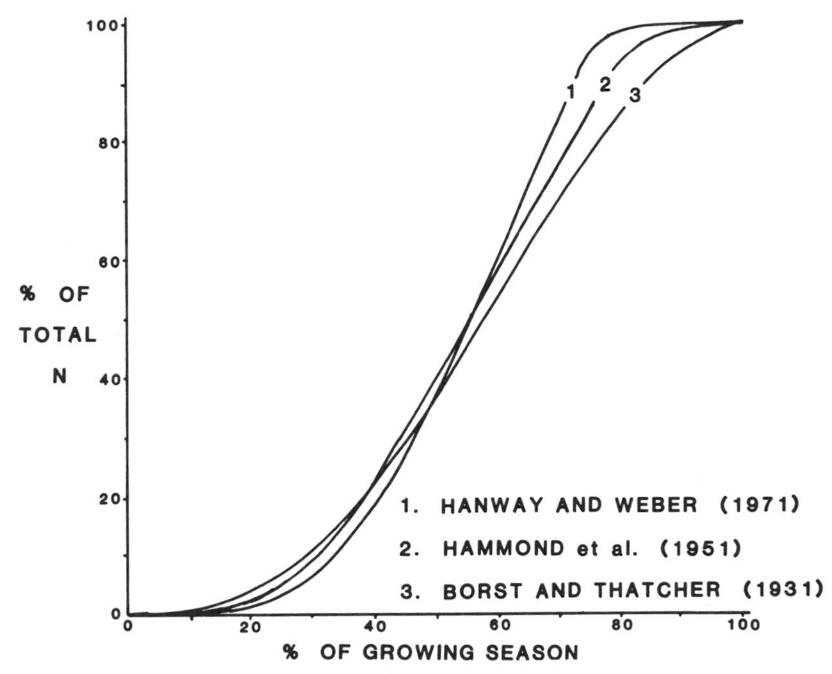

Figure 15.3. Estimated seasonal nitrogen accumulation in field-grown soybean. Calculated from Hanway and Weber (1971a,b,c) using the variety 'Hawkeye' grown in 1961 and 1963. (Data from Hammond et al., 1951, using 'Richland' soybean grown on Webster soil; data from Borst and Thatcher, 1931, using 'Manchu' soybean grown in 1925.)

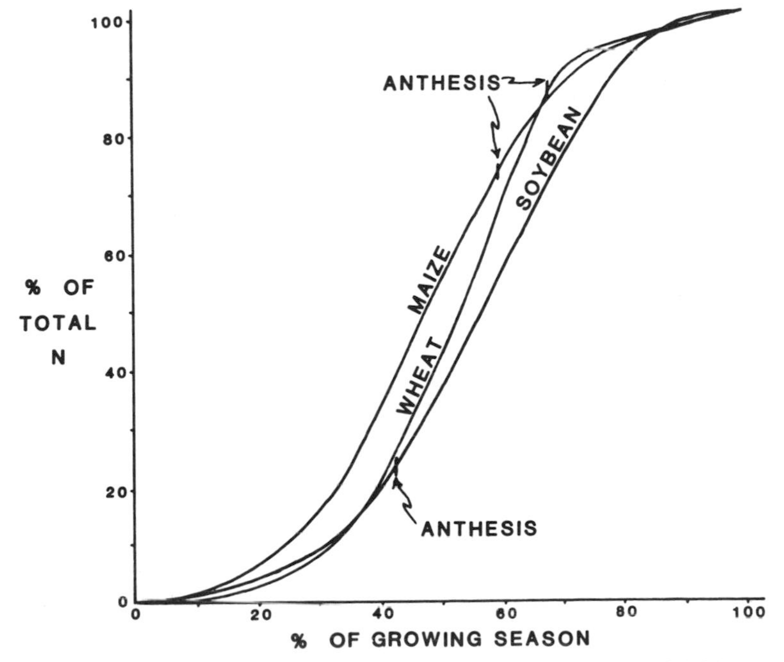

Figure 15.4. Estimates of mean seasonal nitrogen accumulation of field-grown wheat, maize, and soybean derived from Figures 15.1, 15.2, and 15.3, respectively.

three crops (Figure 15.4). The soybean varieties employed in the studies summarized (Figure 15.3) generally possessed the indeterminate growth habit of soybean grown in the midwestern states of the United States.* In the soybean, anthesis occurred when only about 25% of the total N had been accumulated. Vegetative growth was less than half complete at this time (Egli and Leggett 1973; Hanway and Weber 1971a,c), and thus reproductive and vegetative growth compete for carbohydrates and minerals for more than half of the soybean life cycle. A different pattern was apparent in wheat (Figure 15.4). Anthesis occurred after about 85% of the total N had been assimilated. In maize, anthesis occurred after 70% of the total N had been assimilated. The occurrence of anthesis in maize, and particularly in wheat, after the major portion of total N had been assimilated implied that the competition for reduced N between vegetative and reproductive growth in these species was less than in the soybean. Further, a comparison of the Nitrogen Harvest Indices (NHI)* (Cranvin 1976; Cox and Frey 1978) in these crops (Table 15.1) demonstrated that in wheat and soybean approximately 70% of the final plant nitrogen was contained in the grain at harvest. The estimated NHI in maize of 0.63 was somewhat lower. It is important to note that although wheat accumulated 85% of total seasonal N before anthesis, the NHI of this crop was equal to or higher than maize or soybean. The wheat plant must, therefore, be very effective in the remobilization of nitrogen from the vegetative tissue to the grain during the seed-filling process. A measure of the ability to remobilize nitrogen from vegetative tissue (and by inference to the reproductive tissue) was used by McNeal, Berg, and Watson (1966); they did not apply a name to the ratio they calculated, but the term Nitrogen Remobilization Efficiency (NRE) would be descriptive.† The NRE in wheat was 0.70, substantially higher than either maize or soybean (Table 15.1). The high NRE in wheat verified the exceptional capacity of this species to efficiently remobilize and translocate a large proportion of vegetative nitrogen to the developing seed. While the soybean had a lower NRE than wheat, it possessed a slightly higher NHI. The occurrence of anthesis early in the growth cycle in soybean evidently allows a large fraction of the assimilated nitrogen to be utilized directly by the developing seed rather than first undergoing incorporation into leaf or stem proteins. Such incorporation and subsequent remobilization may be energetically inefficient.

*Hammond, Black and Norman (1951) used 'Richland' soybean, which is somewhat more determinate than the typical indeterminate soybean. Indeterminate genotypes were grown on approximately 60% of U.S. hectarage in 1980.

*Nitrogen Harvest Index = the fraction of total above ground nitrogen yield of a crop contained in the grain.

†Nitrogen Remobilization Efficiency = (Maximum leaf + stem nitrogen) − (final leaf + stem nitrogen) ÷ (maximum leaf + stem nitrogen).

$$\text{Nitrogen Remobilization Efficiency} = \frac{(\text{Maximum leaf} + \text{stem nitrogen}) - (\text{final leaf} + \text{stem nitrogen})}{(\text{Maximum leaf} + \text{stem nitrogen})}$$

Table 15.1 Estimates of Nitrogen Harvest Index and Nitrogen Remobilization Efficiency in Field-Grown Wheat, Maize, and Soybean

Crop and source of information	Nitrogen harvest index		Nitrogen remobilization efficiency (%)	
Wheat				
Austin, R. B. et al. (1980)[a]	0.75		—	
Knowles and Watkin (1931)	0.74		72	
Lal and Sharma (1973)	—		78[b]	
McNeal, Berg, and Watson (1966)[c]	0.64		59	
McNeal et al. (1968)[d]	0.71		62	
Boatwright and Haas (1961)	0.68		—	
Mean		0.70		68
Maize				
Hay, Earley, and DeTurk (1953)	0.68		53[e]	
Hanway (1962a,b)[f]	0.62		42	
Jones and Huston (1914)	0.65		41	
Sayre (1948, 1955)	0.56		39	
Mean		0.63		43
Soybean				
Borst and Thatcher (1931)[g]	0.66		47	
Hanway & Weber (1971a,b,c)[h]	0.66		51	
Hammond, Black & Norman (1951)[i]	0.80		—	
Jeppson, Johnson, and Hadley (1978)	0.74		—	
Mean		0.72		49

[a] Mean of 12 genotypes at two locations.
[b] Mean of two varieties for two years.
[c] Mean of 5 spring wheat varieties under dryland conditions.
[d] Mean of 7 varieties under dryland and irrigated conditions.
[e] Calculated from Hay, Earley, and DeTurk (1953), Table II.
[f] Estimated from Plot 1004 (no apparent nutrient deficiencies).
[g] Data from 'Manchu' soybeans grown in 1925.
[h] Data from Hawkeye soybeans grown in 1961 and 1963.
[i] Includes Richland soybean grown on Webster soil only.

Undoubtedly, environment strongly influences nitrogen accumulation (McNeal et al. 1968; Deckard, Lambert, and Hageman 1973; Lal, Reddy, and Modi 1978), Nitrogen Harvest Index (Spratt and Gasser 1970), and nitrogen remobilization (Beauchamp, Kannenberg, and Hunter 1976). Despite the confounding effects of environment, however, seasonal nitrogen flux in wheat, maize, and soybean was apparently dissimilar in a number of respects. Interspecific comparisons should suggest possible approaches to genetic im-

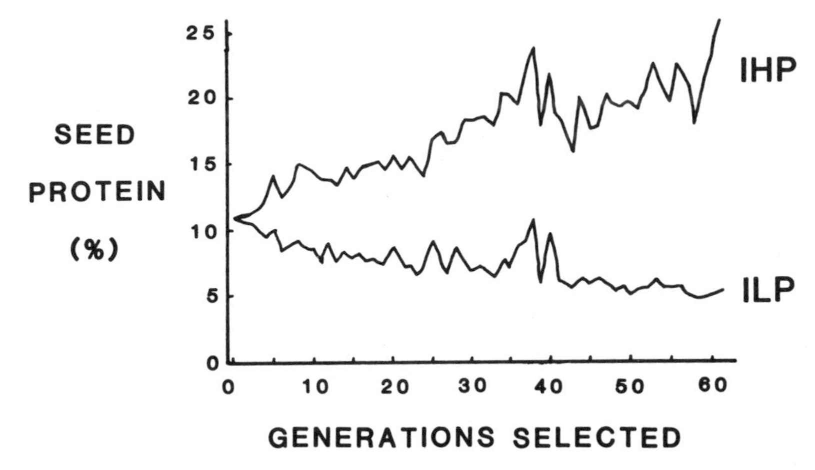

Figure 15.5. Mean protein content of Illinois High Protein (IHP) and Illinois Low Protein (ILP) maize after 61 generations of selection. (Data are from Leng (1962).)

provement in the nitrogen economy of crop plants, but intraspecific genetic variability is the necessary prerequisite for genetic improvement of a particular crop. An outstanding example of such variability is presented in the classical selection experiments begun in 1896 at the Illinois Agricultural Experiment Station involving continuous selection of high and low grain protein (Figure 15.5) from one original maize foundation stock (Winter 1929; Hoener and DeTurk 1938; Woodwroth, Leng, and Jugenheimer 1952). The continuous progress through selection and the presence of residual genetic variability after many generations of selection (Leng 1962; Dudley and Lambert 1969) demonstrate the involvement of a large number of genes in the control of grain protein. It is reasonable to assume that many aspects of nitrogen metabolism were genetically modified during the selection process. This assumption leads to the conclusion that genetic alteration and improvement of individual steps of nitrogen metabolism could be a useful alternative approach to selection based upon an end product such as seed protein percentage.

GENETIC CONTROL OF NITROGEN METABOLISM

Despite intensive interest in the genetics of the quality and quantity of seed proteins—the harvestable end product of nitrogen metabolism—a surprisingly limited amount of information is available on the genetics of nitrogen metabolism itself. The available reports usually consider the genetics of only one aspect of nitrogen metabolism. The limited amount of literature and the necessarily narrow scope of most reports make it difficult to identify those aspects of nitrogen metabolism that would merit inclusion in a scheme directed toward the genetic improvement of the efficiency or productivity of the overall process of nitrogen metabolism. Among the steps in the process of nitrogen metabolism that could be considered for inclusion in a selection procedure are the following:

1. NO_3- or NH_4^+ uptake and/or N_2-fixation
2. Translocation of NO_3- or nitrogen compounds from the root
3. Nitrate reduction
4. Assimilation of reduced nitrogen
5. Protein synthesis in vegetative tissue
6. Protease activity and remobilization of amino acids
7. Phloem loading
8. Translocation of amino acids to the developing seed
9. Phloem unloading at the site of the developing seed
10. Synthesis of seed storage proteins.

There is presently little or no data to suggest the presence of genetic variability in grain crops for items 2, 4, 5, 7, or 9. Genetic variation for the remaining items has been demonstrated. This is particularly true of the genetics of nitrate reduction, which has been extensively studied by Dr. R. H. Hageman at the University of Illinois and others. A review of the current state of knowledge in these areas is a necessary first step in devising an integrated selection system for improved efficiency and productivity of nitrogen metabolism.

Nitrogen uptake. Goodman (1977) stated that the root uptake of soil nitrogen was a function of the size and activity of the root. He cited work by Holmes (1967) that showed differences in root size and nitrogen uptake rate within two *Lolium perenne* cultivars. Pollmer et al. (1979) observed that maize hybrids with high seed protein percentage and protein yield had higher nitrogen uptake than standard hybrids. The authors stated that one possible reason for the "more intensive" nitrogen uptake was a superior root system. Chevalier and Schrader (1977) also demonstrated genotypic variation in maize for nitrate uptake from a nutrient solution. Nitrate absorption was related to root dry weight, but nitrate absorbed per gm dry root weight was not constant. This finding indicated that physiological activity, root affinity for NO_3-, or some other factor had a role in NO_3-absorption and that genetic control was probable.

Other evidence supported the genetic control of root affinity for NO_3^-. The low *Km* (high affinity) of 0.021 mM in maize (Van den Honert and Hooymans 1955) and the high *Km* (low affinity) of 0.11 mM in *Hordeum vulgare* L. (barley) lead Rao and Rains (1976) to speculate that the low affinity in barley might be due to intensive breeding for high nitrogen response with a resultant low efficiency for nitrogen uptake from the soil. This hypothesis might be verified by comparing the feed barley cultivar 'Arivat' used by Rao and Rains (1976) with a malting barley which, because of the restriction upon seed protein content imposed by brewing requirements, is generally grown on soils that have not received nitrogen fertilization. Thus, high response to nitrogen has not been a consideration in the breeding of malting barleys. Such a study would also serve to ascertain the presence of intraspecific genetic variability for root nitrate affinity, which is yet to be established. High affinity for nitrate and the resultant ability of the plant to more effectively scavenge soil nitrogen should have desirable economic and ecological benefits.

Nitrogen fixation. The ability of legumes to fix nitrogen symbiotically with bacteria of the species *Rhizobium* has received much attention in recent years. Intensive research in the biochemistry, physiology, and genetics of the legume host and the Rhizobium microsymbiont has been conducted with the goal of enhancing N_2-fixation. Genetic improvement of fixation requires the presence of exploitable genetic variability. Only recently has such genetic variability been reported in the grain legumes. Wacek and Brill (1976) noted varietal differences in acetylene reduction activity (ARA) in very young (14 days) plants grown in vials. They noted some similarities in the ARA levels measured in vials and those measured on the same genotypes grown in the field. Cregan, Sloger, and Kuykendall (1980) reported that certain soybean lines demonstrated consistently high ARA and nodulation in two distinctly different field environments. Wide differences in total seasonal nitrogen fixation (C_2H_2 reduction) were observed among *Phaseolus vulgaris* L. (bean) cultivars by Westermann and Kolar (1978). Similarly, Zary et al. (1978) found intraspecific variability in *Vigna inguiculata* (L.) Walp (cowpea) for ARA and nodule mass per plant. The apparent presence of genetic variability for nitrogen fixation (C_2H_2 reduction) in soybean, bean, and cowpea indicates that breeding for increased N_2-fixation is possible. Some evidence would suggest, however, that in the soybean greater productivity may be obtained through an increase in the efficiency of N_2-fixation rather than through greater fixation per se. Under low soil nitrate conditions, the soybean is capable of fixing most of the nitrogen required for growth (Vest, Weber, and Sloger 1973). Further, the growth and seed yield response to applied nitrogen has not been consistently positive (deMooy, Pesek, and Spaldon 1973). Evidence suggesting that N_2-fixation is limited by the amount of energy available to the root system and nodule (Lawn and Brun 1974; Hardy, Havelka, and Quebedeaux 1976) further supports the goal of more-efficient nitrogen fixation rather than increased fixation per se. Caldwell and Vest (1970) inoculated Rhizobium-free soil in three environments with 28 *Rhizobium japonicum* (Kirchner) Buchanan strains. The yield of five soybean genotypes was obtained with each of the 28 strains. Significant seed yield differences were associated with certain strains. The lack of significant Rhizobium strain x soybean genotype interaction indicated that particular strains were more-efficient N_2 fixers across a spectrum of soybean genotypes. Thus, more-efficient nitrogen fixation could be attained by providing nodulation with only one or more superior Rhizobium strains. As G. E. Ham (1976) pointed out, hwoever, the establishment of a superior strain in the nodules of plants growing in soils already populated with Rhizobium is a difficult challenge.

Two similar but somewhat different approaches to this problem have been outlined. Devine and Weber (1977) proposed the use of the nonnodulating soybean genotype (rj_1rj_1), which would effectively eliminate nodulation by indigenous *R. japonicum* strains. The system required the discovery or development of a highly efficient Rhizobium genotype with the capacity to effectively nodulate the rj_1rj_1 soybean. Kvien, Ham, and Lambert (1978) screened 1600 soybean genotypes to find lines that were poorly nodulated by indigenous strains but would nodulate well with strains known to be highly effective. The soybean host would thus function to preferentially accept nodulation by superior Rhizobium genotypes. The proposals of Devine and Weber (1977) and Kvien, Ham and Lambert (1978) both appear to be promising genetic approaches to the improved efficiency of N_2-fixation in soybeans and by analogy to other legume species.

Nitrate reductase. Nitrate reductase (NR) is considered to be the rate-limiting step in the assimilation of reduced nitrogen in non-legumes (Hageman, Leng, and Dudley 1967; Beevers and Hageman 1969). Because of the close relationship between available soil nitrate and plant growth, and the substrate inducible nature of NR, it can be argued that soil nitrate supply imposes the real limit upon nitrogen assimilation. Hageman et al. (1976), however, cited evidence from Deckard (1970) indicating that NR was more useful as an index of grain and grain protein production than nitrate tissue content. Hageman et al. (1976) further reported the work of Purvis (1972), in which genotypic differences in NR levels were independent of tissue nitrate content. This evidence suggested that NR would be useful as a tool in selecting plant genotypes with enhanced nitrate assimilation properties. Genotypic variability for NR in maize had been previously established by Zieserl and Hageman (1962), Schrader et al. (1966), and Warner et al. (1969). Croy and Hageman (1970) apparently found no genetic variability in wheat for NR activity but noted that selection based upon NR activity at a point in the season when soil nitrate was in adequate supply could identify genotypes with higher grain protein. Thus, as a result of accumulated physiological and genetic evidence, a number of workers initiated efforts to establish the usefulness of NR as a criterion for the selection of plant genotypes with higher seed protein or seed yield.

Eck, et al. (1975) assayed NR activity in a total of 41 *Sorghum bicolor* (L.) Moench (grain sorghum) hybrids and inbreds grown in the growth chamber. The broad-sense heritability of in vitro NR activity was high ($H = 0.9$, mean of assays made at three plant ages), while narrow-sense heritability was lower, $h^2 = 0.41$. The significant amounts of additive and nonadditive genetic variance noted in this report indicated the potential for genetic advance via pure line or hybrid development. Deckard and Busch (1978) attempted to use NR assays in predicting the grain yield or protein of wheat lines. Mid-parental in vitro NR activity of the second leaf in growth chamber-grown seedlings was correlated with the mean yield of 21F_6 lines from each of 25 crosses ($r = 0.532$, $P \leq 0.01$). Nevertheless, in vitro NR activity could not distinguish the highest from the lowest yielding lines within each cross. In addition, NR activity was not related to total plant nitrogen accumulation. Deckard and Busch (1978) concluded that differing ratios of NR activity to actual nitrogen accumulation or unequal Nitrogen Harvest Indices (NHI) may have been partially responsible for their inability to identify lines with high yield or grain protein using the NR assay.

Previous studies in a number of crops had suggested the occurrence of genotypic differences in reduced nitrogen input estimated from NR activity and actual nitrogen accumulation. Deckard, Lambert, and Hageman (1973) found that the ratio of seasonal input of reduced nitrogen estimated from in vitro NR assays to total reduced N in above-ground tissue in six maize hybrids varied from 2.2 to 4.1. Among 11 grain sorghum hybrids the same ratio varied from 0.55 to 0.93 (Eck, Wilson, and Martinez 1975). Dalling, Halloran, and Wilson (1975) also detected genotypic differences in the ratio of N accumulation estimated by the NR assay to that actually accumulated in five wheat varieties.

Genotypic differences in NHI similar to those found by Deckard and Busch (1978) had also been confounding factors in other efforts to identify plant genotypes with high grain yield or protein using NR assays. Dalling, Halloran, and Wilson (1975) found that wide differences in NHI were responsible for the low correlation between grain nitrogen and total season NR activity ($r = 0.014$) in five wheat varieties.

It seems unreasonable, however, to anticipate a high positive correlation between NR activity and seed yield or seed nitrogen content in a group of plant genotypes. A long sequence of metabolic events occurs between nitrate reduction and the determination of seed yield or N content. Any or all of these events may be affected dissimilarly by the plant genotype or by plant interaction with the environment. An inequality in NHI reduces the correlation between NR activity and percentage or total seed nitrogen. Genotypic differences in the ratio of enzyme-estimated nitrogen accumulation to actual N accumulation indicates a fundamental difficulty in the use of NR activity as a plant selection tool. A clear understanding of the genotypic variability in this ratio will be necessary for the successful use of NR in plant improvement.

Nitrogen remobilization. Miller (1939) reviewed the early literature relating to the movement of nitrogen from vegetation to the developing grain. The first observation of this phenomenon was made by Adorjan (1902), who found that 68.1% of the mature grain nitrogen was translocated from the stems, leaves, and glumes of wheat. While a number of workers have subsequently reviewed the literature regarding nitrogen redistribution or remobilization (Williams 1955; Ohlrogge 1960), Seth, Hebert, and Middleton (1960) were apparently the first to suggest possible genetic variation for Nitrogen Remobilization Efficiency (NRE). Of four wheat varieties they studied, those with high seed protein percentage either absorbed more nitrogen from the soil or translocated a greater proportion of vegetative N to the kernel. Johnson, Schmidt, and Mattern (1968) noted that the higher percent grain protein in wheat lines derived from crosses with 'Atlas 66' was due to more complete translocation of nitrogen from the foilage to the grain. A more recent report in maize (Pollmer et al. 1979) noted good general combining ability for NRE in several inbreds from a breeding program for high grain protein concentration and yield. Jeppson, Johnson, and Hadley (1978) observed no significant differences for NRE among a group of 11 nodulating and nonnodulating soybean genotypes. As a group, however, sets of four lines, previously classified as efficient and inefficient based upon Harvest Nitrogen Index, mobilized 60% and 51%, respectively, of accumulated N from leaflets. The efficient lines remobilized 53% and the inefficient genotypes 43% of petiole nitrogen. Similarly, stem NRE was 80% for the efficient and 73% for the inefficient genotypes. Differences among the NRE of different plant parts were also observed by Lal, Reddy, and Modi (1978). The mean NRE of five wheat varieties ranged from 49.4% for the spike chaff to 82.7% for the lower leaves. Obvious varietal differences were also noted for the NRE of different plant parts.

The presence of genetic variability for nitrogen remobilization has been established in a number of crops, but the use of such variability in a plant improvement effort is as yet unreported. Neither has the heritability or an indication of the type of genetic control been established. This may be due in part to the difficulty in the measurement of nitrogen remobilization, which requires nitrogen determinations at a number of times during the plant growth cycle to obtain an estimate of maximum and final vegetative N. A biochemical procedure for estimating Nitrogen Remobilization Efficiency would be an important step toward facilitating the use of this trait in a selection program. Useful information in this regard was reported by Rao and Croy (1972). They found higher protease (pH 4.0 and pH 7.0) levels after anthesis in the leaf blades of a high, as compared to a low, grain protein wheat. Similarly, in *Oryza sativa* L. (rice) a high grain protein variety, 'IR480', had almost a twofold greater leaf protease (pH 7.8) activity then the lower protein variety 'IR8' (Perez et al. 1973). The protease enzymes are responsible for the degradation of polypeptides into amino acids, and therefore high protease activity would be associated with plant genotypes with a high potential for nitrogen remobilization from the vegetative tissue. The use of protease activity to estimate nitrogen remobilization would require that enzyme activity be the limiting factor in nitrogen movement from the vegetation and would necessitate a high positive correlation of activity and remobilization. Dalling, Boland, and Wilson (1976) used acid proteinase activity (pH 4.2) to estimate nitrogen loss from the flag leaf, first and second leaves below the flag leaf, glumes, and stem of two wheat varieties. They obtained a correlation of $r=0.893$ ($P \leq 0.01$) between estimated and actual nitrogen loss. Dalling, Bolling and Wilson (1976) therefore concluded that acid proteinase was the limiting factor in the loss of nitrogen from vegetative tissue. This work also indicated the quantitative relationship between protease activity and nitrogen movement from the leaf.

Cagampang, Cruz, and Juliano (1971) used a unique approach to ascertain the presence of genetic variability for nitrogen remobilization capacity in seven pairs of rice cultivars. Each pair consisted of a high and low seed protein line selected from the same rice cross. Ten days after flowering, tillers were cut off at the internode below the panicle and bleeding sap was collected from each stem. Over the 12-hr period from 700 to 1900 hrs the bleeding sap of the seven low protein lines averaged 65 μg amino N/10 tillers versus 90.1 μg in the seven high seed protein lines.

Unquestionably, one approach to the increased productivity of protein per hectare is to augment the proportion of vegetative nitrogen translocated to the seed, thus leaving less nitrogen in the crop residue. A paucity of information exists concerning the genetic control of the remobilization process, but the

Table 15.2 Correlation between in Vivo Nitrate Reductase Activity and Grain Yield and Protein Production of 36 Maize Hybrids Grown at Urbana and 45 Hybrids Grown at DeKalb, Illinois

	Correlation (r) of NR with	
Location	Grain yield	Grain protein production
Urbana, Illinois	0.41[a]	0.33
DeKalb, Illinois	0.35	0.04

[a] Significant at the 0.05 level.

Source: Data from Hageman et al. (1976).

evidence that is available offers the expectation of genetic improvement. It seems likely that in the next decade much greater attention will be paid to nitrogen remobilization and methodology useful in the genetic improvement of remobilization.

Protein synthesis in the seed. The conversion of vegetative protein to seed storage protein begins with nitrogen remobilization and terminates with the incorporation of amino acids into seed storage proteins. The strength of the sink provided by the developing seed could thus be of considerable importance in the achievement of maximum N movement from vegetation to seed. Few research reports have dealt with genetic variation in the capacity of developing seeds to serve as a sink for free amino acids. Seth, Herbert, and Middleton (1960) concluded that differences in the seed protein content of wheat varieties were related to the rate of protein synthesis in the developing grain. Whether rate of kernel protein synthesis was a function of the ability of the sink to synthesize protein or merely to differences in amino acid supply was unclear. Brunori et al. (1980) reported different patterns of nitrogen accumulation in the developing kernels of a diverse group of wheats. Wide variation in rate and duration of nitrogen deposition suggested the possibility of using these characteristics as criteria in the selection of genotypes with the potential for high seed protein accumulation.

A detailed study of biochemical differences in the developing rice kernel was conducted by Cruz, Cagampang, and Juliano (1970). They collected grain samples from varieties differing in percent grain protein at four-day intervals from 4 to 32 days after anthesis and assayed for free amino nitrogen, protein, and RNA. Rapid protein accumulation in the seed occurred from 4 to 16 days after flowering with no increase after 16 days. Throughout the 4–16 day period, levels of soluble amino N, protein, and RNA were higher in the variety with high seed protein. In a second experiment, Cruz, Cagampang and Juliano (1970) compared high and low seed protein segregants from single crosses of the low seed protein line 'IR8,' with high protein varieties. In addition to assays of protein, soluble amino N, and RNA, developing kernels were analyzed for L-leucine-U-^{14}C incorporation. In one pair of genotypes which had similar growth duration and seed yield, the high-protein segregant demonstrated higher levels of leucine-U-^{14}C incorporation, soluble amino nitrogen, and protein/kernel than did its low protein counterpart. Free amino acid content and the rate of amino acid incorporation were the two factors that consistently distinguished high from low protein varieties. The authors concluded that the levels of the enzymes involved in storage protein synthesis were probably not induced by high free amino acid content in the developing kernel, but were under separate genetic control. This conclusion suggests that the efficiency with which vegetative N is transferred to the seed can be increased through genetic improvement of the capacity of the developing kernel to incorporate free amino acids.

SELECTION FOR IMPROVED NITROGEN METABOLISM

Surveying the literature of the genetics of nitrogen metabolism in higher plants demonstrates the limited knowledge of the subject. How should one proceed if the goal is the improved efficiency of N metabolism with a resultant increase in crop productivity? Most grain crop improvement efforts with the goal of increased productivity use harvestable end product, i.e., seed yield, seed protein yield, or seed protein percentage, as the major selection criterion. This approach provides a good measure of genetic advance, but offers no knowledge of the physiological and biochemical basis of the improvement. Such knowledge may be important in the identification of factors limiting further advance. Selection based upon the harvestable end product has the additional and extremely important virtue of integrating the effects of genotype, environment, and the interaction of the two, over the entire crop season. The measurement of one physiological trait at one or a few times during the season is unlikely to offer a dependable estimate of the harvestable end product of a particular genotype.

Hageman et al. (1976) provided an interesting insight into this problem using NR data from partial diallel crosses of nine and ten maize inbreds grown, respectively, at Urbana and DeKalb, Illinois, in 1973. In vivo NR was assayed at two postsilking dates separated by seven to ten days. A significant but relatively low correlation between NR and hybrid grain yield was obtained at Urbana (Table 15.2). In vivo NR activity showed no significant relationship, however, with grain protein production at either location. Hageman et al. (1976) suggested a number of reasons for the lack of correlation, including water stress during the plant reproductive period, improper timing of sampling, and the large amount of genetic diversity among the hybrids. When the hybrids grown at DeKalb were divided into groups, each group with one parent in common, the correlation of NR activity with grain yield and grain protein was higher within most groups than it had been without grouping. This finding supported the conclusion that genetic diversity reduced the association between NR activity and grain or protein yield. It can be presumed that limiting the amount of diversity decreases the confounding effects of other genetic differences, such as nitrate uptake, nitrogen remobilization, sink strength, etc. Rather than limiting the genetic variability present in a screening program, however, the objective of more-efficient N metabolism is best served through maximal diversity. The most-appropriate screening procedure is one that can accommodate genetic variability for a number of traits. Such a procedure would approximate an integrated selection system similar to selection based upon the end product, but would have the advantage of describing how the end product was obtained. If such a system were applied to the process of N metabolism, identification of the factor(s) limiting plant nitrogen economy might then be possible. Hageman, Leng, and Dudley (1967) were aware of the necessity of simultaneously observing a number of traits in a biochemical selection program. According to Hageman, Leng, and Dudley, such a selection system would require the determination of the major enzymatic controls involved in the processes of growth and yield, and particularly "the optimum levels of activity of each such enzyme in combinations with specified levels of the other enzymes." This biochemical approach to maize breeding proposed by Hageman, Leng, and Dudley (1967) suggested selection based purely upon enzymatic activity and optimum combinations of activity levels of various enzymes. In the case of nitrogen metabolism, a more-flexible approach in which enzyme assays are used in combination with traits such as Harvest Nitrogen Index, total N accumulation, or Nitrogen Remobilization Efficiency would seem appropriate.

One can only speculate as to what enzymatic or physiological processes would be most profitably included in a selection program. Inclusion of a particular item may depend upon the crop in question. Devising a selection scheme will require the close cooperation of the geneticist and physiologist to

1. determine critical limiting physiological or enzymatic processes,
2. identify those processes for which genetic variability is most likely to exist,
3. select the genetic stocks most likely to possess useful variability, and
4. establish the breeding methodology to facilitate greatest genetic advance.

The decision to include a particular enzymatic and physiological selection criterion in a selection system would be based upon both theoretical consideration and empirical evidence. Studies similar to that by Austin et al. (1980) may provide useful impirical evidence. They attempted to identify physiological changes associated with genetic emprovements in wheat yields. Recently developed experimental lines were compared with a series of varieties grown over the last 70 years. Triats such as dry matter harvest index were important factors associated with yield increases.

The selection of genetic materials for inclusion in a screening effort is a major determinant of the success of such an effort. The limited genetic variability in most major grain crops (National Academy of Sciences 1972) suggests that commonly grown cultivars may be genetically very similar. Thus it is imperative that the use of unique or exotic germplasm sources be considered. The numerous crop germplasm collections make a wide range of genetic material available in most crops.

CONCLUSIONS

The productivity of a grain crop is a function of a number of genetically controlled physiological processes that interact with the environment to yield a particular quantity of final product. In the case of nitrogen metabolism, the genetic control and environmental interaction of the various aspects of the process have been described to varying degrees. Undoubtedly, further physiological and genetic information needs to be accumulated; however, this necessity should not preclude the establishment of plant improvement efforts that rely at least partially upon physiological selection criteria. Such selection programs should not be initiated with the expectation that significantly improved genetic materials will be developed immediately. A physiological approach to breeding necessitates the assembly into one genotype of desirable physiological characteristics from a spectrum of genetic backgrounds. Unanticipated genetic and physiological difficulties will probably be encountered as a result of the juxtapositioning of traits from widely divergent genetic sources. Consideration must also be given to the environment in which physiological selection is applied. Halloran and Lee (1979) suggested that such traits as the translocation of N from vegetation to developing grain in wheat can be significantly affected by growth under artificial conditions. They recommended screening under field plot conditions. Despite the anticipated difficulties, the development of a plant improvement program using physiological selection methods should provide a scientific challenge with the prospect of a valuable and practical end product.

LITERATURE CITED

Adorjan, J. 1902. *Die Nahrstoffaufnahme des Weizens*. Jour. Landw. 50:193–230.

Altschul, Aaron, M. 1976. *The protein-calorie trade-off*. Pages 5–16 in *Genetic Improvement of Seed Proteins*. Proc. Nat. Acad. Sci. USA.

Austin, R. B., J. Bingham, R. D. Blackwell, L. T. Evans, M. A. Ford, C. L. Morgan, and M. Taylor. 1980. *Genetic improvements in winter wheat yields since 1900 and associated physiological changes*. J. Agric. Sci. 94:675–88.

Beauchamp, E. G., L. W. Kannenberg, and R. B. Hunter. 1976. *Nitrogen accumulation and translocation in corn genotypes following silking*. Agron. J. 68:418–22.

Beevers, Leonard, and R. H. Hageman. 1969. *Nitrate reduction in higher plants*. Ann. Rev. Plant Physiol. 20:495–522.

Boatwright, G. O., and H. J. Haas. 1961. *Development and composition of spring wheat as influenced by nitrogen and phosphorus*. Agron. J. 53:33–36.

Borst, H. L., and L. E. Thatcher. 1931. *Life history and composition of the soybean plant*. Ohio Agric. Exp. Sta. Bull. 494.

Brunori, A., H. Axmann, A. Figuoroa, and A. Micke. 1980. *Kinetics of nitrogen and dry matter accumulation in the developing seed of some varieties and mutant lines of Triticum aestivum*. Z. Pflanzenzuecht. 84:201–18.

Cagampang, G. B., L. J. Cruz, and B. O. Juliano. 1971. *Free amino acids in the bleeding sap and developing grain of the rice plant*. Cereal Chem. 48:533–39.

Caldwell, B. E., and Grant Vest. *Effects of Rhizobium japonicum strains on soybean yields*. Crop Sci. 10:19–21.

Chevalier, Peggy, and L. E. Schrader. 1977. *Genotypic differences in nitrate absorption and partitioning of N among plant parts in maize*. Crop Sci. 17:897–901.

Cox, T. S., and K. J. Frey. 1978. *Nitrogen harvest index in oats*. Agron. Abstr. Amer. Soc. Agron.

Cranvin, David, T. 1976. *Interrelationships between carbohydrate and nitrogen metabolism*. Pages 172–91. in *Genetic Improvement of Seed Proteins*. Proc. Nat. Acad. Sci. USA.

Cregan, P. B., C. Sloger, and L. D. Kuykendall. 1980. *Field screening of soybeans for differential N-fixing capacity*. Agron. Abstr. Amer. Soc. Agron.

Croy, Lavoy I., and R. H. Hageman. 1970. *Relationship of nitrate reductase activity to grain protein production in wheat*. Crop Sci. 10:280–85.

Cruz, L. J., G. B. Cagampang, and B. O. Juliano. 1970. *Biochemical factors affecting protein accumulation in the rice grain*. Plant Physiol. 46:743–47.

Dalling, M. J., Geraldine Boland, and J. H. Wilson. 1976. *Relation between acid proteinase activity and redistribution of nitrogen during grain development in wheat*. Austral. J. Plant Physiol. 3:721–30.

Dalling, M. J., G. M. Halloran, and J. H. Wilson. 1975. *The relation between nitrate reductase activity and grain nitrogen productivity in wheat*. Austral. J. Agric. Res. 26:1–10.

Deckard, E. L. 1970. *Nitrate reductase activity and its relationship to yields of grain and grain protein in normal and opaque-2 corn (Zea mays* L). Ph.D. dissertation, Univ. of Illinois, Urbana.

Deckard, E. L., and R. H. Busch. 1978. *Nitrate reductase assays as a prediction test for crosses and lines in spring wheat*. Crop Sci. 18:289–93.

Deckard, E. L., R. J. Lambert, and R. H . Hageman. 1973. *Nitrate reductase activity in corn leaves as related to yields of grain and grain protein*. Crop Sci. 13:343–50.

deMooy, C. J., John Pesek, and Emil Spaldon. 1973. *Mineral nutrition,* Pages 267-352 in B. E. Caldwell, ed. *Soybeans: Improvement, Production, and Uses*. Amer. Soc. of Agron., Madison, Wisconsin.

Devine, T. E., and D. F. Weber. 1977. *Genetic specificity of nodulation*. Euphytica 26:527–35.

Dudley, J. W., and R. J. Lambert. 1969. *Genetic variability after 65 generations of selection in Illinois high oil, low oil, high protein and low protein strains of Zea mays* L. Crop Sci. 9:179–81.

Eck, H. V., E. C. Gilmore, D. B. Ferguson, and G. C. Wilson. 1975. *Heritability of nitrate reductase and cyanide levels in seedlings of grain sorghum cultivars*. Crop Sci. 15:421–24.

Eck, H. V., G. C. Wilson, and Tito Martinez. 1975. *Nitrate reductase activity of grain sorghum leaves as related to yields of grain, dry matter, and nitrogen*. Crop Sci. 15:557–61.

Egli, D. B., and J. E. Leggett. 1973. *Dry matter accumulation patterns in determinate and indeterminate soybeans*. Crop Sci. 13:220–22.

Goodman, P. J. 1977. *Genetic control of inorganic nitrogen assimilation of crop plants*. Pages 165–76. in E. J. Hewitt and C. V. Cutting, ed. *Nitrogen Assimilation of Plants*. Academic Press, London.

Hageman, R. H., and D. Flesher. 1960. *Nitrate reductase activity in corn seedlings as affected by light and nitrate content of nutrient media*. Plant Physiol. 35:700–708.

Hageman, R. H., R. J. Lambert, Dale Loussaert, M. Dalling, and L. A. Klepper. 1976. Pages 103–31 *Nitrate and nitrate reductase as factors limiting protein synthesis*. in *Genetic Improvement of Seed Proteins*. Proc. Nat. Acad. Sci. USA.

Hageman, R. H., E. R. Leng, and J. W. Dudley. 1967. *A biochemical approach to plant breeding*. Adv. Agron. 19:45–86.

Halloran, G. M., and J. W. Lee. 1979. *Plant nitrogen distribution in wheat cultivars*. Austral. J. Agric. Res. 30:779–89.

Ham, G. E. 1976. *Competition among Strains of Rhizobia*. Proc. World Soybean Res. Conf., Urbana, Illinois, Aug. 3–8, 1975.

Hammond, L. C., C. A. Black, and A. G. Norman. 1951. *Nutrient uptake by soybeans on two Iowa soils*. Iowa Agric. Exp. Sta. Res. Bull, 384.

Hanway, J. J. 1962a. *Corn growth and composition in relation to soil fertility. II. Uptake of N, P, and K and their distribution in different plant parts during the growing season*. Agron. J. 54:217–22.

———. 1962b. *Corn growth and composition in relation to soil fertility. III. Percentages of N, P, and K in different plant parts in relation to stage of growth*. Agron. J. 54:222–29.

Hanway, J. J., and C. R. Weber. 1971a. *Dry matter accumulation in soybean (Glycine max (L.) Merrill) plants as influenced by N, P, and K fertilization*. Agron, J. 63:263–66.

———. 1971b. *N, P, and K percentages in soybean (Glycine max (L.) Merrill) plant parts*. Agron. J. 63:286–90.

———. 1971c. *Accumulation of N, P, and K by soybean (Glycine max (L.) Merrill) plants*. Agron. J. 63:406–8.

Hardy, R. W. F., U. D. Havelka, and B. Quebedeaux. 1976. *Opportunities for improved seed yield and protein production: N_2 fixation, CO_2 fixation, and O_2 control of reproductive growth*. Pages 196–228 in *Genetic Improvement of Seed Proteins*. Proc. Nat. Acad. Sci. USA.

Hay, R. E., E. B. Earley, and E. E. DeTurk. 1953. *Concentration and translocation of nitrogen compounds in the corn plant (Zea mays) during grain development*. Plant Physiol. 28:606–21.

Hoener, I. R., and E. E. DeTurk. 1938. *The absorption and utilization of nitrate nitrogen during vegetative growth by Illinois high protein and Illinois low protein corn*. J. Amer. Soc. Agron. 30:232–43.

Holmes, D. P. 1967. *Variation in nitrogen uptake and utilization in Lolium*. Ph.D. dissertation, University of Wales, Aberystwyth.

Jeppson, R. G., R. R. Johnson, and H. H. Hadley. 1978. *Variation in mobilization of plant nitrogen to the grain in nodulating and non-nodulating soybean genotypes*. Crop Sci. 18:1058–62.

Johnson, V. A., J. W. Schmidt, and P. J. Mattern. 1968. *Cereal breeding for better protein impact*. Econ. Bot. 22:16–25.

Jones, W. J., Jr., and H. A. Huston, 1914. *Composition of maize at various stages of growth*. Indiana Agr. Exp. Sta. Bull. 174.

Knowles, F., and J. E. Watkin. 1931. *The assimilation and translocation of plant nutrients in wheat during growth*. J. Agric. Sci. 21:612–37.

Kvien, C., G. E. Ham, and J. W. Lambert. 1978. *Improved recovery of introduced Rhizobium japonicum strains by field-grown soybeans*. Agron. Abstr. Amer. Soc. Agron.

Lal, Pyare, G. G. Reddy, and M. S. Modi. 1978. *Accumulation and redistribution of dry matter and N in triticale and wheat varieties under water stress conditions*. Agron. J. 70:623–26.

Lal, P., and K. C. Sharma. 1973. *Accumulation and redistribution of nitrogen and dry matter in dwarf wheat as influenced by soil moisture and nitrogen fertilization*. Ind. J. Agric. Sci. 43:486–92.

Lawn, R. J., and W. A. Brun. 1974. *Symbiotic nitrogen fixation in soybeans. I. Effect of photosynthetic source-sink manipulations*. Crop Sci. 14:11–16.

Leng, E. R. 1962. *Results of long-term selection for chemical composition in maize and their significance in evaluating breeding systems*. Zeitschrift für Pflanzenzüchtung 47:67–91.

McKee, H. S. 1962. *Nitrogen metabolism in plants*. Clarendon Press, Oxford.

McNeal, F. H., M. A. Berg, and C. A. Watson. 1966. *Nitrogen and dry matter in five spring wheat varieties at successive stages of development*. Agron. J. 58:605–8.

McNeal, F. H., G. O. Boatwright, M. A. Berg, and C. A. Watson. 1968. *Nitrogen in plant parts of seven spring wheat varieties at successive stages of development*. Crop Sci. 8:535–37.

Miller, Edwin C. 1939. *A physiological study of the winter wheat plant at different stages of its development*. Kansas Tech. Bull. 47.

National Academy of Sciences. 1972. *Genetic Vulnerability of Major Crops*. GPO, Washington, D.C.

Ohlrogge, A. J. 1960. *Mineral nutrition of soybeans*. Adv. Agron. 12:229–68.

Perez, C. M., G. B. Cagampang, B. V. Esmama, R. V. Monserrate, and B. O. Juliano. 1973. *Protein metabolism in leaves and developing grains of rices differing in grain protein content*. Plant Physiol. 51:537–42.

Pollmer, W. G., D. Eberhard, D. Klein, and B. S. Dhillon. 1979. *Genetic control of nitrogen uptake and translocation in maize*. Crop Sci. 19:82–86.

Purvis, A. C. 1972. *Effects of varying levels of carbon dioxide on the induction of nitrate reductase and accumlation of nitrate in corn (Zea mays L.) seedlings*. Ph.D. dissertation, University of Illinois, Urbana.

Rao, S. C., and L. I. Croy. 1972. *Protease and nitrate reductase seasonal patterns and their relation to grain production of 'high' versus 'low' protein wheat varieties*. J. Agric. Food Chem. 20:1138–41.

Rao, S. C., and D. W. Rains. 1976. *Nitrate absorption by barley. I. Kenetics and energetics*. Plant Physiol. 57:55–58.

Sayre, J. D. 1948. *Mineral accumulation in corn*. Plant Physiol. 23:267–81.

———. 1955. *Mineral nutrition in corn*. Pages 293–314 in G. F. Sprague, ed. *Corn and Corn Improvement*. Academic Press, New York.

Schrader, L. E., D. M. Peterson, E. R. Leng, and R. H. Hageman. 1966. *Nitrate reductase activity of maize hybrids and their parental inbred*. Crop Sci. 6:169–73.

Seth, J., T. T. Hebert, and G. K. Middleton. 1960. *Nitrogen utilization in high and low protein wheat varieties*. Agron. J. 52:207–9.

Spratt, E. D., and J.K.R. Gasser. 1970. *Effects of fertilizer-nitrogen and water supply on distribution of dry matter and nitrogen between different parts of wheat*. Can. J. Plant Sci. 50:613–25.

Van den Honert, T. H., and J. J. M. Hooymans. 1955. *On the absorption of nitrate by maize in water culture*. Acta Bot. Neerl. 4:376–84.

Vest, Grant, D. F. Weber, and C. Sloger. 1973. *Nodulation and nitrogen fixation*. Pages. 353–90 in B. E. Caldwell, ed. *Soybeans: Improvement, Production, and Uses*. Amer. Soc. of Agron., Madison, Wisconsin.

Wacek, T. J., and Winston J. Brill. 1976. *Simple, rapid assay for screening nitrogen-fixing ability in soybean*. Crop Sci. 16:519–23.

Warner, R. W., R. H. Hageman, J. W. Dudley, and R. J. Lambert. 1969. *Inheritance of nitrate reductase activity in Zea mays (L)*. Proc. National Acad. Sci. USA 62:785–92.

Westermann, D. T., and J. J. Kolar. 1978. *Symbiotic N_2 (C_2H_2) fixation by bean*. Crop Sci. 18:986–90.

Williams, R. F. 1955. *Redistribution of mineral elements during development*. Ann. Rev. Plant Physiol. 6:25–42.

Winter, F. L. 1929. *The mean and variability as affected by continuous selection for composition in corn*. J. Agric. Res. 39:451–76.

Woodworth, C. M., E. R. Leng, and R. W. Jugenheimer. 1952. *Fifty generations of selection for oil and protein content in corn*. Agron. J. 44:60–65.

Zary, K. W., J. C. Miller, Jr., R. W. Weaver, and L. W. Barnes. 1978. *Intraspecific variability for nitrogen fixation in southern pea (Vigna unguiculata (L) Walp*. J. Amer. Soc. Hort. Sci. 103:806–8.

Zieserl, John F., and R. H. Hageman. 1962. *Effect of genetic composition on nitrate reductase activity in maize*. Crop Sci. 2:512–15.

16] Source-Sink Relationships and Flowering

by R. M. SACHS* and
W. P. HACKETT*

ABSTRACT

Studies of phloem transport and partitioning of assimilates are the basis for source-sink reasoning. Mature leaves are the major sources (net exporters of assimilates), whereas all regions of relatively intense meristematic activity are sinks (net importers of assimilates). Source strength is a function of all factors affecting net photosynthesis (leaf area, irradiance, CO_2 concentration, stomatal aperture, respiratory costs); sink demand is a function of factors that enhance or depress metabolic activity (namely hormones, temperature, and oxygen concentration). Assimilate partitioning, or nutrient diversion, from sources to sinks is then a function of the parameters cited. Physiologists and geneticists have used source-sink reasoning to explain fruit set and abscission, root:shoot ratios, axillary bud growth, and harvest indices. We have used a nutrient diversion hypothesis to account for many, varied experimental results as they relate to the control of flowering. We suggest that this hypothesis is as yet indistinguishable experimentally from one that proposes that flowering is regulated by specific translocatable morphogens. We believe that sink activiation, e.g., through increased assimilate flow to an apical meristem and synthesis of substrate-limited enzymes, would appear in traditional analyses as a specific florigen. Also, activation of competitive sinks, e.g., roots, young leaves, or nonreceptive shoot apical meristems, would appear as a translocatable inhibitor of floral initiation.

INTRODUCTION

Mason and Maskell (1928) were probably the first to use the terms *source* and *sink* in describing the observed movement of sugars in cotton plants. Since

*Department of Environmental Horticulture, University of California, Davis, CA 95616.

STRATEGIES OF PLANT REPRODUCTION (BARC Symposium number 6—Werner J. Meudt, ed.)
Allanheld, Osmun, Totowa

then there has been general agreement that sources are exporters and sinks are importers of assimilates. Mature leaves are the major net exporters, whereas all regions of relatively intense metabolic activity, e.g., root and shoot apical meristems, young leaves, fruit and storage tissues, etc., are net importers of assimilates. Source strength is a function of all factors affecting net photosynthesis, e.g., leaf area, irradiance, CO_2 level, stomatal aperture, and respiratory costs, and sink demand is a function of factors that enhance or depress metabolic activity, e.g., temperature, oxygen levels, and hormones. Resource allocation, or nutrient diversion, in plants is then an accounting of assimilate movement from sources to sinks as a function of the parameters cited. Source-sink reasoning is used to explain fruit set and abscission, root:shoot ratios, and axillary bud growth (Phillips 1975). Crop performances have been analyzed to assess whether economic yield is limited by source strength or sink demand or perhaps some combination of the two. Useful accounts of some source-sink reasoning as they apply to crop yield and plant development are found in Warren-Wilson (1972), Cook and Evans (1976), Wardell (1976), Wareing (1978), and by Woolhouse (see Chapter 13). On several occasions we have employed these concepts in a nutrient diversion hypothesis to account for many experimental results relating to the control of flowering (Sachs 1977; Sachs and Hackett 1977; Sachs et. al. 1979). We have suggested that this hypothesis provides a valuable contrast to, but has so far been experimentally indistinguishable from, the hypothesis that proposes that flowering is regulated by specific translocatable morphogens.

PHOTOSYNTHESIS AND THE CONTROL OF FLOWERING—MODIFYING SOURCE STRENGTH

Photosynthesis has been assigned an important role in flowering, dating back to the first studies of Georg Klebs in 1913. High irradiances often override and replace photoperiodic signals. In both daylength-sensitive and day-neutral species, flower initiation and development to anthesis have higher irradiance (and presumably photosynthetic) requirements than continued vegetative development (see extensive review by Sachs et al. 1979). Particularly impressive studies of floral initiation as a function of irradiance have been performed by Wardell (1976) with tobacco, by Friend, Deputy, and Quedado (1978) with *Brassica,* by Bodson et. al. (1977) with *Sinapis,* by Hackett and Sachs (1966) with *Bougainvillea,* and by Cockshull (1972) with chrysanthemum in inductive and non-inductive conditions. Studies with chrysanthemum have led to the development of a comprehensive photosynthesis-based model of floral initiation related to growth rate of the apical meristem (Charles-Edwards et al. 1979). In this species, apical dome size is critical for the transition from vegetative to reproductive development (Horridge and Cockshull 1979); according to the model, flowering is governed almost entirely by photosynthesis, which promotes apical dome growth, and an inhibitor, which controls leaf initiation, not flowering per se. An important consequence of the studies with chrysanthemum is that one understands better why the ability to control flowering by chemical or environmental parameters is often a function of irradiance and temperature and other factors affecting net photosynthesis. For example, a recent study with chrysanthemum shows that the night interruption irradiance required for preventing floral initiation in chrysanthemum is a direct function of daytime irradiance (Sachs, Kofranek, and Kubota 1980). If the apical dome size photosynthesis model is correct, then light interruption of the dark period diverts nutrients from the apical dome so that it does not grow as rapidly as it would in short days.

The idea that source strength is important is indicated by the fact that flowering in *Bougainvillea* is a function of leaf area (Ramina, Hackett, and Sachs 1979) and that sugar levels can be adjusted to replace or to reduce daylength or low-temperature requirements for flower initiation in tissue explants of chicory (Nitsch et al. 1967). Direct evidence for increased ^{14}C transport to apical buds suggests that increased assimilate flow (T'se et al. 1974) to or sugar concentration at receptor meristems (Bodson 1977) is an important component of induction, evocation, and development to anthesis.

Proponents of a specific floral hormone argue that increased photosynthesis enhances phloem loading and transport of the hormone with photosynthetic assimilates from induced leaves to receptor meristems. Photosynthetic assimilates are, in this view, assigned a supportive but not determining role in floral evocation and development. Great weight must be given to the findings for species such as *Xanthium* and *Pharbitis* in which plant development is clearly not limited by photosynthesis and in which one inductive cycle promotes or a short exposure to photomorphogenetically active light inhibits flowering. What could be translocated from leaf to meristem (other than some specific morphogen) that would cause a rapid, dramatic change in developmental program of the latter? We argue here that one must consider agents that (a) reduce sink demand of young leaves surrounding the apical meristem and/or the subapical meristematic tissues just below the apical meristem, and/or (b) agents that promote sink activity of the apical meristem proper. In the first case, these agents are most likely to appear as floral stimuli in timed defoliation or grafting experiments. Conversely, in the second case agents that promote activity of competing sinks will appear as inhibitors of floral initiation in traditional analyses. With either interpretation, however, induction becomes a means of modifying sink strength and, hence, resource allocation.

TIMING OF ASSIMILATE SUPPLY

Timing of photosynthetic input is critical and particularly perplexing in daylength-sensitive plants in which floral initiation is promoted by a single inductive cycle. High-intensity light of ca 31000 lux (96 J m^{-2} s^{-1}) applied during the second 8-hr period of an inductive long day for *Sinapis alba* inhibits flowering, compared to an 8-hr exposure to 8000 lux (Bodson et al. 1977). If CO_2 is removed during this period, high-intensity light does not inhibit flowering, indicating that photosynthetic assimilates are involved. An exposure of 31000 lux during the first 8-hr period of an inductive long day does not inhibit and may in fact promote flowering, compared to irradiance with 8000 lux. Thus photosynthesis at high rates *at the wrong time* inhibits flowering. Is there a coordination of quantity of assimilate transport with events required for evocation in the apical meristem? Fontaine (1972) and Miginiac (1972) have evidence for *Anagallis* and *Scrofularia*, respectively, that the apical meristem is more receptive to stimuli for floral evocation at certain plastochron stages than at others. If this "morphogenetic window" for floral initiation and primordial development is related to apical dome diameter, there is then additional circumstantial evidence for the nonspecific inhibitor of flowering proposed by Charles-Edwards et al. (1979) for chrysanthemum, a species that requires several short days (up to 14 in some cultivars) for most rapid floral initiation and development. The significance of timing of assimilate supply to the apical meristematic region may be accounted for in two as yet indistinguishable ways: assimilate supply during one part of the plastochron may promote leaf initiation and leaf primordia development (competing sink-type inhibition) or, at another stage, promote growth of the apical meristem proper. This latter idea suggests the notion of a *meristeme d'attente*, a quiescent region of the apical meristem that must be activated for initiation of flowers, and fits a nutritional hypothesis very well (Sachs 1977). A slight modification of this idea to include as well minimum apical volume and competing roles of leaf primordia (Charles-Edwards et al. 1979) also fits data for timing of initiation and assimilate flux. Lyndon (1981 personal communication) has proposed that the relative growth rate of the apical dome vis-à-vis new lateral primordia is the key parameter in floral initiation, rather than the absolute size of the dome.

SINK DEMAND—RESOURCE ALLOCATION—NUTRIENT DIVERSION

Sink demand is often increased by exogenously applied auxins, gibberellins and, even more ubiquitously, by cytokinins (Patrick 1976; Letham, Goodwin, and Higgins 1978). There is evidence, too, for endogenous transport of these substances in the phloem. Hence, it is reasonable to adopt hypotheses that suggest that sink demand may be increased by some of the known hormones at the same time that source strength increases. In fact, in any organism one might suggest that the signal for increased metabolism at growth centers (sinks) ought to precede or accompany increased metabolite supply. Cytokinin-induced promotion of flowering in several species—e.g., chrysanthemum (Pharis 1972), W-38 tobacco (Sachs et al. 1979), *Pharbitis* (Ogawa and King 1980) *Bougainvillea* (T'se et al. 1974), *Vitis* (Srinivasan and Mullins 1979)—may fit the hypothesis that increased sink demand constitutes a part of the inductive signal. Ogawa and King (1979, 1980) interpret their findings of cytokinin-induced promotion or inhibition of floral initiation in *Pharbitis* as a result of nonspecific promotion or interference with transport of the floral stimulus that is carried (passively) with assimilates in the phloem. This is a floral stimulus diversion rather than a nutrient diversion hypothesis. The inhibitory role of non-induced leaves, inserted between the receptor meristem and the leaf, is a classic example of floral stimulus diversion (King and Zeevaart 1973; Zeevaart, Brede, and Cetas 1977).

There is evidence from studies on metabolism and distribution of naturally occurring growth regulators that photoperiodic and low-temperature induction may operate at the growth regulator level (Van Staden and Wareing 1972; Henson and Wareing 1974; Metzger and Zeevaart 1980). Owing to improved methods of isolation, detection, and identification of growth substances and their derivatives, further studies are required on compartmentation and derivatization of hormones before the results of earlier papers can be fully interpreted (see discussion in Bernier, Kinet, and Sachs 1981, Vol. 2). Nevertheless, it seems likely that distribution of sink promoters and inhibitors is affected by environmental signals that affect flower initiation.

Sink activation. What is causal, at the biochemical level, to increased sink demand of target meristems (or decreased sink demand in competing meristematic tissues)? Should one look for increased synthesis of macromolecules in general, or are there perhaps key, regulatory, enzymatic steps that must be activated to enable subsequent increased syntheses? One energy transduction system, the adenylate energy charge, has received increasing attention in many physiological systems, including those pertaining to the regulation of flowering in *Chenopodium rubrum*, SDP (Wagner 1976, 1977). In this species the compartmentation of adenylate kinase (AK) isozymes is modified by short-day induction such that mitochondrial AK is decreased relative to chloroplast AK. Although these studies were with whole seedlings and were not designed to determine AK compartmentation in meristems as distinct from cotyledons, they show that changes in the adenylate energy charge are likely to occur as a result of photoperiodic induction; thus reactions regulated by energy charge will be altered as well.

Another suggestive study by Auderset et al. (1980) shows increased glucose-6-phosphate dehydrogenase (GDH) activity in apices of *Spinacia oleraceae*, a LDP, during the early hours of long-day induction. GDH increases of up to 60% were observed after 7 to 9 hr of the supplemental light period of the first LD. This increase in GDH activity is an indicator of increased metabolism via the pentose phosphate pathway (PPP), of great significance owing to resulting enhanced capacity for reduction of NADP to NADPH. NADPH is the key reductant in many biosynthetic pathways, including those for the synthesis of long-chain fatty acids and steroids, both required in relatively larger amounts for synthesis of membranes, which in turn do increase in evoked meristems (Havelange, Bernier, and Jacqmard 1974; Auderset and Greppin 1977; Havelange 1980). Thus, GDH activity as a marker for the PPP may serve as an indicator of "sinkness" as well.

Of particular significance for our purposes are the findings that GDH activity may be regulated in part by substrate availability. Glucose and ribose will promote GDH activity in *Wolffia* by as much as 50%, whereas galactose, rhamnose, and lactose may depress activity (Eichorn and Augsten 1977). Thus, in *Spinacia* and other species as well, the supply of assimilates, particularly certain sugars, to the shoot apical meristem may constitute a fundamental signal for regulation of the PPP. Through substrate-induced enzyme synthesis, quantitative variations in assimilate supply can be amplified in effect and appear to "cause" qualitative changes.

Competing sinks. As noted earlier, the nutrient diversion hypothesis predicts that assimilate limitations at a target meristem can be induced by enhanced activity of competing sinks, thereby inhibiting floral initiation. Young leaf removal in tomato (DeZeeuw 1956), *Bougainvillea* (T'se et al. 1974), W38 tobacco (Wardell 1976), and *Sinapsis* (G. Bernier, personal communication) promote floral initiation and development in conditions not optimal for flowering. Young leaves are net importers of assimilates from mature leaves (Larson, Isebrand, and Dickson, 1980; Turgeon 1980), and rank high as competitive sinks to the target meristems (T'se et al. 1974). We suggest that GA-induced inhibition of inflorescence differentiation in *Bougainvillea* may result from GA-induced activation of competing meristems, particularly of the rib meristems (e.g., those active in stem elongation and leaf blade elongation). It is likely that young leaves are sources of gibberellins, auxins, and perhaps other promoters of competing sinks. Henson and Wareing (1977) show, too, that leaf removal increases cytokinin activity in buds of *Xanthium;* indicating that leaves, regardless of age, may be considered competing sinks for root-derived cytokinins.

Root initiation and particularly root expansion inhibits floral initiation in the LDP *Anagallis arvensis,* particularly if the onset of rapid root elongation coincides with LD induction (Bismuth et al. 1979). These recent studies fit well with Miginiac's long-held views of the importance of correlations for floral initiation (Miginiac 1978) and those expressed more fully in a review by Krekule et al. (1979). Recent studies by McDaniel (1980) with W38 tobacco, showing that root induction delays flowering, can be interpreted as floral inhibition due to assimilate diversion to the root system and not necessarily due to a direct influence of something derived from the roots. Irradiances are not given in McDaniel's paper; hence it is not possible to estimate accurately source limitations to flowering. But it may be reasonable to assume from the minimum nodes to flower that the plants were source-limited. Our research with W38 tobacco (Sachs et al. 1979) under 4500 lux continuous irradiance shows flower initiation at approximately 35 nodes, essentially the same as in McDaniel's studies, whereas at higher irradiance flower initiation occurred at node 22. In our studies, young leaves were removed continuously after four fully expanded leaves had formed, and leaf area was fixed at about 1800 cm^2 per plant. This young leaf removal procedure, as Wardell (1976) found, is a significant factor in influencing the node to flowering in W38 grown at low irradiances. In McDaniel's studies young leaves were not removed from the developing shoots; hence, if the young leaves act as competing sinks with the apical meristem proper, it is possible that McDaniel's plants were source-limited with respect to flower initiation. Under such circumstances, it is not surprising that root initiation and rapid root growth will further deny the shoot apical meristematic region of assimilates adequate for floral initiation.

CONCLUSIONS

Our intent in this chapter has been to emphasize the physiological significance of intraplant competition for assimilates in matters of floral initiation and development. Thus, we attempted to draw close parallels with source: sink studies related to fruit set and fruit thinning, branching and apical dominance, and root: shoot ratios. The value of source: sink reasoning in studies of reproductive development is that it presents an alternate hypothesis to the specific floral hormone(s) hypothesis. It is at present impossible to distinguish between these two hypotheses, but nutrient diversion accounts better for correlative inhibition and the apparently important, nonspecific, indirect roles of many hormones and other substances. Nutrient diversion may well be floral hormone diversion, but in the absence of direct evidence for organ-specific hormones in plants, or of specific gene derepressors, it appears more likely that the signal for changed morphogenetic patterns will be found in the response of shoot and root apical meristems to general substrate levels (source strength) and growth promoters or inhibitors (sink activators/inhibitors). The proposition that terrestial autotrophic plants modify their behavior to optimize assimilation relative to evapotranspiration (Cowan and Farquhar 1977) suggests that reproductive development, too, will be strongly linked to these fundamental parameters which are in turn of paramount importance in source: sink relationships.

At the core of source-sink reasoning as it applies to problems of morphogenesis is the notion that a morphogenetic event will occur if the tissues receive the substrates or cofactors for growth. In this view genetic information will be expressed in most tissues if nutrients, in the broadest sense, are ample. Thus, hormonal action in plants is not at the depressor level but at the growth-modulating level, from whence gene regulation follows. In the case of floral evocation we propose that some or all of the apical meristem is relatively starved for assimilates (or ATP or NADH) and hence cannot go through the accelerated growth stages required for reproductive development. Further research on transport and distribution of ^{14}C-labeled assimilates would aid in detecting source: sink relationships between donor leaves and receptor buds in cases of graft-transmissible inhibitors and promoters. By similar techniques the action of competing sinks, e.g., root systems, or assimilate movement can be detected. Radioautographic studies are required to detect partitioning among meristematic tissues, such as newly initiated leaves and subapical rib meristems on the one hand and the apical dome on the other. Little is known about the qualitative nature of assimilates arriving at reproductive meristems, e.g., whether sugars, amino acids, ATP, or some complex mixture of the known hormones and other compounds, which occur in low concentration as well, must be considered as part of the nutritional fare of meristems. Assessment of the physiological significance of the various components can most readily be made by in vitro feeding of potentially reproductive meristems (or other responsive tissue).

LITERATURE CITED

Auderset, G., and H. Greppin. 1977. *Effet de l'induction florale sur l'evolution ultrastructurale de l'apex caulinaire de Spinacia oleracea Nobel.* Protoplasma 91:281–301.

Auderset, G., P. B. Gahan, A. L. Dawson, and H. Greppin. 1980. *Glucose-6-phosphate dehydrogenase as an easy marker of floral induction in shoot apices of Spincia oleracea var. Nobel.* Plant Sci. Let.

Bernier, G., J. M. Kinet, and R. M. Sachs. 1981. *The Physiology of Flowering,* vols 1 and 2. CRC Press, Boca Raton, Florida.

Bismuth, F., J. Brulfert, and E. Miginiac. 1979. Mise à fleurs de l'*Anagallis arvensis* L. en cours de rhizogenèse. Physiol. Vég. 17:477–82.

Bodson, M. 1977. *Changes in the carbohydrate content of the leaf and the apical bud during transition to flowering.* Planta 135:19–23.

Bodson, M, R. W. King, L. T. Evans, and G. Bernier. 1977. *The role of photosynthesis in flowering of the long day plant Sinapis alba.* Austral. J. Plant Physiol. 4:467–78.

Charles-Edwards, D. A., K. E. Cockskull, J. S. Horridge, and J. H. M. Thornley. 1979. *A model of flowering in Chrysanthemum.* Ann. Bot. 44:557–66.

Cockshull, K. E. 1972. *Photoperiodic control of flowering in the chrysanthemum.* Pages 235–50. *Crop Processes in Controlled Environments,* A. R. Rees, K. E. Cockshull, D. W. Hand, and R. G. Hurd; eds. Academic Press, New York.

Cook, M. G., and L. T. Evans. 1976. *Effects of sink size, geometry and distance from source on the distribution of assimilates in wheat.* Pages 393-400 in *Transport and Transfer Processes in Plants;* I. F. Wardlaw and J. B. Passioura; eds. Academic Press, New York.

Cowan, I. R., and G. D. Farquhar. 1977. *Stomatal function in relation to leaf metabolism and environment.* Symp. Soc. Exp. Biol. 31:471–505.

DeZeeuw, D. 1956. *Leaf-induced inhibition of flowering in tomato.* Proc. Kon. Ned. Adad. Wetensch. Amsterdam 59:535–40.

Eichhorn, M., and H. Augsten. 1977. *Der Einfluss löslicher Kohlenhydrate auf die Aktivität der glucose-6-phosphat-Dehydrogenase verschiedenaltriger Wolffia—Populationer unter Berücksichtiging von energy charge, 0_2-Austach und Pyruvat-Gehalt.* Z. Pflanzen physiol 84:37-48.

Fontaine, D. 1972. *Incidence du stade de developpement de point vegetatif de l'Anagallis arvensis L. au cours du plastochrone, sur la mise a fleurs de plantes soumises a des conditions limitantes d'induction.* C. R. Acad. Sci. Paris 274:2984–87.

Friend, D. F. G., J. Deputy, and R. Quedado. 1978. *Photosynehetic and photomorphogenetic effects of high photon flux densities on the flowering of two long day plants, Anagallis arv̌ensis and Brassica campestris.* Pages 59–72 *Photosynthesis and Plant Development;* R. Marcelle, H. Clijsters, and M. Van Poncke, eds. W. Junk, The Hague.

Hackett, W. P., and R. M. Sachs. 1966. *Flowering in Bougainvillea "San Diego Red."* Proc. Amer. Soc. Hort. Sci. 88:606–12.

Havelange, A. 1980. *The quantitative ultrastructure of the meristematic cells of Xanthium strumarium during the transition to flowering.* Amer. J. Bot. 67:1171–78.

Havelange, A., G. Bernier, and A. Jacqmard. 1974. *Descriptive and quantitative study of ultrastructural changes in the apical meristem of mustard in transition to flowering. II. The cytoplasm, mitochondria and proplastids.* J. Cell Sci. 16:431–32.

Henson, I. E., and P. F. Wareing. 1974. *Cytokinins in Xanthium strumarium: A rapid response to short day treatment.* Physiol. Plant. 32:185–88.

Henson, I. E., and P. F. Wareing. 1977a. *Cytokinins in Xanthium strumarium L; Some aspects of the photoperiodic control of endogenous levels.* New Phytol. 78:35–45.

Henson, I. E., and P. F. Wareing. 1977b. *An effect of defoliation on the cytokinin content of buds of Xanthium strumarium*. Plant Sci. Let 9:27–31.

Horridge, J. S., and K. E. Cockskull. 1979. *Size of the Chrysanthemum shoot apex in relation to inflorescence initiation and development*. Ann. Bot. 44:547–56.

King, R. W., and J. A. D. Zeevaart. 1973. *Floral stimulus movement in Perilla and flower inhibition caused by non-induced leaves*. Plant Physiol. 51:727–38.

Krekule, J. 1979. *Stimulation and inhibition of flowering: Morphological and physiological studies*. Pages 19–57 in *La Physiologie de la floraison*, R. Jacques and P. Champagnat, eds. CNRS No. 285. Paris.

Larson, P. R., J. G. Isebrands, and R. E. Dickson. 1980. *Sink to source transition of Populus leaves*. Ber. Deutsch. Bot. Ges. 93:79–90.

Letham, D. S., P. B. Goodwin, and T. J. V. Higgins. 1978. *Phytohormones and Related Compounds. A Comprehensive Treatise*. Vol 1: *The Biochemistry of Phytohormones and Related Compounds*. Elsevier, Amsterdam.

McDaniel, C. N. 1980. *Influence of leaves and roots on meristem development in Nicotiana tabacum L. cv. Wisconsin* 38. Planta 148:462–67.

Mason, T. G., and E. J. Maskell. 1928. *Studies on the transport of carbohydrates in the cotton plant*. II. *The factors determining the rate and the direction of movement of sugars*. Ann. Bot. O.S. 42:571–636.

Metzger, J. D., and J. A. D. Zeevaart. 1980. *Comparison of the levels of 6 endogenous gibberellins in roots and shoots of spinach in relation to photoperiod*. Plant Physiol. 66:679–83.

Miginiac, E. 1972. *Cinetique d'action comparée des racines et de la kinetine sur le developpement floral desbourgeons cotyledonaires chez le Scrofularia arguta Sol.* Physiol. Veg. 10:627–36.

———. 1978. *Some aspects of regulation of flowering: role of correlative factors in photoperiodic plants*. Bot Mag. Tokyo. Special Issue 1:159–73.

Nitsch, J. P., C. Nitsch, C. Rossini, F. Ringe, and H. Harada. 1967. *L'induction florale in vitro*. Pages 369–82. in *Cellular and Molecular Aspects of Floral Induction*, G. Bernier, ed. Longman Group L&D, London.

Ogawa, Y., and R. W. King. 1979. *Indirect action of benzyladenine and other chemicals on flowering of Pharbitis nil Chois. Action by interference with assimilate translocation from induced cotyledons*. Plant Physiol. 63:643–49.

———. 1980. *Flowering in seedlings of Pharbitis nil induced by benzyladenine applied under a non-inductive daylength*. Plant and Cell Physiol. 21:1109–16.

Patrick, J. W. 1976. *Hormone-directed transport of metabolites*. Pages 433–46 in *Transport and Transfer Processes in Plants;* I. F. Wardlaw and J. B. Passioura, eds. Academic Press, New York.

Pharis, R. P. 1972. *Flowering of chrysanthemum under non-inductive long days by gibberellins and by N^6-benzyladenine*. Planta 105:205–12.

Phillips, I. D. J. 1975. *Apical dominance*. Ann. Rev. Plant Physiol. 26:341–67.

Ramina, A., W. P. Hackett, and R. M. Sachs. 1979. *Flowering in Bougainvillea: a function of assimilate supply and nutrient diversion*. Plant Physiol. 64:810–13.

Sachs, R. M. 1977. *Nutrient diversion: an hypothesis to explain chemical control of flowering*. Hort. Sci. 12:220–22.

Sachs, R. M., 1979. *Metabolism and energetics in flowering*. Pages 168–208 in *Physiologie de la floraison*, R. Jacques and P. Champagnat, eds. CNRS No. 285. Paris

Sachs, R. M., and W. P. Hackett. 1969. *Control of vegetative and reproductive development in seed plants*. Hort. Sci. 4:103–7.

———. 1977. *Chemical control of flowering*. Acta Horticulturac 68:29–49.

Sachs, R. M., W. P. Hackett, A. Ramina, and C. Maloof. 1979. *Photosynthetic assimilation and nutrient diversion as controlling factors in flower initiation in Bougainvillea 'San Diego Red' and Nicotiana tabacum cv. Wis. 38*. Pages 95–102 in *Photosynthesis and Plant Development*, R. Marcelle, H. Clijsters, and M. Van Poucke, eds. Dr. W. Junk. The Hague.

Sachs, R. M., A. M. Kofranek, and J. Kubota. 1980. *Radiant energy required for the night-break inhibition of floral initiation is a function of daytime light input in Chrysanthemum*. Hort. Sci. 15:609–10.

Srinivasan, C., and M. G. Mullins. 1979. *Flowering in Vitis: Conversion of Tendrils into inflorescences and bunches of grapes*. Planta 145:187–92.

T'se, A., A. Ramina, W. P. Hackett, and R. M. Sachs. 1974. *Enhanced inflorescence development in Bougainvillea 'San Diego Red' by removal of young leaves and cytokinin treatments*. Plant Physiol. 54:404–7.

Turgeon, R. 1980. *The import to export transition; experiments on Coleus blumei*. Ber. Deutsch Bot. ges. 93:91–97.

Van Staden, J., and P. F. Wareing. 1972. *The effect of photoperiod on levels of endogenous cytokinins in Xanthium strumarium*. Physiol. Plant. 27:331–37.

Wagner, E. 1976. *The nature of photoperiodic time measurement: energy transduction and phytochrome action in seedlings of Chenopodium rubrum*. Pages 419–43 in *Light and Plant Development;* H. Smith, ed. Butterworth, London.

———. 1977. *Molecular basis of physiological rhythms*. Pages 33–72 in *Integration of Activity in the Higher Plant,* D. H. Jennings, ed. Soc. Exptl. Biol. Symp. 31.

Wardell, W. C. 1976. *Floral activity in solutions of deoxyribonucleic acid extracted from tobacco stems*. Plant Physiol. 57:855–61.

Wareing, P. F. 1978. *Plant development and crop yield*. Pages 1–17 in *Photosynthesis and Plant Development,* R. Marcelle, H. Clijsters, and M. Van Poucke, eds. Dr. W. Junk, The Hague.

Warren-Wilson, Jr. 1972. *Control of crop processes*. Pages 7-30 in *Crop Processes in Controlled Environments,* A. R. Rees, K. E. Cockskull, D. W. Hand, and R. G. Hurd, eds. Academic Press, New York.

Zeevaart, J. A. D., J. M. Brede, and C. B. Cetas. 1977. *Translocation patterns in Xanthium in relation to long day inhibition of flowering*. Plant Physiol. 60:747–53.

17] Managing Flowering, Fruit Set, and Seed Development in Apple with Chemical Growth Regulators

by MAX W. WILLIAMS*

ABSTRACT

The management of apple fruit production is an effort to control the amount of flowering, fruit set, and seed development to maintain high yields of quality fruit. Sophisticated chemical methods have been developed to prevent alternate year or biennial bearing in fruit trees. These chemical bioregulants affect the endogenous hormone balance by altering biochemical pathways and creating stress in the tree, which shifts the balance between vegetative growth and reproductive growth in favor of flowering. In the process, seed numbers and the amount of fruit set are either reduced or increased according to treatment. There is an antagonistic relationship between flowering and seed development. Trees carrying fruit with a high complement of seeds have less tendency to initiate flowers than those with fewer seeds. Seeds produce gibberellins (GA) and auxins (IAA), which promote vegetative and fruit growth at the expense of flowering. Application of chemical growth retardants reduces GA and IAA production, which suppresses vegetative growth and increases flower initiation.

INTRODUCTION

Horticulturists use a number of cultural and chemical practices to influence the balance between the vegetative and reproductive growth of fruit trees. The challenge is to regulate tree growth so that high, sustained, annual production of quality fruit is maintained.

The need for consistent annual flowering is apparent. From 30% to 40% of the growing points (spurs) should flower each season. Most of the flower initiation in apple trees occurs 3 to 6 weeks after bloom. A small amount,

*USDA-ARS, Tree Fruit Research Laboratory, Wenatchee, Washington 98801.

STRATEGIES OF PLANT REPRODUCTION (BARC Symposium number 6—Werner J. Meudt, ed.)
Allanheld, Osmun, Totowa

especially on the current season's growth, occurs throughout the season after terminal growth has ceased. On most bearing trees, 90% of the bloom is born on spurs two years of age and older.

Just before the early flower initiation period, much cell division occurs as new growth in many parts of the tree. Final flower devlopment from the previous year's floral initiation is at its peak, followed by pollination and fertilization of the developed flowers, which stimulates rapid cell division in the new fruitlets where ovule and seed development is proceeding. At the same time, new leaves are forming and active shoot growth begins. These closely related and concurrent events are a vitalizing influence at the flowering site, but are collectively devitalizing on a whole tree basis because of the competitive "sink" effect. This early period preceding flower initiation is therefore a very important and "critical time" (Williams 1973). Strong, active, vegetative growth and early fruit development during this period interfere with the endogenous growth regulator balance needed for flower initiation.

Seed development also has a profound influence on the delicate balance between vegetative growth and flower initiation. The presence of seeds in apple fruit favors vegetative growth. Spur-type Golden Delicious trees with a moderate crop of fully seeded fruit produce a considerable amount of terminal shoot growth. Similar trees without a seeded crop form very little terminal shoot extension growth. There is much evidence that gibberellic acid (GA) production, stimulated by seed formation, favors shoot elongation (Dennis and Nitsch 1966; Luckwill, Weaver, and MacMillan 1969; Luckwill 1974; Grochowska and Karaszewska 1978; and Hoad 1978). Profuse seed development and GA production not only stimulate vegetative growth but also prevent flower formation, which leads to biennial bearing in apple trees.

BIENNIAL BEARING

Biennial or alternate year bearing is a major problem in apple orchards throughout the world. The cyclic series of events in biennial bearing are shown in Figure 17.1. A "snowball" or 100% bloom leads to a heavy fruit set that reduces the number of resting spurs and prevents flower initiation for the next year. The absence of flowers means no fruit are set, and the resting spurs form flowers and a "snowball" bloom for the following year. These events continue to be repeated, unless the initial fruit set is partly reduced by adverse weather or by artificial means, such as early hand or chemical thinning.

Before chemical thinning was practiced, an entire production area was often in a biennial condition, especially when a general spring frost severely reduced the fruit set. The pattern thus established prevailed until changed by another damaging frost, or by adequate early hand thinning. The cyclic pattern of apple production is shown by comparison of yields of fruit in the USA between 1935 and 1973 (Fig. 17.2). Chemical thinning began commercially about 1949, and the yields after 1949 are more consistent than in earlier years because of the use of chemicals to stimulate abscission of the excess fruit. The use of bioregulants to control alternation has become a standard orchard practice,

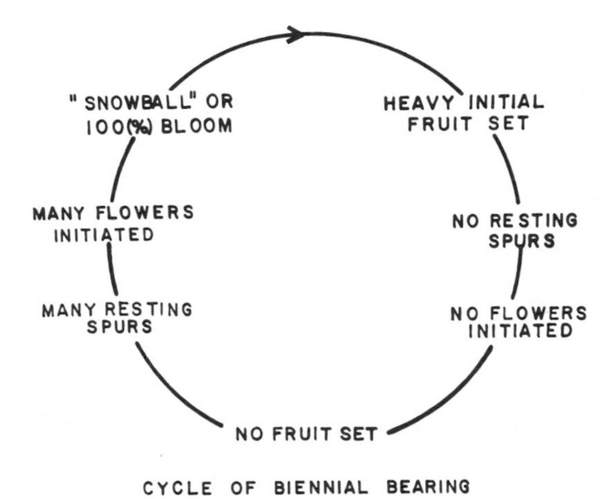

Figure 17.1. Cycle of biennial bearing

and progressive fruit growers utilize chemicals to reduce labor costs and improve production.

Tree vigor is a major factor in biennial bearing. Trees deficient in N tend to follow an alternate bearing habit more than those in good vigor. Yearly application of moderate amounts of N keeps trees vigorous and helps offset biennial bearing. In Washington State, the ideal N level for Golden Delicious leaves sampled in mid-July to mid-August is about 2.0% on a dry weight basis (Williams and Billingsley 1974). Red Delicious leaves should contain about 2.5% N. The ideal N level varies with the cultivar, the rootstock, the soil, the climate, and the quality of fruit desired. Other cultural management tools such as pruning, limb girdling, and irrigation affect annual cropping but are difficult to control and are insufficient to offset the effect of major spring frosts on biennial bearing.

The number of flowering spurs and the amount of initial fruit set remaining on the tree until the "June drop" period influence annual bearing. The number of seeds formed in the fruit also has a pronounced affect on flower initiation. Chan and Cain (1967) found a highly significant correlation between seed content and the number of spurs initiating flowers. Golden Delicious fruit often contain a full complement of seeds, especially when cross-pollinated. Seeds produce large quantities of gibberellins that favor vegetative growth and interfere with flower formation (Dennis and Nitch 1966; Luckwill 1974). The same is true for pears; Huet (1973) reported that spurs with seedless fruit had a greater tendency to flower the next year than spurs with seeded fruit.

The time of fruit removal is very important for flower initiation and a good bloom the following year. The amount of return bloom is increased by removing fruit early, especially with cultivars with high seed numbers in the fruit. Harley et al. (1942) thinned Yellow Newtown trees to a ratio of 70 leaves per fruit at various times after bloom and found a close correlation between the time of thinning and the amount of return bloom (Fig. 17.3). Yellow Newtown

Figure 17.2. Percentage of growing points forming blossoms on main leaders of a biennial-bearing 'Yellow Newtown' tree following periodic fruit thinning to one per 70 leaves at 33, 38, 54, 66, and 76 days after full bloom in a bearing year at Wenatchee, Washington. (From Harley et al. 1942.)

is subject to biennial bearing, and usually the fruits have a high number of seeds. With most cultivars, thinning to 70 leaves per fruit constitutes overthinning. Even with overthinning, Harley et al. (1942) obtained less than 50% bloom return on the limbs thinned at 33 days from full bloom.

Golden Delicious trees also require early thinning to assure a good return bloom. If maximum yields are obtained, a bearing Golden Delicious tree grown under irrigation will annually bear from 40 to 50 fruit per 100 blossoming clusters. The leaf-to-fruit ratio, established early in the season, should be about 30 to 40 per fruit. The leaf efficiency appears to vary greatly with the cultivar; Winesap, for instance, can be kept annual with a lower leaf-to-fruit ratio. As the trees age and lose vigor, the excess fruit must be removed soon after bloom, and the leaf-to-fruit ratio should be increased. Harley et al. (1942) concluded that if thinning is delayed until after the June "drop," very few blossom buds are initiated.

Early removal of excess fruit (a) reduces the number of seeds producing gibberellins and other endogenous hormones, (b) reduces competition for photosynthates and nutrients, (c) improves growth and ultimate size of the remaining fruit, and (d) removes small, blemished, and poorly shaped fruit. Thinning not only affects the quality of fruit harvested, but more importantly determines the amount of crop the next year. The value of early thinning in overcoming biennial bearing is often underestimated; it is very significant because it affects the crop for two seasons.

Early fruit removal is the main way to overcome biennial bearing. All fruit should be removed from about half of the fruiting spurs, rather than reduce the fruit load to one per spur. Golden Delicious fruit grown under irrigation in the

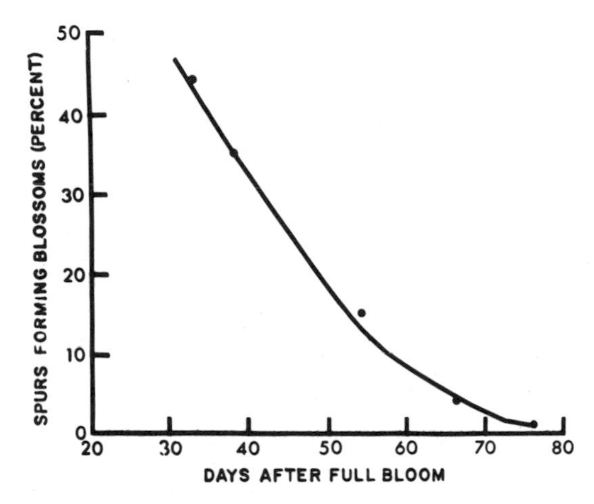

Figure 17.3. Total apple production in United States from 1935 to 1973. Note alternate-year bearing before 1949.

western part of the USA have better quality if two or three fruit remain on a spur. When only one fruit is on a spur, it usually is too large and does not keep well. With more fruit per cluster, it is possible to obtain high yields and still have many resting spurs for the next year's crop.

When hand labor is available and trees are small, entire flower clusters can be removed at bloom time. Spacing the clusters about 25cm apart in all directions will result in many resting spurs. Fruit set on the remaining flower clusters can later, at the grower's convenience, be reduced to two or three per cluster without affecting the following year's crop. The early cluster thinning will have already assured a return bloom. Cluster thinning is the most certain method of off-setting biennial bearing, but its use is limited by the lack of and high cost of labor at bloom time.

BIOREGULATION OF FLOWERING AND FRUIT SET

Considerable progress has been made in chemically regulating flowering and fruit set. Chemical thinning effectively and inexpensively prevents biennial bearing. Its advantage is that it can be done early during either the bloom or postbloom period. Chemicals can be applied as single or multiple sprays. Specific suggestions on the concentration and timing of sprays are reported by Williams (1979) and Williams and Edgerton (1981).

In fruit-growing areas where the bloom is uniform and the weather is dry, a 4,6-dinitro-ortho-cresol (DNOC) blossom-thinning spray can be used. The effect of a single DNOC spray applied to Golden Delicious trees annually for 4 years is presented in Table 17.1. The untreated control trees became biennial,

Table 17.1 Bloom and Yield of Golden Delicious Chemically Thinned for Four Years at Full Bloom with Dinitro-O-cresol

	Control		Sprayed	
Year	Bloom (%)	Yield (kg)	Bloom (%)	Yield (kg)
1	96[a]	720	93[a]	570
2	30	345	58	525
3	80	585	61	435
4	20	180	60	390

[a] Average of 10 trees (from Batjer 1965).

Note: DNOC concentration was 500 ppm.

whereas the treated ones consistently had a 50% to 60% return bloom and a full crop of fruit. Total yield differed little after 4 years, but fruit quality and size were superior on the treated trees.

To assure adequate thinning and return bloom on most biennial cultivars, including Golden Delicious, a second and possibly a third chemical thinning spray may be necessary. In the Northwest, 1-naphthaleneacetamide (NAAm) and 1-naphthyl N-methylcarbamate (carbaryl) are used on Golden Delicious as postbloom thinners. NAAm is applied 7 to 15 days from bloom (2 mm fruit diameter) and, if necessary, followed by carbaryl at 15 to 25 days from bloom (10 mm fruit diameter). NAAm removes the small fruit on one-year-old wood in the tops of the trees. NAA is more erratic than NAAm and may overthin when applied after a DNOC spray; thus NAAm and carbaryl are most often used for thinning Golden Delicious in the western USA. Recently, NAA has been combined with carbaryl for more consistent thinning of Golden Delicious and Spur Delicious.

Combinations of chemical thinning agents and growth regulators, such as succinic acid-2,2-dimethylhydrazide (daminozide) or (2-chloroethyl)-phosphonic acid (ethephon), are used successfully to assist in overcoming biennial bearing (Williams 1972, 1973). The combinations are particularly effective on spur-type trees that have the greatest tendency for biennial bearing. Ethephon acts as a growth inhibitor, a flower promoter, and a chemical thinning agent. Daminozide acts only as a growth inhibitor and flower promoter, and it often increases fruit set when applied at or shortly after bloom. A side benefit of daminozide is increased firmness and quality of the fruit. Some of these spray combinations are shown in Table 17.2. Treatments 2 through 8 are used commercially on Golden Delicious in Washington State. The choice of treatment depends on the amount of bloom present and the amount of fruit removal desired. If bloom is 30% or less and there are an adequate number of resting spurs, daminozide is omitted.

The effect of daminozide and ethephon on flowering of moderate- and high-vigor Starkrimson Delicious trees grafted on seedling rootstocks is shown in

Table 17.2 Fruit Set on Ten-Year-Old "Snowball" Bloom Golden Delicious Trees after Treatment with Combination Sprays of Chemical Thinning Agents and Growth Regulators

Treatment[a]	Number of fruit per 100 blossom clusters	Return bloom potential
1. Control	95	None
2. DNOC	67	Light (25%)
3. DNOC + NAAm + carbaryl	56	Moderate (40%)
4. NAAm + carbaryl	75	Light (30%)
5. DNOC + NAAm + daminozide	59	Moderate (40%)
6. NAAm + daminozide	71	Light (25%)
7. DNOC + NAAm + ethephon	43	Heavy (60%)
8. NAAm + ethephon	66	Moderate (40%)

[a] DNOC was applied at full bloom; the other chemicals were applied as a tank mix at 15 days after full bloom. Application rates were DNOC, 375 ppm; NAAm, 34 ppm; carbaryl, 600 ppm; ethephon, 450 ppm.

Source: Williams (1973).

Table 17.3. All the chemical combinations were affective on the moderate-vigor trees, but flower induction on the high-vigor trees was more difficult. Heavy pruning and overfertilization were responsible for the high vigor.

Ethephon cannot be used at high rates on bearing trees because of its fruit-thinning effect and slight reduction of fruit size. Complete defruiting may occur when ethephon is applied earlier than 4 weeks from bloom. When used between 5 and 7 weeks from bloom at low concentration or in combination with daminozide, little fruit thinning occurs, and fruit size on young trees under irrigated conditions is acceptable.

INFLUENCE OF POSTBLOOM CHEMICAL THINNING AGENTS ON SEED ABORTION AND FRUIT SIZE

Postbloom thinning agents such as NAA and carbaryl are known to cause seed abortion. Carbaryl causes the most seed abortion in Delicious fruit. Usually, warm weather following the application of the chemical causes stress in the tree, and most of the seedless fruit abscise. In some seasons, especially in cooler growing areas, Delicious trees set parthenocarpic fruit. When postbloom thinners are used in cool seasons, the percentage of parthenocarpic fruit is increased. When weather conditions favor parthenocarpic fruit set, the number of seedless or low seed-count fruit persisting on the tree is increased and the average fruit size is noticeably reduced. Because of the tendency to abort seeds and increase flower initiation, carbaryl is favored over other

Table 17.3 Effect of Growth Retardants on Flower Initiation of High- and Moderate-Vigor Apple Trees

Treatment[a] and concentration	Time applied (days after bloom)	Terminal-shoot growth 1971	Percent bloom 1972	
			Moderate vigor	High vigor
Control		100	7	1
Daminozide 1000 ppm	14	82	5	1
Ethephon 500 ppm	14	73	31	4
Ethephon 500 ppm	28	76	37	—
Ethephon 500 ppm	42	82	12	—
Daminozide 1000 ppm + ethephon 500 ppm	14	58	28	10
Daminozide 1000 ppm + ethephon 500 ppm	28	67	35	31
Daminozide 1000 ppm + ethephon 500 ppm	42	87	25	1
Daminozide 1000 ppm + TIBA 25 ppm	14	79	27	7

[a]Surfactant used with all treatments contained 236 ml of Spray Modifier/378 liters of spray solution.

Source: Williams (1973).

postbloom thinning agents to promote consistent return bloom. NAA causes slightly less seed abortion, but return bloom is less than with carbaryl. Interestingly, the greater the percentage of parthenocarpic fruit on the tree, the higher the amount of flower initiation. This observation agrees with Chan and Cain (1967), who showed that trees with seeded fruit initiate fewer flowers than those with seedless fruit (Table 17.4).

The effect of seed number on the size of Delicious fruit was reported by Williams (1977). At least 7 seeds per fruit are necessary for maximum fruit size (Fig. 17.4). To obtain maximum fruit size in the later blooming, cooler districts, it is advisable to rely more on the blossom and early postbloom thinners, such as DNOC and ethephon, and less on the late postbloom chemicals. This is true even with the greater risk of overthinning with blossom thinners. An alternative is to use the postbloom chemicals at as late an effective date as possible to take advantage of the warmer days, which will increase stress and subsequent thinning.

EVENTS LEADING TO FLOWER INITIATION

To be active in promoting flower initiation, a chemical application or cultural practice must temporarily alter the normal growth processes during the

Table 17.4 Effect of Seeds in Apple ('Spencer Seedless') Fruits on Flower Initiation

	Percent spurs flowering		
Spurs bearing	1964	1965	1966
Seedless fruits	96	100	90
Seeded fruits	0	7	27
No fruits	98	—	—

Source: Chan and Cain (1967).

"critical period" or the period before flower initiation (Williams 1973). This period for most apple trees in Washington State is from 3 to 6 weeks after bloom, but will vary in other parts of the world because of cultural practices and climatic effects. Barnard (1938) indicated that the time of flower bud differentiation is more closely related to the size of the current season crop and the time of cessation of shoot elongation than to the weather conditions, and that differentiation occurs latest in cultivars that are the most vigorous and vegetative.

Before a young nonbearing tree can be chemically induced to flower, it must have developed beyond the late juvenile stage into what Zimmerman (1973) referred to as the transition phase. The conditions or events I consider important for the initiation of flower primordia in apple buds are presented in Figure 17.5.

In general, flower induction in young, vigorous fruit trees is associated with a temporary reduction of terminal-shoot growth, but not complete cessation of lateral growth. Arguments to the contrary have been given by Monselise, Goren, and Halevy (1966), Luckwill (1970), and Tromp (1972). Nevertheless, others (including the author) believe a change in rate of vegetative growth is necessary for flower initiation to occur. Davis (1957) strongly contends that flowering follows the cessation of shoot growth. Seidlova and Krekule (1973) concluded that "the inhibition of vegetative growth during the period preceding flower induction favors flowering and that inhibition of vegetative growth is a prerequisite for flowering." Krekule (1978, p. 46) further states that "cessation of growth in trees has often been associated with stimulation of flowering." Bukovac (1981) also refers to the competitive aspects between vegetative and reproductive development in fruit trees.

With trees close to flowering naturally, total growth at the end of the season may be about the same as with nonflowering trees. But the rate of growth during the "critical period" must be altered either to allow utilization of available metabolites in the flower induction process or to reduce the levels of gibberellin GA and auxins IAA in potential flowering meristems, which allows flower initiation to proceed. Harley et al. (1942) suggested that roots and other structural tissues make the first demand on photosynthates, and when these demands are satisfied, blossom buds are formed. Liverman and Bonner (1953)

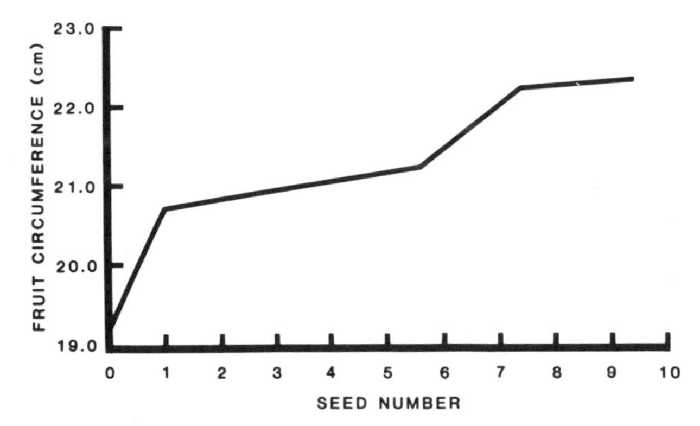

Figure 17.4. Relation between number of seeds per fruit and size of 'Delicious' fruit at harvest, 1976. (From Williams 1977.)

and Priestly (1970) support the premise that flower initiation depends on endogenous growth regulators that require the presence of adequate carbohydrate reserves.

The defoliation experiments of Magness (1917) and Harley et al. (1942) showed that removal of leaves from nonbearing spurs prevented flower formation. The general interpretation given to the response was that the leaves produced photosynthates and a flowering stimulus, which was transported to the bud to induce flowering. Zeevaart (1978) presents a number of options concerning the composition of the flowering stimulus, and Sachs (1979) summarizes what I believe is the situation in fruit trees. He considers the flowering stimulus to be hormonal and suspects that it is a mixture of compounds that differ by plant species and changes in time during floral evocation, initiation, and development. He also states that "the signal coming from leaves cannot be readily separated from bulk assimilate transport."

MODE OF ACTION OF BIOREGULANTS IN FLOWERING

All of my experimental observations suggest an indirect rather than a direct effect of exogenous growth-regulating chemicals on flowering. It is proposed that daminozide, ethephon, and most other growth retardants are effective, in varying degrees, in repressing IAA and GA production and distribution in actively growing shoots. The production of auxin is believed by Kato and Ito (1962) and Kuraishi and Muir (1962) to be stimulated by GA production in young leaves, and both chemicals stimulate internode elongation. In most plants, IAA is responsible for control of apical dominance (Sachs and Thimann 1964). Application of ethephon to nonbearing Delicious apple trees causes a temporary cessation of terminal growth and a tendency to increase lateral growth, giving the appearance of a release of the spurs and lateral buds from apical dominance. Aminoethoxyvinylglycine (AVG), an endogenous ethylene

ACCUMULATION OF CARBOHYDRATES
IN ROOTS, TRUNK, LIMBS, AND SPURS

INTERRUPTED VEGETATIVE GROWTH

REDUCED GIBBERELLIN AND AUXIN
IN GROWING TERMINALS

INCREASE IN ABSCISIC ACID, PHORETIN,
OR OTHER INHIBITORS

ACCUMULATION OF HORMONES
IN POTENTIAL FLOWERING MERISTEMS

INCREASED SRNA PRODUCTION, GENE
ACTIVATION AND PROTEIN SYNTHESIS

ACTIVE CELL DIVISION IN MERISTEMS

APPEARANCE OF FLOWER PRIMORDIA

Figure 17.5. Sequence of events necessary for flower initiation in apple trees.

inhibitor, also suppresses apical dominance in apple by interfering with auxin utilization (Williams 1980).

The basic mode of action of 2,3,5-triiodobenzoic acid (TIBA), daminozide, and ethephon in flower promotion appears to involve a common endogenous growth regulator system. TIBA reduces apical dominance and shoot elongation through its effect on auxin transport and production (Goldschmidt 1968). Daminozide reduces apical dominance by blocking auxin and possible GA synthesis (Ryugo and Sachs 1969; Williams and Stahly 1970; Luckwill and Silva 1979). Ethephon is broken down to ethylene by the plant tissues and inhibits auxin production and movement, thus destroying apical dominance (Morgan and Gausman 1966; Burg and Burg 1967). TIBA, daminozide, ethephon, and other growth retardants are active in the control of auxin (IAA) and GA metabolism and transport, which in turn reduces shoot elongation and allows a temporary increase in cytokinins and other metabolites, favoring active protein synthesis, cell division, and the initiation of flower primordia.

The role of natural inhibitors, such as ABA, phloretin, and other phenols, in the flowering process is little understood. It is possible that these inhibitors play a vital role in the flower initiation process. The role of IAA, GA, and inhibitors in flowering are discussed by Buban and Faust (1981), but more studies are necessary to determine their effect on the level of the natural growth regulators and the process of flower initiation.

NEW COMPOUNDS FOR REGULATING FLOWERING, FRUIT SET, AND SEED PRODUCTION IN APPLE

Endogenous ethylene suppressants are being tested on fruit trees. One of the compounds, aminoethoxyvinylglycine (AVG), an analog of rhizobitoxine discovered by Owens, Lieberman, and Kunishi (1971), can be applied either pre- or postharvest.* Ethylene production is suppressed throughout the dormant period, and apical dominance of apple and pear trees is counteracted (Williams 1980). There is a marked increase in lateral growth and total leaf surface (Williams 1980). This response is useful in increasing branching and the early bearing surface of apple and pear trees.

Spring applications of AVG drastically reduce "June drop" and increase fruit set. AVG is a useful tool to study the mode of action of the chemical thinning agents. When AVG is applied immediately before or after an application of NAA, NAAm, or carbaryl, endogenous ethylene production, which is normally stimulated by the chemical thinners, is supressed and fruit abscission is prevented (Williams 1980).

The average number of seeds in AVG-treated and control fruit is the same at harvest time, but the increased number of apples means there are more seeds on the tree, which prevents flowering and increases the tendency for alternate-year bearing. Therefore, excessive use of the chemical must be avoided.

A number of new, vegetative growth-retarding compounds are being tested for controlling terminal shoot growth in mature apple trees. Many fruit trees continue to grow excessively after their allotted space in the row is filled, and the result is shading and lowered productivity. Growers attempt to correct the problem by making heavy pruning cuts, which further aggrevates the problem. The compound ICI pp333* and its analogs have been applied to apple trees either by foliar spray or ground application. The amount of terminal growth control ranges from 10% to 90%, depending on the concentration and method of application. No adverse effects have been observed on the size or quality of the fruit. Trees receiving ground applications require little or no dormant pruning. Light penetration into the treated trees is much better as evidenced by a higher percentage of red color development on the fruit. The compound ICI pp333 has a wide range of activity on other crops and is being tested on all types of fruit trees.

Complete control of the vegetative and reproductive growth of fruit trees with chemical bioregulants will make it possible to fully manage fruit production. Maximum efficiency can be obtained and high yields of quality fruit will be provided for the consumer on an annual basis.

AVG and ICI pp333 are experimental compounds and are not registered by EPA for use in commercial orchards.

*Aminoethoxyvinylglycine is available from Fluka Ag. Chemische Fabrik, CH-9470 Buchs, Switzerland.

*Supplied by ICI America, P.O. Box 208, Goldsboro, NC 27530.

LITERATURE CITED

Barnard, C. 1938. *Studies of growth and fruit bud formation. VI. A summary of observations during the season 1930-31 to 1934-35*. Austral. Council Sci. and Ind. Res. J. 11:61–70.

Batjer, L. P. 1965. *Fruit thinning with chemicals*. U.S. Dept. Agric. Info. Bull. 289.

Buban, T. and M. Faust. 1981. *Flower bud induction in apple trees: Internal control and differentiation*. In *Horticultural Reviews*, vol. 4.

Bukovac, M. J. 1981. *Interrelationships between vegetative and reproductive development of fruit trees*. Proc. Wash. state Univ. Short Course on Use of Plant Growth Regulators on Fruit Trees.

Burg, S. P., and E. A. Burg. 1967. *Inhibition of polar auxin transport by ethylene*. Plant Physiol., 42:1224–28.

Chan, B. G., and J. C. Cain. 1967. *Effect of seed formation of subsequent flowering in apple*. Proc. Amer. Soc. Hort. Sci. 91:63–68.

Davis, L. D. 1957. *Flowering and alternate bearing*. Proc. Amer. Soc. Hort. Sci. 70:545–56.

Dennis, F. G., Jr., and J. P. Nitsch. 1966. *Identification of gibberellins A_4 and A_7 in immature apple seeds*. Nature. 211:781–82.

Goldschmidt, M. H. M. 1968. *The transport of auxin*. Ann. Rev. Plant Physiol., 19:347–57.

Grochowska, M., and A. Karaszewska. 1978. *A possible role of hormones in growth and development of apple trees and a suggestion on how to modify their action*. Acta Hort. 80:457–64.

Harley, C. P., J. R. Magness, M. P. Masure, L. A. Fletcher, and E. S. Degman. 1942. *Investigations on the cause and control of biennial bearing of apples*. U.S. Dept. Agric. Tech. Bull. 792.

Hoad, G. V., 1978. *The role of seed derived hormones in the control of flowering in apple*. Acta Hort. 80:93–103.

Huet, J., 1973. *Floral initiation in pear trees*. Symposium on Growth Regulators in Fruit Production. Acta Hort. 1(34):193–98.

Kato, T., and H. Ito. 1962. *Physiological factors associated with the shoot growth of apple trees*. Tohoku J. Agric. Res. 13:1–21.

Krekule, J., 1979. *Stimulation and inhibition of flowering*. In *La Physiologie de la floraison* [pages 19–58], (R. Jacques and P. Champagnat, eds.) CNRS No. 285. Paris.

Kuraishi, S., and R. M. Muir. 1962. *Increase in diffusible auxin after treatment with gibberellin*. Science 137:760–61.

———. 1963. *Mode of action of growth retarding chemicals*. Plant Physiol. 38:19–24.

Liverman, J. L., and J. Bonner. 1953. *Biochemistry of the photoperiodic response. The high intensity light reaction*. Bot. Gaz. 115:121–28.

Luckwill, L. C., 1970. *The control of growth and fruitfulness of apple trees*. Pages 237–54 *in* L. C. Luckwill and C. V. Cutting eds. *Physiology of Tree Crops*. Academic Press, London and New York.

———. 1974. *A new look at the process of fruit bud formation in apple*. Proc. 19th Int. Hort. Congr. 3:237–45.

Luckwill, L. C., and J. M. Silva. 1979. *The effects of daminozide and gibberellic acid on flower initiation, growth, and fruiting of apple cv Golden Delicious*. J. Hort. Sci. 54:217–23.

Luckwill, L. C., P. Weaver, and J. MacMillan. 1969. *Gibberellins and other growth hormones in apple seeds*. J. Hort. Sci. 44:413–24.

Magness, J. R., 1917. *Studies in fruit bud formation*. Ore. Agric. Exp. Sta. Bull. 146:3–27.

Monselise, S. P., R. Goren, and A. H. Halevy. 1966. *Effects of B nine, cycocel, and benzothiazole oxyacetate on flower bud induction of lemon trees*. Proc. Amer. Soc. Hort. Sci. 89:195–200.

Morgan, P. W., and H. W. Gausman. 1966. *Effects of ethylene on auxin transport*. Plant Physiol. 41:45–52.

Owens, L. D., M. Lieberman, and A. Kunishi. 1971. *Inhibition of ethylene production by Rhizobitoxine*. Plant Physiol. 48:1–4.

Priestly, C. A., 1970. *Carbohydrate storage and utilization*. In *Physiology of Tree Crops*. L. C. Luckwill and C. V. Cutting, eds. Academic Press, London and New York.

Ryugo, K., and R. M. Sachs. 1969. *In vitro and in vivo studies of Alar (1,1-dimethylaminosuccinamic acid, B-995) and related substances*. J. Amer. Soc. Hort. Sci. 94:529–33.

Sachs, T., and K. O. Thimann. 1964. *Release of lateral buds from apical dominance*. Nature 201:939–40.

Sachs, R., 1979. *Metabolism and energetics in flowering*. Pages 168–208 in La Physiologie de la floraison, [R. Jacques and P. Champagnat, eds.] CNRS No 285. Paris

Seidlova, F., and J. Krekule. 1973. *The negative response of photoperiod floral induction in Chenopodium rubrum L. to preceeding growth*. Ann. Bot. 37:605–14.

Tromp, P. J., 1972. *The interaction of growth regulators and tree orientation on fruit-bud formation in apple*. ISHS Symp. on Use of Growth Regulators in Fruit Production, Acta Hort 34:185–188.

Williams, M. W., 1972. *Induction of spur and flower bud formation in young apple trees with chemical growth retardants*. J. Amer. Soc. Hort. Sci. 97(2):210–12.

———. 1973. *Chemical control of vegetative growth and flowering of apple trees*. Acta Hort. 2 (34), Symposium on growth regulators in fruit production. pp. 167–74.

———. 1977. *Weather and chemicals affect size of apples*. Proc. Wash. State Hort. Assoc. vol. 73:157–61.

———. 1979. *Chemical thinning of apples*. In *Horticultural Reviews*, vol 1:270–300.

———. 1980. *Retention of fruit firmness and increase in vegetative growth and fruit set of apples with aminoethoxyvinylglycine*. Hort. Sci. 15(1):76–77.

Williams, M. W., and H. D. Billingsley. 1974. *Effect of nitrogen fertilizer on yield, size, and color of 'Golden Delicious' apple*. J. Amer. Soc. Hort. Sci. 99(2):144–45.

Williams, M. W., and L. J. Edgerton. 1974. *Biennial bearing of apple trees*. Proc. 19th Int. Hort. Congress, vol. 3:343–52.

———. 1981. *Fruit thinning of apples and pears with chemicals*. U.S. Dept. Agric. Info. Bull. 289 (revised).

Williams, M.W., and E. A. Stahly. 1970. *N-malonyl-D-tryptophan in apple fruits treated with succinic acid 2,2-dimethylhydrazide*. Plant Physiol. 46:123–215.

Zeevaart, J. A. D. 1979. *Perception, nature and complexity of transmitted signals*. Pages 60–90 in *La Physiologie de la floraison,* R. Jacques and P. Champagnat, eds. CNRS No. 285, Paris.

Zimmerman, R. H. 1973. *Juvenility and flowering of fruit trees*. Acta Hort. 34:139–42.

18] Strategies and Specifications for Management of in Vitro Plant Propagation

by W. R. SHARP,* D. A. EVANS,* C. E. FLICK* and H. E. SOMMER†

INTRODUCTION

Large scale, in vitro propagation (IVP) strategies are important for cloning, tissue culture breeding, and disease irradication. IVP can be accomplished through either the in vitro culture of meristematic explants or nonmeristematic explants, depending on the propagation or crop improvement strategy. Meristematic explants include isolated meristems, shoot tips, or axillary buds, while nonmeristematic explants include any nonmeristematic tissue explant, e.g., leaf, petiole, root, etc. The difference between the two explant sources is that the meristematic explants consist of cells with a developmental commitment to the plantlet pattern of development, while the nonmeristematic explant cells are committed to alternative developmental pathways. These latter cells, whether of sporophytic or gametophytic origin, can redifferentiate and become committed to the plantlet pattern of development only when subjected to an appropriate culture regime.

Apomixis, asexual reproduction, and vegetative propagation are general terms often used to define IVP from either sporophytic or gametophytic tissues in the absence of fertilization. Diploid plants are obtainable from sporophytic tissues, whereas haploid plants may be obtained from tissues of gametophytic origin. Organogenesis and embryogenesis are the two developmental processes through which IVP occurs. Organogenesis is limited to the initiation of organ primordia, while an embryo, an initial stage in plant development, is characterized by a bipolar structure bearing shoot and root poles and following a series of developmental sequences. Organogencsis may occur through either

*DNA Plant Technology Corporation, 2611 Branch Pike, Cinnaminson, New Jersey and
†School of Forest Resources, University of Georgia, Athens, Georgia.

STRATEGIES OF PLANT REPRODUCTION (BARC Symposium number 6—Werner J. Meudt, ed.)
Allanheld, Osmun, Totowa

the culture of meristems, e.g., isolated meristems, shoot tips, or axillary buds, or through adventitious shoot development in cultures of nonmeristematic tissue. Several categories of embryogenesis may be defined: androgenesis implies that the embryos are of microspore origin, parthenogenesis pertains to embryos of oospore or ovum origin, gynogenesis is restricted to embryos derived from incompletely fertilized oospores and, finally, apometry describes the origin of embryos from either synergids or antipodals.

PATTERNS OF IVP

Shoot apex axillary bud propagation. Certain species, or unique genetic variants thereof, are often difficult or time consuming to propagate using established horticultural methods. The use of organized structures, e.g., mass clonal propagation via IVP, shoot apex/axillary bud or meristem for production of genetically identical plants is important on a commercial basis (Murashige 1974). The apical shoot tip or axillary bud is surface sterilized and placed onto a simplified culture medium for production of multiple shoots that are subsequently separated and subcultured onto a rooting medium. Plantlets are then transplanted to soil or peat pots and, thereafter, transferred to the greenhouse or field. This mode of IVP represents the state of the art for species of horticultural importance and is the most widely used procedure in the commercial tissue culture propagation industry, because plants produced by this method are as a rule genetically uniform. This is because such plants result from pre-existing or newly formed meristems without an intervening callus stage.

Murashige (1974) has described a general three-stage procedure for production of plants from meristematic explants that normally requires alteration of culture medium or growth conditions between stages. Stage I pertains to the establishment of tissue in vitro. Stage II is most important as it requires production of multiple shoots. Stage III must result in root formation and conditioning of propagules prior to transfer to the greenhouse. High light intensity is important in Stage III (Murashige 1978). The media and culture conditions are not altered for many species between Stages I and II (Murashige 1978).

Frequency of plant regeneration occuring in tissue culture using explants from apical meristems, axillary buds, or shoot tips is determined by the growth regulator and concentrations thereof used in the culture medium. The growth regulators 6-BA, NAA, and GA_3 are especially important for high-frequency regeneration. Some mass propagation schemes recycle in vitro propagated shoots to decrease the time for achieving production goals.

Adventitious shoot development from either sporophytic or gametophytic tissues. De novo formation of meristematic loci in cell cultures or induced-organogenic determined cells (IODC), leading to the organization of a well-defined shoot and/or root meristem, has been thoroughly described in the literature (Evans, Sharp, and Flick 1981). The primordia originate from small groups of cells with a developmental commitment to either shoot and/or root organogenesis. The redetermination of cells comprising mature tissues to shoot/root formation is often dependent on the relative cytokinin/auxin ratio in the nutrient medium (Skoog and Miller 1957; Evans, Sharp, and Flick 1981). High concentrations of cytokinin favor shoot formation, whereas high concentrations of auxin induce the formation of roots. Factors important to determining the frequency of IODC development pertain to explant age, seasonal variation, explant source, culture medium, growth regulators, and environmental conditions.

Induced-organogenic determined cells (IODC)—one step. Shoot induction can occur from nonorganogenic determined cells cultured on a medium containing appropriate concentrations of specific growth regulators. The growth regulators promote cellular proliferation and redetermination of certain cellular populations to the shoot morphogenesis pattern of development.

The most-frequent growth regulator additives to the culture medium consist of IAA + Kin, 6-BA, IAA + 6-BA, or NAA + 6-BA. Cytokinin is removed from the culture medium for root development.

Tobacco has been used as a model system for investigations of the IODC mode of IVP studies on adventitious shoot development. Totipotency was first demonstrated with *Nicotiana tabacum* by regeneration of mature plants from single cells (Vasil and Hildebrandt 1965). Some of the basic correlates for organogenesis have been elucidated using tobacco. Plant regeneration from isolated protoplasts was first accomplished with *N. tabacum* (Takebe, Labib, and Melchers 1971), and the first somatic hybrid plant was obtained between two *Nicotiana* species, *N. glauca* and *N. langsdorfii* (Carlson, Smith and Dearing 1972). The most-frequently used culture medium regimes used for plant regeneration in tobacco and other taxa undergoing induced organogenic determination are reviewed in Evans (1982) and Evans, Sharp, and Flick (1981).

The two subfamilies of the *Gramineae, Poacoideae* (grasses), and *Panicoideae* (cereals), have been cultivated in vitro with limited success. The primary limitation has been explant source, as meristematic tissues have been used to initiate callus cultures capable of plant regeneration from each cereal species and most grass species cultured in vitro. Sugarcane is certainly the most malleable of the grass species examined in vitro. Although the immature inflorerescence is the most-successful explant, a variety of explants may be used for plant regeneration, including apical meristems, young leaves, and pith parenchyma. Plants have been regenerated from long-term callus cultures in sugarcane (Nadar and Heinz 1977).

Induced-organogenic determined cells (IODC)—two step. Shoot induction occurs for many of the monocotyledons via a two-step culture medium regime that mimics the one used for Induced embryogenic determined cells (IEDC). The primary culture medium usually contains 2,4-D, while 2,4-D is either deleted in the secondary culture medium, added at a reduced concentration, or replaced by a weaker auxin.

Species of numerous families of dicotyledonous plants can readily be induced to undergo plant regeneration in vitro, but in general monocotyledons have been more difficult to culture. This is unfortunate, as *graminaceous* species are extremely important sources of nutrition. Because of the large number of agriculturally important species, most investigators using *graminaceous* species for studies on plant regeneration have restricted themselves to cultivated crops.

Somatic embryogenesis. Two general patterns of embryogenic development of in vitro embryogenesis are discernible: (a) direct embryogenesis: embryos originate directly from tissues in the absence of callus proliferation (i.e., nucellar cells of polyembryonic varieties of citrus; epidermal cells of hypocotyl in wild carrot and *Ranunculus sceleratus*), and (b) indirect embryogenesis: callus proliferation is a prerequisite to embryo development (i.e., secondary phloem or domestic carrot; inner hypocotyl tissues of wild carrot; leaf tissue explants of coffee; pollen of rice, etc). An understanding of these two different patterns of development depends upon consideration of the determinative events of cytodifferentiation during the mitotic cell cycle. It is well known that the fate of determined daughter cells following mitosis occurs at least one mitotic cell prior to differentiation (Yeoman 1970). In other words, cells that undergo embryogenesis directly are the daughters of a prior determinative cell division. Such determined cells may undergo a postmitotic arrestment until environmental conditions are favorable for commencement of the mitotic developmental sequence characteristic of embryogenesis. The two general patterns of in vitro embryogenic development, direct and indirect, may be further characterized by their relative times of determination and differentiation into embryogenic cells. Direct embryogenesis proceeds from preembryogenic determined cells (PEDC) (Konar and Natajara 1965; Kato and Takeuchi 1966), while indirect embryogenesis requires the redetermination of differentiated cells, callus proliferation, and differentiation of induced-embryogenic determined cells (IEDC).

Apparently, PEDCs require either synthesis of an inducer substance or removal of an inhibitory substance prior to resumption of mitotic activity and embryogenic development. Conversely, cells undergoing IEDC differentiation require a mitogenic substance to reenter the mitotic cell cycle and/or exposure to specific concentrations of growth regulators. Cyto-differentiation and the emergence of multicellular organization are multistep processes in which each step leads to the establishment of a particular pattern of gene activation, resulting in transition to the next essential state of development (Street 1978). Arrestment may occur at any step in this process.

The view that external applications of growth regulators can be permissive or inhibitive of differentiation but not determinative has been suggested (Street 1978). Tisserat, Esan, and Murashige (1979) state that the explant and certain of its associated physiological qualities are the most-significant determinants of embryo intiation, while the "in vitro" environment acts primarily to enhance or repress the embryogenic process. That is, the cells that undergo embryo initiation are predetermined, and their subsequent exposure to exogenous growth regulators simply allows embryogenesis to occur (Tisserat, Esan, and

Murashige 1979). Street (1978) believes that growth regulators may best be regarded as activating agents toward previously induced cells that are preconditioned to respond in specific ways.

We are in agreement with Street (1978) and Tisserat, Esan, and Murashige (1979) if their definition is restricted to PEDC. Their concept of embryogenesis is limited to predetermined embryogenic cells, where growth regulators serve only to initiate embryo development from PEDCs and/or the cloning of these PEDCs. An alternative concept must be developed to explain how induced embryogenic determined cells (IEDCs) become committed to embryogenic development, since the occurrence of such cells requires redetermination and commitment to embryogenesis.

Pre-embryogenic determined cells (PEDC). Citrus nucellus. The natural occurrence of polyembryony in many species of *Citrus* (Bacchi 1943) and its economic importance have long been realized (Webber 1931). Thus, not only a zygotic embryo, but also several adventive embryos may be found within a single seed. These adventive embryos have been shown to originate in single cells of the nucellus, near the micropyle of the ovule, and appear to be initiated after fertilization, soon before or after the first zygotic division (Bacchi 1943).

In addition to the natural occurrence of nucellar polyembryony, in vitro cultures of nucellar explants may give rise to embryos and eventually to fully developed plants (Kochba, Spiegel-Roy, and Safran 1972). Nucellar tissues from both unfertilized and fertilized ovules may undergo embryogenesis in vitro (Button and Bornman 1971; Mitra and Chaturvedi 1972). Cultures of monoembryonic cultivars like the Shamouti orange also have been shown to be embryogenic (Button, Kochba, and Bornman 1974). Early characterization of embryogenesis in nucellar cultures consists of an initial proliferation of callus and subsequent development of pseudobulbils in the absence of exogenous plant growth regulators (RangaSwamy 1958). Some of these pseudobulbils continue embryogenic development and eventually become entire plantlets. In other experiments with several cultivars of normally monoembryonic species, however, callus and pseudobulbil formation were not prerequisite to embryo development, and embryos arose directly from nucellar tissue (Button and Bornman 1971). Mitra and Chaturvedi (1972) reported that embryos may arise directly from the nucellus, or indirectly from nucellar callus. Many of these experiments include growth regulators such as NAA and Kinetin, as well as complex addenda like malt or yeast extract and coconut milk. These may be beneficial in increasing the frequency of observed embryos. It must be stressed, however, that embryogenesis may be observed in the absence of these components.

Button, Kochba, and Bornman (1974) characterized embryogenesis in habituated nucellar callus cultures. This callus was found to be composed not of unorganized parenchymatous tissue, but solely of numerous proembryos. Embryogenesis has been observed to occur from single cells in the periphery of the callus, as well as from existing proembryos. Some of these developing embryos may enlarge only to a globular stage, commonly referred to as pseudobulbils. These rarely develop into plants. Other proembryos follow the developmental sequence characteristic of zygotic embryogenesis and eventually develop into plants. The fact that this callus was habituated, i.e., autonomous for exogenous growth regulators, in no way decreased its embryogenic potential. That the presence of exogenous growth regulators actually depressed embryogenesis lends further support to the concept that exogenous growth regulators may be viewed best as inductive agents for determination.

It is our view that embryogenesis from nucellar tissues, both in vivo and in vitro, may be considered best as cases of PEDC-mediated embryogenesis. The cells of the nucellus are actually PEDCs, and their proliferation as a callus mass and subsequent embryogenesis may be viewed as the cloning of PEDCs. Thus, it appears that embryogenesis in nucellar tissue can accur in the absence of exogenous growth regulators. Although low concentrations of kinetin and NAA have been shown to be beneficial, they are not absolute requirements for embryogenic determination. Exogenous growth regulators probably contribute to the cloning of these PEDCs, thus increasing the relative number of embryogenic cells. Of course, it is also possible that an additional population of predetermined cells (PEDCs), requiring exogenously supplied growth regulators for development, further contribute to the relative number of embryos.

Induced-embryogenic determined cells (IEDC). The concept of IEDC explains the redetermination of differentiated cells to the embryogenic pattern of development. Evidence exists that the auxin or auxin/cytokinin concentration in the primary culture medium or conditioning medium is not only critical to the onset of mitotic activity in nonmitotic differentiated cells, but to the epigenetic redetermination of these cells to the embryogenic state of development.

A statement characterizing the role of growth regulators in gene expression as direct or indirect cannot be made. Regardless of how growth regulators control gene expression, evidence exists that the auxin, 2,4-D elicits a response at the transcriptional and translational levels during primary culture. Subsequently, an additional response at the transcriptional and translation levels occurs shortly after subculture onto a secondary or induction medium (Sengupta 1978; Sengupta and Raghavan 1980a, b).

Numerous examples of IEDC have been reported in the literature in which embryogenic cells, resulting from cellular redetermination, proceed through the various stages of embryo development and form plants. Development of the embryogenic determined cells is usually restricted during culture on the primary or conditioning medium. Thereafter, cells need to be subcultured onto a secondary culture medium or induction medium for continued development in the embryogenic determined cells. Documented examples of IEDC have been reported in the literature for the following crops: endive (Vasil and Hildebrandt 1966), carrot (Reinert, Tazawa, and Semenoff 1967), pumpkin (Jelaska 1974), celery (Williams and Collin 1976), grape (Krul and Worley 1977), caraway (Ammirato 1977), cotton (Price and Smith 1979), cacao (Pence, Hasegawa, and Janick 1979), eggplant (Matsuoka and Hinata 1979), date palm (Reynolds and Murashige 1979), and coffee (Sondahl, Spahlinger, and Sharp 1979), alfalfa (Kao and Michayluk 1980), and pearl millet (Vasil and Vasil 1980).

Androgenesis and gynogenesis involve the development of IEDCs. In these instances, either the microspore or megaspore must undergo a quantal mitotic division, resulting in an embryogenic determined cell. In the former, a quantal mitotic division results in a generative cell and a vegetative cell. The vegetative cell, or in some instances the generative cell (Raghavan 1976) or a fusion cell, is then determined as an embryogenic mother cell (Sunderland 1974, 1977).

OTHER MODES OF IVP

Protoplasts. The plant cell wall interferes with the ability to genetically modify plant cells. The wall can be digested with enzymatic treatment, but the resulting protoplasts have reduced viability. Nonetheless, it is possible to genetically alter protoplasts with mutagen treatment, cell fusion, or uptake of exogenous DNA. Genetically modified protoplasts must then be regenerated into intact plants. Application of protoplast technology for crop improvement is limited to the species in a small number of plant genera in which plants can be regenerated from protoplasts.

Protoplasts can be isolated following treatment of plant tissue with mixtures of cellulase, pectinase, Driselase, and Rhozyme (Gamborg and Wetter 1975). Procedures have been devised for the release of protoplasts from leaf tissue, stem tissue, flower petals, or cell suspension or callus cultures. Isolation of protoplasts is usually achieved with enzymes dissolved in a solution containing osmotic stabilizers to prevent bursting and swelling of the protoplasts. Following the release of the protoplasts, the enzyme solution is removed by filtration and centrifugation. Isolated protoplasts are amenable to genetic modification following enzyme removal. Following chemical or physical treatment, modified protoplasts are cultured in nutrient medium. In some cases, protoplast culture medium may be extremely complex (Kao and Michayluk 1975). Viable protoplasts regenerate a cell wall and undergo mitosis in the culture medium. First mitosis is usually observed in 2 to 7 days after removal of digestive enzymes. As the protoplast-derived cells continue to reproduce, new culture medium must be added, or culture medium sufficient for regeneration must be substituted for the protoplast culture medium.

While protoplasts can be isolated from any plant tissue (Vasil and Vasil 1980), the condition of the donor plant prior to protoplast isolation is extremely important if plants are to be regenerated (Shepard and Totten 1977). In many cases, greatest success in plant regeneration has been achieved with protoplasts isolated from cultured cells. Cultured cells are already grown under aseptic and controlled environmental conditions and in many cases are already programmed for rapid cell devision. If leaf tissue is to be used for protoplast isolation, optimum growth conditions of plants must be ascertained.

Protoplast-derived cell colonies must usually be transferred to fresh medium for plant regeneration. In most cases, where organogenesis occurs, a sequence of culture media must be used to achieve shoot, then root, formation. If embryogenesis is possible, only a single transfer of culture medium is required to recover plants from cultured cells. Consequently, regeneration of plants from cells derived from protoplasts is achieved using the same hormone requirements as plant regeneration from callus cells produced from intact plant tissue.

Plants have been regenerated from protoplasts isolated from at least 35 plant species (Vasil and Vasil 1980). In at least some cases regenerated plants remain chromosomally stable (Evans 1979). Most of the species capable of regeneration, though, are *Solanaceous* plants, including species in the *Datura, Hyoscyamus, Lycopersicon, Nicotiana, Petunia,* and *Solanum* genera (Vasil and Vasil 1980). Very few economically important crop species have been regenerated from protoplasts. Successful regeneration protocols have been published for asparagus (Bui-Dang-Ha and Mackenzie 1973), carrot (Dudits et al. 1976), rapeseed (Kartha et al. 1974), tobacco (Nagata and Takebe 1971), potato (Shepard and Totten 1977), millet (Vasil and Vasil 1980), and alfalfa (Kao and Michayluk 1980). Extension of this list to other economically important crops, particularly cereals and legumes, is essential for application of protoplast technology for crop improvement.

Cells of two species can be combined by fusion of protoplasts. Methodology for the isolation and fusion of protoplasts has been described by Gamborg and Wetter (1975). Following fusion treatment, protoplasts regenerate cell walls and undergo mitosis. Hybrid cells must be identified in mixtures of parental, homokaryotic, and hybrid cells. Identification of hybrids is usually accomplished using genetic markers (e.g., White and Vasil 1979). Somatic hybrid plants have been recovered for interspecies combinations not possible using conventional genetics (Evans 1982). In most cases, the species combinations in interspecific hybridization represent *solanaceous* crop plants and include only tomato, potato, tobacco, and carrot among cultivated crops.

Haploids. Haploid plants can be recovered using the IVP schemes of either androgenesis or from callus mediated plantlet development. Sharp, Reed, and Evans (1982) have reported that most species capable of haploid production proceed via direct androgenesis. Taxonomic differences are apparent, though, between androgenesis and callus mediated regeneration. The majority of *solanaceous* species capable of producing haploid plants undergo direct embryogenesis, while the majority of *graminaceous* species undergo callus-mediated haploid plant regeneration (Sharp, Reed, and Evans 1982). Crop species capable of haploid plant production from cultured anthers include rapeseed, tobacco, rye, potato, corn, asparagus, pepper, strawberry, rubber, barley, sweet potato, wheat, grapes, and clover.

Embryos. In vitro pollination and embryo rescue is an important mode of IVP for the development of interspecific and intergeneric hybridizations, which are rare in nature. Success in this technique depends upon two basic considerations (Yeung, Thorpe, and Jensen 1981). These are (a) using pollen grains and ovules at the proper stage, and (b) defining nutrient media that will support pollen germination, pollen tube growth, and embryo development.

Embryo abortion, which occurs quite frequently as a result of unsuccessful crosses in breeding, can be eliminated. This is accomplished through the aseptic culture of the embryo in a nutrient medium. This approach has been successful for interspecific crosses in cotton, barley, tomato, rice, and jute. Success has also been achieved with intergeneric hybrids, e.g., *Hordeum* and *Secale, Hordeum* and *Hordelymus, Triticum* and *Secale,* and *Tripsacum* and *Zea*. Another novel use of embryo culture has been in the production of monoploids and doubled monoploids of barley.

GENETIC BASIS FOR IVP

Progress in the application of in vitro embryogenesis and organogenesis using nonmeristematic explants to agriculture is dependent upon an understanding of the genetic basis for these phenomena. Such a basis has been established for the frequency of organogenesis in alfalfa and tomato. Genetic lines of alfalfa with 67% regeneration were selected from hypocotyl tissue culture regenerated plants with an initial plant regeneration frequency of only 12% (Bingham et al 1975). Shoot morphogenesis in tomato germplasm, originating from cultured leaf discs with different regeneration potentials, were examined by Frankenberger and Tigchelaar (1980). An analysis of diallele F_1 hybrids was clearly supportive of a genetic basis for shoot morphogenesis. Work accomplished by Jacobsen and Sopory (1978), using selections obtained from potato clones with low frequency androgenesis, have resulted in identification or recombinants with a 30% to 40% frequency of androgenesis.

STRATEGIES FOR APPLICATION OF IVP

Cloning. Plants produced asexually through shoot apex/axillary bud culture in vitro are of uniform quality and can be produced more rapidly than is possible through sexual reproduction or by conventional modes of propagation (Murashige 1974). A millionfold increase per year in the ratio of clonal multiplication over conventional methods is not unrealistic.

Clonally propagated plants produce uniformly superior seeds, improve progeny evaluation of breeding experiments, and show improved vigor and quality. Moreover, clonal propagation technology can be used for the elimination of pathogens and for propagation of disease-free germplasm for international shipment.

The advantages of IVP cloning strategies over conventional propagation techniques pertain to a reduction in time for accomplishment of large-scale propagation. Meristematic explants are generally used for clonal propagation because such cultures have a lesser degree of genetic variability, as compared to plants regenerated from nonmeristematic explant material. Clonal propagation strategies accomplished through IVP include: increase of plants with poor seed set and low germinability (e.g., hybrid carrot seed), rootstock increase (e.g., M9 apple rootstock), increase of parental lines for breeding (e.g., asparagus), propagation of hybrids (e.g., broccoli), virus-free plant multiplication (e.g., potato), and forest tree propagation (e.g., Douglas fir).

An IVP cloning strategy would be quite useful in propagation of hybrids. This is especially true in crops where male steriles are not readily available, or where continuous commercial production of hybrid seed is expensive. In some species, e.g., greenhouse cucumber, hybrid seeds are quite expensive. Hybrid plants produced by IVP procedures may be less expensive to produce than hybrids originating from seed produced by hand pollination.

Genetic variability and varietal development. IVP procedures can be used for production of genetic variability using (a) nonmeristematic explants for plantlet regeneration, (b) protoplast technology, (c) mutagenesis, or (d) the uptake of foreign DNA via a vector. Factors that appear to be associated with the degree of genetic variability or stability of plants obtained from IVP include, (a) explant source, (b) explant age, (c) culture environment, (d) culture duration, (e) chemical additives, and (f) growth stimulants or regulators. Categories of genetic variability observed in IVP plants consist of: (a) chromosome number and structure, (b) simple gene mutation, (c) cytoplasmic mutation, and (d) epigenetic change. Genetic analysis must be made to determine the basis for genetic variability.

Exciting possibilities exist for taking advantage of the wide use of genetic variability associated in certain IVP plants for development of new cultivars. A determination has not been made as to whether the phenotypic variability observed in IVP sugarcane and potato plants (see Shepard, Bidney, and Shahin 1980) is a result of genetic or epigenetic variability.

Application of genetic variability among regenerated plants. Phenotypic variation has been reported in a number of plant species regenerated via adventitious shoot development. Such genetic variability may be agriculturally useful when integrated into existing breeding programs. The phenotypic variability recovered in regenerated plants reflects either preexisting cellular genetic differences or tissue culture–induced variability. Variability has been observed in both asexually and sexually propagated crops. Plants regenerated from callus of sugarcane have been examined in detail. A wider range of chromosome numbers ($2n = 71 - 300$) has been observed in plants regenerated from sugarcane callus (Heinz, et al. 1977) than in parent lines. These clones of regenerated sugarcane were examined for disease resistance. Resistance to three diseases (eyespot disease, Fiji disease, and downy mildew) was observed in some plants regenerated from a susceptible sugarcane clone (Heinz, et al. 1977). One regenerated line was simultaneously resistant to both Fiji disease and downy mildew. These lines are being incorporated into a conventional breeding program. More recently, plants regenerated from mesophyll protoplasts of potato have been analyzed. As in sugarcane, lines have been identified that are resistant to diseases (late blight and early blight) to which the parent line, Russet Burbank, is susceptible (Shepard, Bidney, and Shahin 1980). Similarly, Sibi (1978) has observed sexually transmitted variability in plants regenerated from lettuce. Pigment and morphological variants were reported most frequently. Sibi (1978) has suggested that the most useful variants may be cytoplasmically inherited. It is hoped that new varieties will soon be developed using these techniques.

Application of protoplast technology. Phenotypic variability has been observed in plants regenerated from mesophyll protoplasts (Shepard, Bidney, and Shahin 1980). Experimenters attempting to apply protoplast regeneration to crop improvement should be cognizant of this potential variability. Separate clones should be monitored and screened for agriculturally useful variability.

Protoplast fusion offers the opportunity to introduce unique genetic variability into crop plants. Novel nuclear-cytoplasmic combinations can be accomplished that are not possible using conventional sexual hybridization. In most plant species, the male cytoplasm is excluded during fertilization. During fusion of protoplasts, the two cytoplasms are mixed. Following fusion, mixed cytoplasms normally segregate to one of the two parental types, thereby permitting effective cytoplasmic transfer. Recently, recombination between species of both mitochondrion (Belliard, Vedel, and Pelletier 1979) and chloroplast (Conde 1981) DNA has been observed in somatic hybrid plants. Agriculaturally important traits known to be cytoplasmically encoded include male sterility and herbicide and disease resistance. Therefore, protoplast fusion may be uniquely applied to the transfer of cytoplasmic traits between breeding lines and to create unique cytoplasmic recombinants that incorporate multiple cytoplasmic traits in a single breeding line.

Uptake of foreign DNA and treatment of protoplasts with chemical or physical mutagens can be used to isolate single genetic changes. As knowledge of plant molecular genetics involving isolation and synthesis of specific genes and manipulation of plasmid vectors accumulates, it will be possible to transfer single genes into plant protoplasts with subsequent plant regeneration.

Application of haploid plants. Haploid plants are important in plant breeding since haploids greatly reduce the time for obtaining homozygous lines following diploidization in allogamous populations. Doubled haploids, when integrated into a conventional breeding program, have resulted in rapid development of new varieties. Diploidized haploid lines can be produced using a colchicine treatment or, alternately, may be produced using leaf midrib culture (Kasperbauer and Collins 1972). Selection from a population of doubled haploids, particularly an F_2 population, can result in rapid fixation of desired characteristics. Kasha and Reinbergs (1980) have reported that use of doubled haploids reduced the time for release of a new variety of barley from 12 years to 5. Doubled haploids have also been proposed for use in tobacco, Asparagus, Brassica, rice, corn, and potato.

Application of embryo culture. Work accomplished by Zenkteller, Misiura, and Guzowska (1975) pertains to the production of interspecific and intergeneric hybrids via in vitro pollination and embryo culture. Another application pertains to overcoming self-incompatibility. Self-incompatibility is usually controlled by a physiological barrier that prevents fusion of otherwise fertile gametes. Placental in vitro pollination has been especially successful for

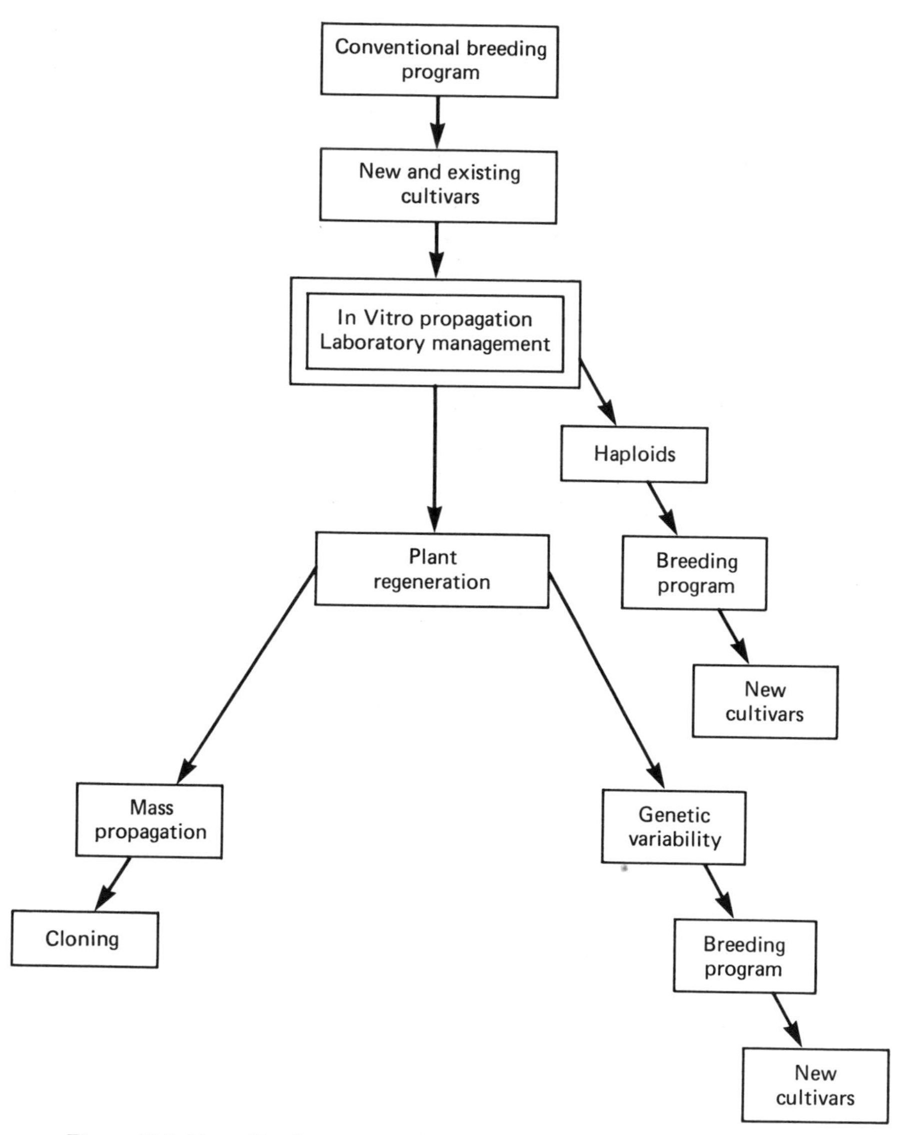

Figure 18.1. Use of in vitro propagation laboratory to complement conventional breeding program.

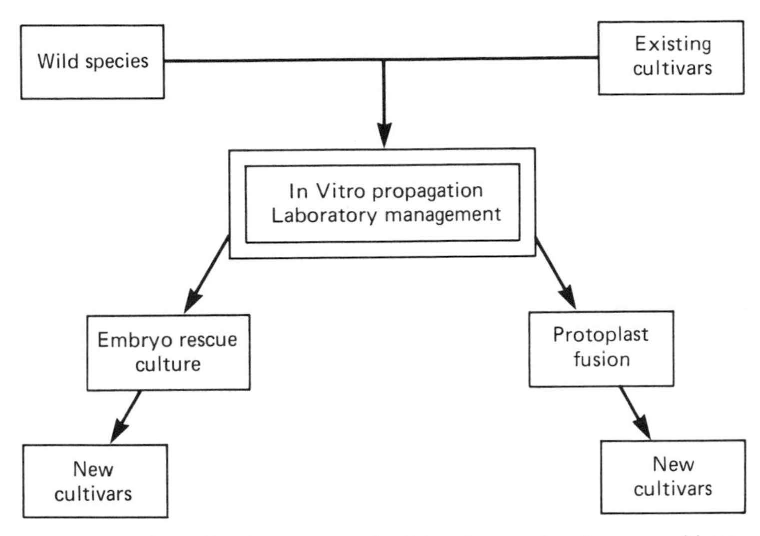

Figure 18.2. Use of in vitro propagation laboratory to develop new cultivars.

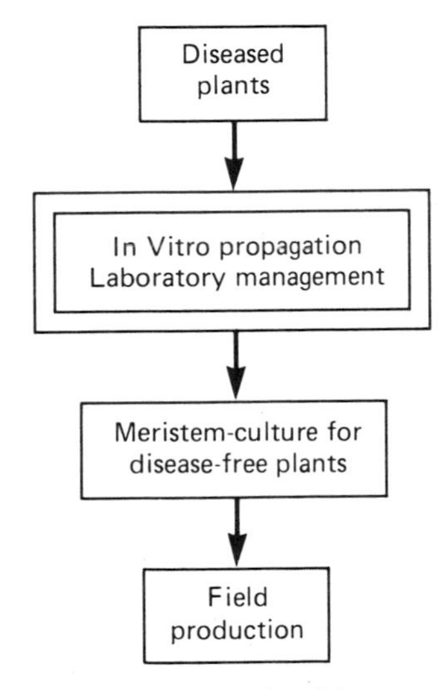

Figure 18.3. Use of in vitro propagation laboratory for production of disease-free plants for crop improvement.

overcoming self-incompatibility. The entire ovule mass of an ovary that is still intact on the placenta is cultured following pollination to permit embryo development (Yeung, Thorpe, and Jensen 1981).

IVP MANAGEMENT STRATEGIES

The establishment of a successful In Vitro Propagation Program requires coordination of effort between personnel in tissue culture and cellular genetics, plant pathology, and breeding. It is imperative for scientists in these key areas to have a general appreciation of tissue culture and cellular genetics in addition to a thorough understanding of crop improvement and breeding goals.

An In Vitro Propagation Program consists of these key areas: (a) complementation of a conventional breeding program through cloning new or existing varieties, or development of haploid or genetic variants from such germplasm (Fig. 18.1)., (b) development of new cultivars from existing cultivars and wild species using embryo rescue culture or protoplast fusion (Fig. 18.2), or (c) eradication of disease through meristem culture (Fig. 18.3).

LITERATURE CITED

Ammirato, P. V. 1977. *Hormonal control of somatic embryo development from cultured cells of caraway*. Plant. Physiol. 59:579–86.

Bacchi, O. 1943. *Cytological observations in Citrus*. III. *Megasporogenesis, fertilization, and polyembryony*. Bot. Gaz. 105:221–25.

Belliard, G., F. Vedel, and G. Pelletier. 1979. *Mitochondrial recombination in cytoplasmic hybrids of Nicotiana tabacum by protoplast fusion*. Nature 218:401–3.

Bingham, E. T., L. V. Hurley, D. M. Kaatz, and J. W. Saunders. 1975. *Breeding alfalfa which regenerates from callus tissue in culture*. Crop Sci. 15:719–21.

Bui-Dang-Ha, D., and I. A. Mackenzie. 1973. *The division of protoplasts from Asparagus offincinalis L. and their growth and differentiation*. Protoplasma 78:215–21.

Button, J., and C. H. Bornman. 1971. *Development of nucellar plants from unpollinated and unfertilized ovules of the Washington navel orange in vitro*. So. Afr. J. Bot. 37:127–34.

Button, Jr., J. Kochba, and C. H. Bornman. 1974. *Fine structure of and embryoid development from embryogenic ovular callus of "Shamouti" orange (Citrus sinensis osb.)*. J. Exp. Bot. 25:446–57.

Carlson, P. S., H. H. Smith and R. D. Dearing. 1972. *Parasexual interspecific plant hybridization*. Prod. Nat. Acad. Sci. USA 69:2292–94.

Conde, M. R. 1981. *Chloroplast DNA recombination in Nicotiana somatic parasexual hybrids*. Genetics 97:s26.

Dudits, D., K. N. Kao, F. Constabel, and O. L. Gamborg. 1976. *Embryogenesis and formation of tetraploid and hexaploid plants from carrot protoplasts*. Can. J. Bot. 54:1063–67.

Evans, D. A. 1979. *Chromosome stability of plants regenerated from mesophyll protoplasts of Nicotiana species*. Z. Pflanzenphysiol. 95:459–63.

———. 1982. *Plant regeneration and genetic analysis of somatic hybrid plants*. In E. D. Earle, ed. *Plant Regeneration and Genetic Variability*. Praeger, New York (forthcoming).

Evans, D. A., W. R. Sharp, and C. E. Flick. 1981. *Plant regeneration from cell cultures*. Pages 214–314 in J. Janick, ed. *Horticultural Reviews*, vol. 3. AVI Press, Westport, Conn.
Frankenberger, E. A., and E. C. Tigchelaar. 1980. *Genetic analysis of shoot formation from excised leaf discs of tomato*. In Vitro 16:217.
Gamborg, O. L., and L. R. Wetter. 1975. *Plant Tissue Culture Methods*. Nat. Res. Council, Saskatoon, Canada.
Heinz, D. J., M. Krishanmurthi, L. G. Nickell, and A. Maretzki. 1977. *Cell, tissue, and organ culture in sugarcane*. Pages 3–17 in J. Reinert and Y. P. S. eds. *Plant Cell, Tissue and Organ Culture*. Springer-Verlag, Berlin.
Jacobsen, E., Bajaj, and S. K. Sopory. 1978. *The influence and possible recombination of genotypes on the product of microspore embryoids in anther culture of Solanum tuberosum dihaploid hybrids*. Theoret. Appl. Genet. 52:119–23.
Jelaska, S. 1974. *Embryogenesis and organogenesis in pumpkin explants*. Physiol. Plant. 31:257–61.
Kao, K. N., and M. R. Michayluk. 1975. *Nutritional requirements for growth of Vicia hajastana cells and protoplasts at a very low population density in liqud media*. Planta 126:105–10.
———. 1980. *Plant regeneration from mesophyll protoplasts of alfalfa*. Z. Pflanzenphysiol. 96:135–41.
Kartha, K. K., M. R. Michayluk, K. N. Kao, O. L. Gamborg, and F. Constabel. 1974. *Callus formation and plant regeneration from mesophyll protoplasts of rape plants (Brassica napus L. cv. Zephyr)*. Plant Sci. Lett. 3:265–71.
Kasha, K. J., and E. Reinbergs. 1980. *Achievements with haploids in barley reserach and breeding*. Pages 215–30 in D. R. Davies, D. A. Hopwood, eds. *The Plant Genome*. The John Innes Charity, Norwich.
Kasperbauer, M. J., and G. B. Collins. 1972. *Reconstitution of diploids from anther-derived haploids in tobacco*. Crop Sci. 12:98–101.
Kato, H., and M. Takeuchi. 1966. *Embryogenesis from the epidermal cells of carrot hypocotyl*. Sci. Papers College Gen. Educ. Univ. Tokyo 16:245–54.
Kochba, J., P. Spiegel-Roy, Y. Safran. 1972. *Adventure plants from ovules and nucelli in citrus*. Planta 106:237–45.
Konar, R. N., and K. Natajara. 1965. *Experimental studies in Rununculus sceleratus L. Development of embryos from the stem epidermis*. Phytomorphology 15:132–37.
Krul, W. R., and J. F. Worley. 1977. *Formation of adventitious embryos in callus cultures of "Seyval", a French hybrid grape*. J. Amer. Soc. Hort. Sci. 102:360–63.
Matsuoka, H., and K. Hinata. 1979. *NAA-induced organogenesis and embryogenesis in hypocotyl callus of Solanum melogena L*. J. Exp. Bot. 30:363–70.
Mitra, G. C., and H. C. Chaturvedi. 1972. *Embryoids and complete plants from unpollinated ovaries and from ovules of in vivo-grown emasculated flower buds of Citrus spp*. Bull. Torr. Bot. Club. 99:184–89.
Murashige, T. 1974. *Plant propagation through tissue cultures*. Ann. Rev. Plant Phys. 25:135–66.
———. 1978. *The impact of tissue culture on agriculture*. Pages 15–26 in T. A. Thorpe, ed. *Frontiers of Plant Tissue Culture*. Univ. Calgary Press, Calgary.
Nadar, H. M., and D. J. Heinz. 1977. *Root and shoot development from sugarcane callus tissue*. Crop Sci. 17:814–16.
Nagata, T., and I. Takebe. 1971. *Plating of isolated tobacco mesophyll protoplasts on agar medium*. Planta 99:12–20.
Pence, V. C., P. M. Hasegawa, and J. Janick. 1979. *Asexual embryogenesis in Theobroma cacao L*. J. Amer. Soc. Hort. Sci. 104:145–48.
Price, H. J., and R. H. Smith. 1979. *Somatic embryogenesis in suspension cultures of Gossypium klotzschianum Anderss*. Planta 145:305–7.

Raghavan, V. 1976. *Role of the generative cell in androgenesis in henbane*. Science 191:388–39.

RangaSwamy, N. S. 1958. *Culture of nucellar tissue of Citrus in vitro*. Experientia 14:111–12.

Reinert, J., M. Tazawa, and S. Semenoff. 1967. *Nitrogen compounds as factors of embryogenesis in vitro*. Nature 216:1215–16.

Reynolds, J. F. and T. Murashige. 1979. *Asexual embryogenesis in callus cultures of palms*. In Vitro 15:383–87.

Sengupta, C. 1978. *Protein and nucleic acid metabolism during somatic embryogenesis in carrot*. Ph.D. dissertation, Ohio State Univ., Columbus.

Sengupta, C., and V. Raghavan, 1980a. *Somatic embryogenesis in carrot cell suspension*. I. *Pattern of protein and nucleic acid synthesis*. J. Exp. Bot. 31:247–58.

———. 1980b. *Somatic embryogenesis in carrot cell suspension*. II. *Synthesis of ribosomla RNA and Poly (A) + RNA*. J. Exp. Bot. 31:259–68.

Sharp, W. R., S. M. Reed, and D. A. Evans. 1982. *Production and application of haploid plants* Chap. 13 *in* P. B. Vose and S. Blixt, eds. *Contemporary Bases for Crop Breeding*. Pergamon Press (forthcoming).

Shepard, J. F., D. Bidney, and E. Shahin. 1980. *Potato protoplasts in crop improvement*. Science 208:17–24.

Shepard, J. F., and R. E. Totten. 1977. *Mesophyll cell protoplasts of potato*. Plant. Physiol. 60:313–16.

Sibi, M. 1978. *Multiplication conforme, non conforme*. Le Selectionneur Français 26:9–18.

Skoog, F., and C. O. Miller, 1957. *Chemical regulation of growth and organ formation in plant tissues cultured in vitro*. Pages 118–31 in *Symposium of Society of Experimental Biology*, vol. 11. Academic Press, New York.

Sondahl, M. R., D. A. Spahlinger, and W. R. Sharp. 1979. *A histological study of high frequency and low frequency induction of somatic embryos in cultured leaf explants of Coffea arabica L*. Z. Pflanzenphysiol. 94:101–8.

Street, Y. E. 1978. *Differentiation in cell and tissue cultures—regulation at the molecular level*. Pages 192–218 in H. R. Schutte, and D. Gross, eds. *Regulation of Developmental Processes in Plants*. VEB Kongress—und Werbedruck, Oberlungwitz.

Sunderland, N. 1974. *Anther culture as a means of haploid production*. Pages 91–122 in K. J. Kasha, ed. *Haploids in Higher Plants, Advances and Potential*. Univ. of Guelph, Guelph, Ontario.

———. 1977. *Observations on anther culture of ornamental plants*. In R. Gautheret, ed. *G. Morel Memorial Volume*. Masson Lie, Paris.

Takebe, I., G. Labib, and G. Melchers. 1971. *Regeneration of whole plants from isolated mesophyll protoplasts of tobacco*. Naturwissen. 58:318–20.

Tisserat, B., E. B. Esan, and T. Murashige. 1979. *Somatic embryogenesis in angiosperms*. Pages 1–78 in J. Janick, ed. *Horticultural Reviews*, vol. 1. AVI Publishing Co., Westport, Conn.

Vasil, I. K., and A. C. Hildebrandt. 1965. *Differentiaion of tobacco plants from single, isolated cells in microcultures*. Science 150:889–92.

———. 1966. *Variation of morphogenetic behavior in plant tissue culture*. I. *Cichorium endiva*. Amer. J. Bot. 53:860–69.

Vasil, I. K., and V. Vasil. 1980. *Isolation and culture of protoplasts*. Pages 1–19 in I. K. Vasil, ed. *International Rev. of Cytology*, vol. 11B. Acad. Press.

Vasil, V. and I. K. Vasil. 1980. *Isolation and culture of cereal protoplasts. Part 2: Embryogenesis and plantlet formation from protoplasts of Pennisetum americanum*. Theoret. Appl. Genet. 56:97–99.

Webber, H. J. 1931. *The economic importance of apogamy in Citrus and Mangifera*. Proc. Amer. Soc. Hort. Sci. 28:57–61.

White, D. W. R., and I. K. Vasil. 1979. *Use of amino acid analogue-resistant cell lines for selection of Nicotiana sylvestris somatic cell hybrids*. Theoret. Appl. Genet. 55:107–12.

Williams, L., and H. A. Collin. 1976. *Embryogenesis and plantlet formation in tissue cultures of celery*. Ann. Bot. 40:325–32.

Yeoman, M. M. 1970. *Early development in callus cultures*. Intern. Rev. Cytol. 29:383–409.

Yeung, E. C., T. A. Thorpe, and C. J. Jensen. 1981. *In vitro fertilization and embryo culture*. Pages 253–71 in T. A. Thorpe, ed. *Plant Tissue Culture Methods and Application in Agriculture*. Academic Press, New York.

Zenkteller, M., E. Misiura, and I. Guzowska. 1975. Page 180 in *Form, Structure and Function in Plants*. B. M. Johri Commemoration Vol., Sarati Prakashan, Meerut.

six

ENVIRONMENTAL AND STRESS FACTORS

19] Simulation of Environmental Impacts on Productivity

by WAYNE L. DECKER* and
CLARENCE M. SAKAMOTO*

ABSTRACT

Using modern simulation and statistical analyses, the impacts of environmental factors (atmospheric and edaphic) can be mathematically expressed. The method of simulation involves the development of mathematical equations that mimic the biological processes. Considerable success with this type of modeling has been reported for photosynthesis, respiration, and translocation. To be useful the simulation of all processes must be combined into an appraisal of the complete plant system. These mathematical expressions provide the modeler with information about how the biophysical processes works and identify the limiting environmental factors. The biophysical model also provides insights into the impacts of management options (irrigation, tillage, etc.) on production.

A second method for mathematically assessing the environmental impacts is through the use of regression statistics. This method provides maximum information concerning total production or grain yields. Regression analyses have been used for a long time in developing models of crop yields, but the recent development of computers and the adoption of newer statistical techniques (principle components and ridge regression) have encouraged regression analyses. The statistical method is useful in obtaining timely estimates of crop production in both deficit and surplus grain areas. The technique also provides a method for determining the risk of famine in food deficit-developing countries.

ENVIRONMENTAL LIMITATIONS ON AGRICULTURAL PRODUCTION

Climate is one of the environmental resources that limits the abundance of agricultural production. This limitation impacts on food supply in both import-

*Department of Atmospheric Science, University of Missouri-Columbia, Columbia, Missouri 65211

STRATEGIES OF PLANT REPRODUCTION (BARC Symposium number 6—Werner J. Meudt, ed.)
Allanheld, Osmun, Totowa

ing and exporting countries, the profits of commercial agricultural enterprises, and ultimately the price of food at the marketplace. Climatic variability impacts grain production in the United States. For example, Runge and Benci (1975) reported that the year-to-year variations in the weather during this century depressed corn yields to 56% below the mean yield in some years and improved yields in other years to values 43% above the average. The impact of climate variations is not confined to temperate and subhumid agriculture. Kreasuwun (1980) reported rice production data for central Thailand that indicated average yields between 1955 and 1977 to have varied from 1.2 to 2.0 kgm/ha. Kreasuwun noted that most of the variation was due to climatic fluctuation with only small improvements in overall yields of rice through fertilization, genetic improvement, and irrigation.

Crop production results from a complex interaction between plant potential, management factors, and natural resources. These resources include weather, availability of irrigation water, and soil productivity. A complete mathematical relationship showing the impact of one part of the system on agricultural yields and profitability without considering interactions with the other components oversimplifies the complex system. An appraisal of production potentials that focuses only on soil productivity and agronomic or economic considerations, but does not account for variations in weather, is incomplete. Similarly, an analysis of weather effects that "average over" genetic, technological, and edaphic considerations is oversimplified and also inadequate.

It is possible to describe the impacts of weather and climate variations on yield of grain or stover through the development of mathematical models. There are three types of mathematical models: (a) the biophysical, which attempts to define equations for each plant process, (b) the statistical, employing least squares analyses, or (c) combinations of the biophysical and statistical. There are many models of the second and third type but only a few of the first.

Mathematical models of plant growth and yields of agricultural crops are developed for three purposes. First, mathematical simulations of plant growth allow for a more-complete understanding of the biological system, how it works, and the limitations imposed by the environment on growth or yield. Second, a mathematical expression of the relationship between the environment and yield permits the prediction of final yields or assessment of crop conditions during the growing season at a fixed point or over a region. Third, a complete mathematical expression can be used to determine the optimal management or production system for a given set of weather or resources inputs. The type of model chosen depends, to a large degree, on its expected use.

SIMULATION OF BIOPHYSICAL PROCESSES IN PLANTS

Basis of simulations. Simulation of plant development is accomplished by defining mathematical expressions that mimic biological processes. The speed of such processes as photosynthesis, respiration, and translocation is determined by environmental conditions. The rate of each process can be mathematically expressed by differential equations. These equations must have initial conditions defined, coefficients experimentally derived, and feedback relationships or interactions described.

There are at least two important features in every system for simulation of plant growth. First, the simulation must account for the input (and loss) of mass. That is, the analysis must obey the law of conservation of mass. Second, the simulation must be driven by energy inputs (usually solar radiation) and must account for all the energy, i.e., there must be a conservation of energy. The second requirement is just as important as the first, because the excesses in energy warm the plant tissue, soil, and air surrounding the plant.

The bookkeeping for the conservation of mass focuses on water exchange between the elements of the soil-plant-atmosphere continuum and on the fate of CO_2 as it is transformed through photosynthesis and translocated through the plant (see Nobel 1975). The successful model also treats the nitrogen balance as the element is translocated from the roots to the growth sites.

The soil system must be realistically tied to the simulation of plant growth, as the soil is an important buffer for water, nitrogen, and other essential elements. Physical barriers, such as clay pans or depth of soil, can limit root penetration and water movement. The textural properties of the soil, which determine the water-holding capacity, must be considered. The native fertility of the soil and fertilizer applications must be incorporated in defining the amount of growth and yield. Thus, the systems approach to the simulation of growth requires quantitative information about the soil; to be realistic, the simulated growth must vary with the soil properties.

The simulation analysis for crop growth must also respond to atmospheric conditions. Stomatal behavior responds to the atmospheric demand as noted by Benci (1974), who showed that the size of the stomates of soybeans decreases as the atmospheric demand for water increases.

The biological processes. Plant growth and the subsequent harvest of grain, fruit, or forage occurs because of biophysical processes occurring within the plant. These processes, some of which are photochemical, determine the production potential of a given set of conditions. According to de Wit (1978), simulation of these processes depends on the ability to define quantitatively the state of the system at a particular time, the driving or forcing variables, the auxiliary variables, and the rate parameters (or statistics). It is important that the interactions between variables not be neglected. When a variable (driving or auxiliary) hastens or slows biological processes, changes are induced in the canopy structure, the environment, or in the biophysical processes elsewhere in the plant system.

The basic processes to be simulated are well known in plant growth and production. Photosynthesis (see de Wit 1978) is driven by radiant energy. To account mathematically for the sum total of solar energy, the interception of radiation by existing leaf area and the efficiency of the photochemical reaction must be known. Of course, the light reaction will be different in C_3 and C_4 plants. Photosynthesis is also driven by the availability of carbon dioxide, with

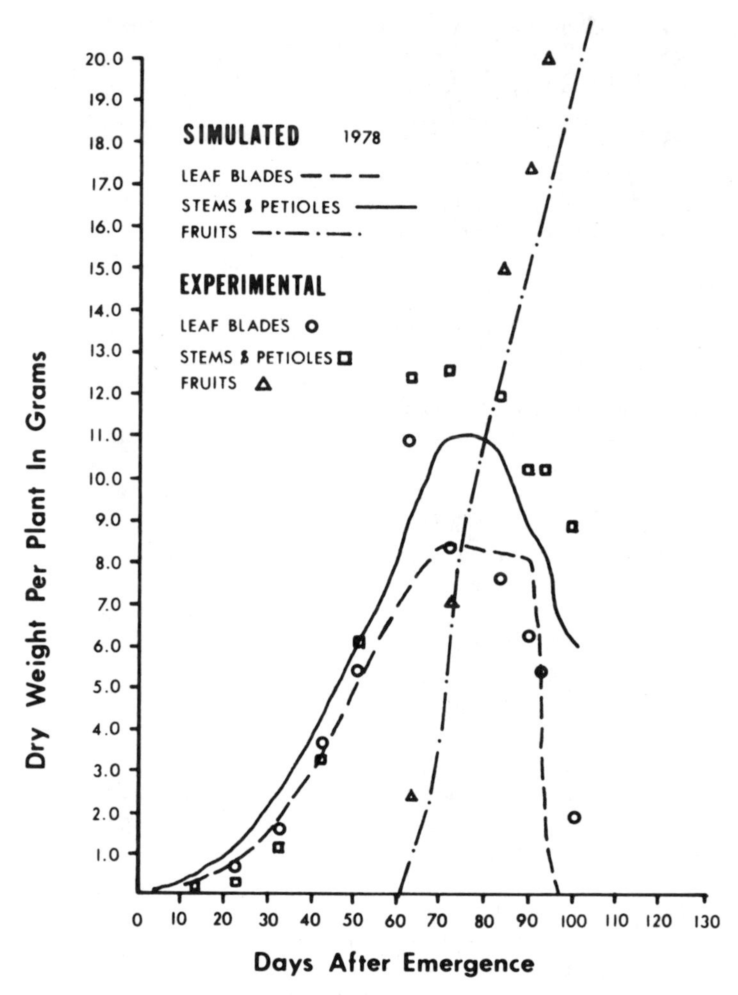

Figure 19.1. Simulated and actual growth and fruit development for soybeans during 1979 at Wooster, Ohio. (After Curry et al. 1980.)

the transport of carbon dioxide to the site of photosynthesis being determined by atmospheric conditions and the entry of carbon dioxide into the plant by the stomatal behavior. The stomatal behavior is a function of the plant's water status.

Dark respiration is driven by temperature. McCree (1974) has presented details on using temperature (a quadratic relationship) for simulating respiration. The temperature and thus the respiration of the individual plant parts are determined by the ambient temperature, radiation balance, and internal water status of the crop.

Translocation must also be simulated in the crop models in order to move photosynate to roots, floral parts, and seeds. Translocation is linked to important concentration gradients within plants. The complex nature of such simulation models can be seen from examples suggested by Curry (1979).

Examples of simulation of plant growth. There have been many attempts to use a systems approach to simulate plant growth and the final yield of crops. The most successful of these focuses on the growth of a single plant and allows for the total production of a plot or field through adjustments for the plant population. One example of the simulation of soybean response to the environment has been developed by Curry et al. (1980). In SOYMOD, Curry and his collaborators use basic meteorological data to drive the biophysical processes of the soybean plant. The weather variables were maximum and minimum temperatures, dew point temperature (assumed to be constant for the day), solar radiation, rainfall, and wind. The system also requires the astrophysical information of day length. Photosynthesis and respiration are estimated by mathematical expressions involving light and temperature. Translocation occurs in response to realistic mechanisms involving concentration gradients and temperature, and the internal water supply simulated from a soil water balance and atmospheric demand. Finally, each plant part is allowed to gain (or lose) dry weight.

In Figure 19.1, the validation of the soybean model is shown for one year of weather data, 1978. These data were not used by Curry in the development of the coefficients for the driving mechanisms in his model. One is impressed by the similarity between the simulated growth and actual growth. Of particular interest for soybeans is the manner by which seed weight is simulated. An even better approximation occurred for the simulation in 1976, when the simulated fruit weights never varied more than six percent from the mean sample yields.

A second example for cotton has been reported by Baker, Landivar, and Reddy (1979). Again, weather data are allowed to drive the system. Photosynthesis occurs with translocation to growing points in the cotton plant. In this model each process has a genetically determined maximum rate, with the actual rates adjusted by inputs of weather and other environmental variables. The validation of the model was again made through comparisons of simulated yields with actual data. The data used for the validation were from Bruce and Romken's (1965) experiment conducted earlier. Figure 19.2 shows the validation of the model for several significant growth and development indicators.

Utility of simulations. Perhaps the greatest utility of simulations is what the models tell the scientist concerning the physiological processes within plants, including identification of limiting factors to growth, the importance of sources and sinks, and the determination of the rates of biophysical processes. These scientific reasons are probably sufficient for conducting simulations.

An additional reason for conducting simulations concerns practical applications in the selection of agronomic practices or farm management options. Using the simulation model for soybeans, Curry et al. (1980) concluded that irrigation would not materially increase soybean yields, especially with high

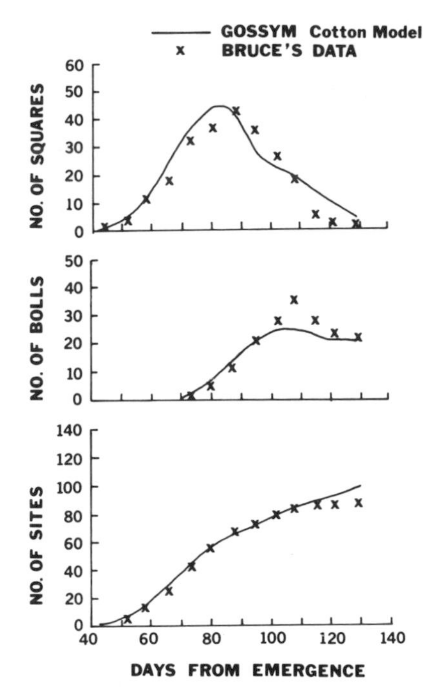

Figure 19.2. Simulated and actual development of reproductive sites for cotton. (After Baker et al. 1979.)

plant populations. Actually, this conclusion should be expected to apply only to the weather for the four years used in the study (1974–1978 at Wooster, Ohio), and may be restricted to the variety used in developing the model (Beeson). The simulator indicated soybean yields increased with plant density (i.e., rate of planting), with the average yield of the four years being nearly four times greater for the plant density of 62 plants/m^2 than with 7 plants/m^2.

Using the model for cotton, Baker, Landivar, and Reddy (1979) also tested the impacts of irrigation treatments. Under the simulated applications the irrigated crop was taller, matured later, and yielded one-third more than the nonirrigated system. Further, Baker simulated the impact of deep tillage through the removal of a clay pan located at between 10 and 15 cm depth in the soil. The yields were increased by one-third through the tillage treatment. The simulation model allowed the researcher to look for the reasons associated with the increased yield. The simulated tillage treatment produced differences in the timing of the set of squares, number of bolls, and size of bolls. These differences were attributed to direct and interactive factors associated with light interception, response to soil water, and nitrogen nutrition.

In addition, simulation can be used to develop systematic risk analysis of moisture supply. This analysis aids in estimating the impact of climatic change, in this case a change in precipitation, which reflects on soil moisture and yield. Richie (1979) simulated the available soil moisture in wheat fields as a function of planting time. His analysis taken over the period of weather records provided information on risk probabilities associated with different dates of planting.

The simulation of the impacts of management options on growth and yields does not necessarily "make it so" in the real world. Field tests and experimental trials are still necessary. Simulation is a tool to identify the important options to be tested in field experimentation. It should reduce the cost of experimentation and the time required for the adoption of a technology.

STATISTICAL METHODS FOR ESTIMATING CROP PRODUCTION

Techniques of analysis. In theory, biophysical models can be used to predict final yields and to make assessments of crop conditions during the growing season. But many of these simulation models are site-specific, i.e., refer to a single soil type and fertility condition, and are genetically restricted to a particular variety of crop. Furthermore, the many processes involved in the soil-plant-atmosphere continuum have yet to be precisely described by biophysical models, and sufficient experimental work has not been completed quantitatively to relate the interactions between the driving and auxiliary variables. With the above limitations, the rational is to short-circuit the system with the intent of maximizing the information content obtained from the available data. This procedure is the premise under which the statistical-empirical approach has been used to simulate the impact of climate variability on food production. Large-area yield estimation and status assessments will probably be obtained from statistically based models for a long time to come. Although considered a black-box approach, these models have already proven useful in estimating commodity production.

Statistically based models have a long history. In the United States, Henry Wallace was one of the first to use crop-weather models. Using corn, Wallace (1920) provided a practical application of the principle of "law of universal regression" first proposed by Galton in 1889. The popularity of these methods subsided until the drought years of the 1930s. In the late 1950s, Dr. Louis Thompson revived interest in statistical modeling through the commodity yield models he developed for the grain belt of the US (see Thompson 1962, 1963, 1969).

The advent of computers permitted analysts to review many additional derived variables, including those that indicate plant stress from temperature and moisture. Figure 19.3 represents an example of a data source for a regression model for estimating Kansas winter wheat yield. The statistical model for predicting yield from these data contained a constant, a trend evaluation assumed to be the technological effect, and the weather variables as predictants of departure in yield from trend. Models of this type have been the basis for evaluating the quantitative impact of weather in the United States and in other countries. Although the regression principles have not been much altered, the kinds of environmental variables used to predict yields have improved. For example, the ratio of evapotranspiration (ET) to potential

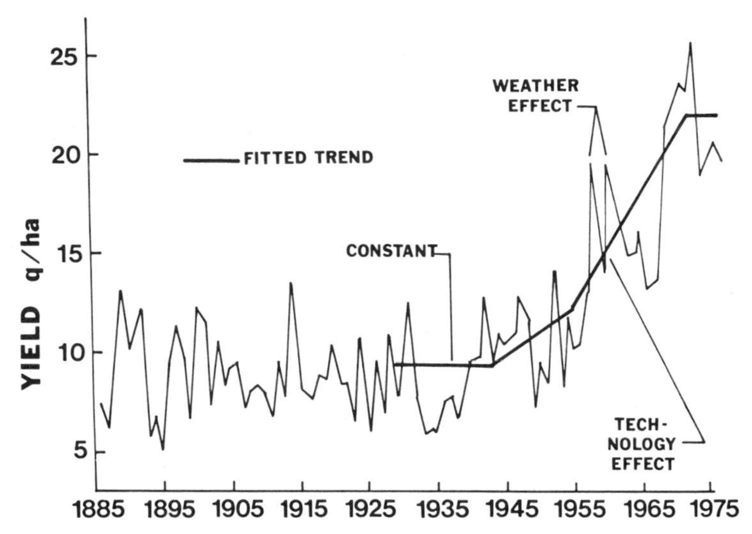

Figure 19.3. Time series of average yields of winter wheat for Kansas with the suggested trend in yields identified.

evapotranspiration (PET) has been successfully used as a variable to indicate the water supply/demand ratio and a measure of plant moisture stress. In a recent report, Kanemasu et al. (1980) suggested that improvements in all types of yield modeling could be achieved with better models of the soil-water balance. Consistent with the state of the art, relatively simple and empirical soil-water relationships were suggested. The goal of using these derived variables is to reduce the multicollinearity problems associated with using multiple meteorological variables in time.

Other statistical methods have been proposed to deal with the collinearity problem in correlation analysis. One is to develop a regression analysis using principal components. An illustration of this procedure for estimating crop productivity is given by Steyaert, LeDuc, and McQuigg (1978). Principal components of atmospheric pressure were used to estimate wheat yield in the US and the Soviet Union. Patterns of synoptic scale atmospheric pressure over large areas, which is correlated with the regional precipitation and temperature patterns, were used in the crop yield model. Another statistical approach involving regression modeling is ridge regression (Marquardt and Snee 1975). In this procedure, a bias is introduced to decrease the variance of the estimated regression coefficient.

There are problems associated with each of the above methods. The researcher should be familiar with the advantages and disadvantages of each statistical model to maximize the information content from these methods.

Problems associated with the use of statistical models. One of the major limiting factors in the application of any model is the availability of quality data

Table 19.1 Station Density of Climatic and Synoptic Network

State	Area/Station (KM^2)	
	Climatic	Synoptic
Indiana	854	23498
Kansas	774	17755
Nebraska	851	33336
North Dakota	1143	26146
Oklahoma	978	22636

within a short time frame. Data also need to represent a finite area, so there are spatial requirements for sufficient climate observations to represent the region. The availability of data affects the accuracy of the models and, ultimately, the cost of the analysis. In the United States, the spatial distribution of reporting weather stations is generally considered inadequate. Table 19.1 shows the station density of the climatological network in five states. The table indicates that the climatological network is about 27 times more dense than the snyoptic network in this area. The data from the synoptic network are available daily (even hourly) from the telecommunications network. The climatological network provides daily reports that are not available to users for two or more months because of communication and publication delays. According to LeDuc et al. (1980) the cost of acquiring and quality controlling these data increases in orders of magnitude as one increases the spatial and temporal scales (Figure 19.4).

One of the difficult problems of statistical regression models is the need to specify the trend or trends in a yield time series. This adjustment is usually done by assuming a linear trend, or some other mathematical form, as a surrogate for technological improvements. Although this assumption may in general be valid, the estimation of crop production in any particular year may be obscured by the possible interactions of technology with weather. In fact, Figure 19.5 shows that climate, as represented by crop season rainfall, may be decreasing through time while the yield is increasing.

The impact of technology, including varietal changes, fertilizer, insect and disease control, management by farmers, and government policies, all impact on the degree to which yield manifests itself in response to the variability of weather. The partial solution to this difficult problem lies in the ability to obtain sufficient data for separate assessments of the contribution of each of these factors to crop production.

Timeliness of the assessment is a major characteristic of a desirable model. The assessment must provide the desired information with sufficient lead time to be considered in decisions relative to management or policy options. The longer the lead time, however, the less accurate is the estimate of crop yield and production. To deal with this problem, analysis of weather and crop yield

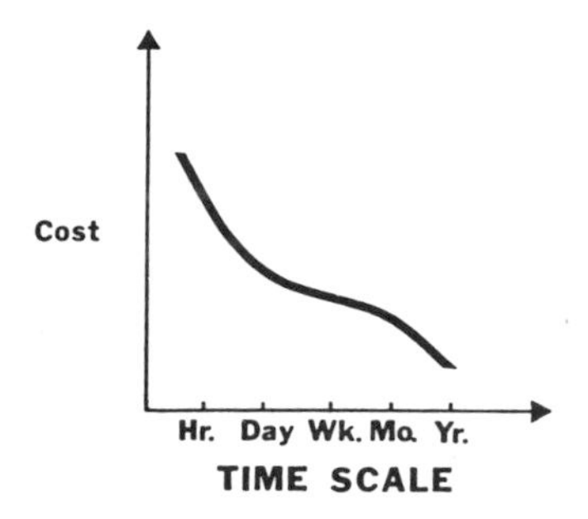

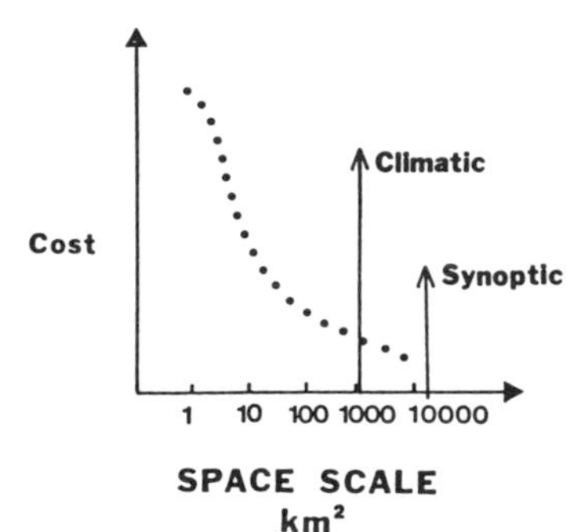

Figure 19.4. Scheme for showing how the cost varies for using climatic data from different time and spatial scales. (After LeDuc et al. 1980.)

data based on conditional probability was used by Strommen et al. (1979) to estimate expected yields at different harvest periods, given the actual weather conditions prior to the estimate. Walther and Hayes (1977) have utilized a Monte Carlo procedure to simulate the probability distribution of yield, given the inputs of temperature and precipitation, in a regression crop yield model up to the forecast period.

The utility of assessments from statistical models. The most-obvious use of statistical models for crop production is the estimation of food and feed grain production as the harvest approaches. This information is vital to both exporting and importing countries to plan orderly marketing strategies. The estimates of crop production are also widely used by commodity traders.

A second use of statistical models is associated with the anticipation of needs in food deficit in developing countries. In these countries, data for preparing crop models are more severely limited. Nevertheless, these are the areas where information on food production can provide early warning on food security to deal with the issues of potential famine. The University of Missouri, in cooperation with NOAA, has been developing such an assessment system using synoptic World Meteorological Organization data from the Global Telecommunication System, regression crop models, and agroclimatic analysis. Using these tools and resources, operational assessments are being

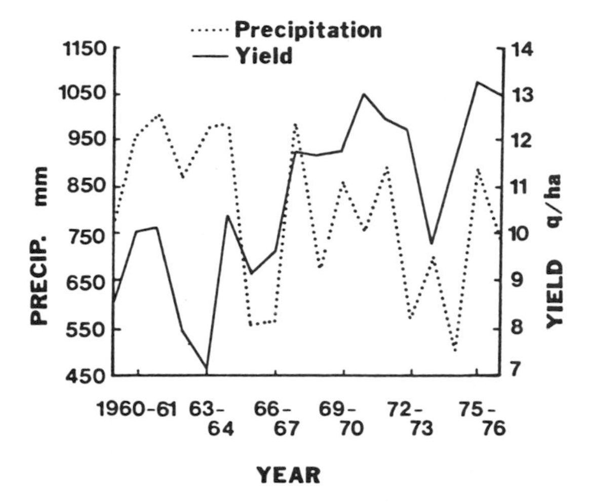

Figure 19.5. Time series of average wheat yield and crop season rainfall for the state of Uttar Pradesh, India.

issued to the Agency for International Development one to two months prior to the internal awareness of a potential shortfall. Even in these developing countries, simple analysis and models can provide useful information for the decisionmakers at the policy-making level of governments.

Figure 19.6 is an illustration of the analysis of precipitation (P) and potential evapotranspiration (PET) to give a Crop Moisture Ratio (P/PET) for a time series of data. In those years where the ratio was low, below the 20th percentile, serious food famine was documented in that country. Using current data, this method allows the development of information on potential food shortages and will be useful in the development of plans to mitigate the impact of climatic variability.

RELATED MODEL SIMULATION PROBLEMS

Although scientists have accomplished much in simulating the environmental impacts on crop productivity, there are opportunities for improving the models and the distribution of the results. The improvements of the models and more-rapid distribution of the assessments from these models will be associated with obtaining real-time environmental data and information. There are four needs for data which need additional development.

1. Rapid data acquisition. There must be an improvement in the availability of basic data for model development and subsequent utilization. This improvement must include both agricultural and meteorological data. A recent significant breakthrough has led to testing of a method for obtaining solar radiation

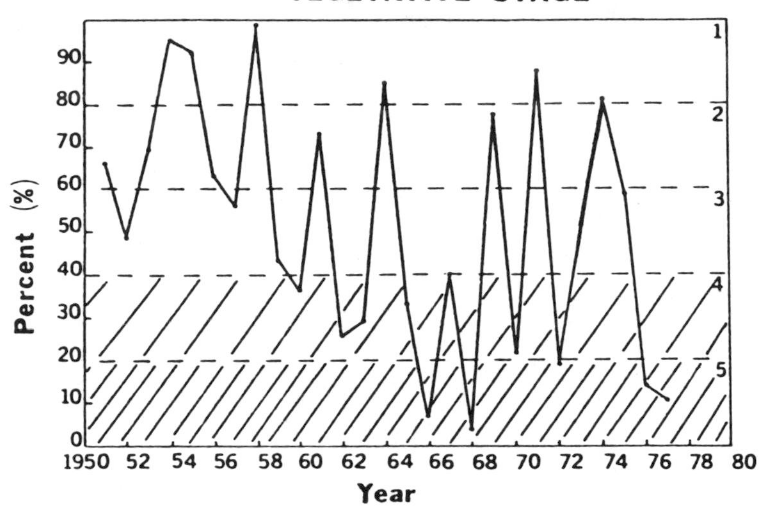

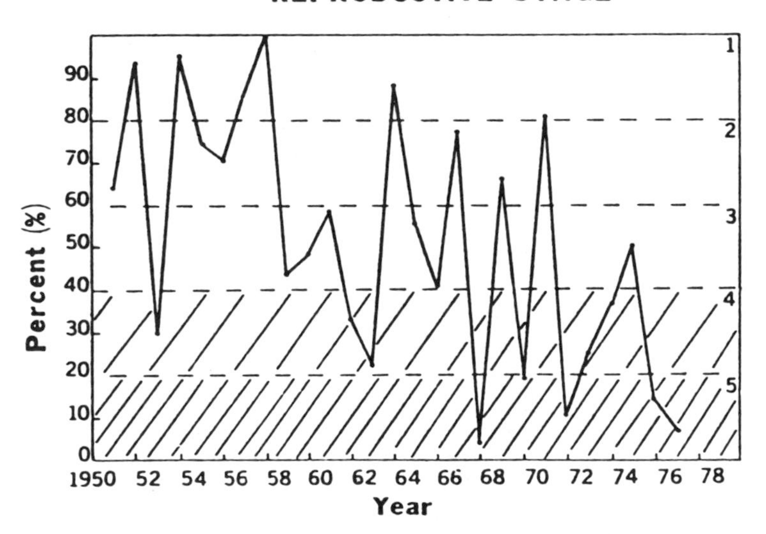

Figure 19.6. Time series of the percent ranking of the crop moisture ratio (P/PET). The numbers on the left are the quintals.

data from weather satellites for areas in all regions of the world (Tarpley, personal communication). This technique suggests that solar radiation may be available for use as a variable in large-area assessment models and may lead to the improvement in the use of the biophysical models that use solar energy as a driving variable.

Weather and climate data can be made available much more quickly with current technologies. Automated data systems, touch-tone telephone communication, and satellite communication will all be used in the future to obtain real-time data for assessments.

2. *Episodic events*. The simulations of crop response must include the occurrences of events that are sudden, unanticipated, and injurious to crop production. These episodic events are due to natural physical occurrences, such as hail, and to biological pests, i.e., insects and disease. For example, a freeze of May 1976 in Kansas at flowering of wheat caused a yield reduction. A corn blight infestation greatly reduced the total crop for 1970. Field experiments together with laboratory studies could help improve the modeling of these events at various periods of growth and development. Riesink (1981), for example, has successfully modeled peak larval feeding for alfalfa and corn rootworm using degree days. Production estimation requires a systems approach to these problems. As with data needs, remote-sensing studies can provide useful information on the spatial extent of physical characteristics and crop abnormalities.

3. *Remote sensing to augment regular observations*. There is a need to increase the precision of precipitation estimates for both field size or large-area assessments. The regular rainfall observations from synoptic and climatic stations must be supplemental to estimates derived from radar and satellite coverage. A multiagency federal program called AgRISTARS (Agriculture Resources Inventory Surveys Through Aerospace Remote Sensing) is currently addressing this issue. Both LANDSAT and meteorological satellites have shown significant potential for estimating crop productivity in an area.

4. *Technology impact*. The impact of management in terms of technological inputs into crop production will be a major factor in describing productivity over large areas. This suggests that field experiments should be carefully planned so that environmental data, including soil conditions, are concurrently observed on site to permit the evaluation of technology and the environment.

CONCLUDING REMARKS

The advances in computer technology have greatly aided the goals of simulating the impact of the environment on crop production systems. These simulations have provided estimates of biophysical processes and large-area productivity. The progress of the past decade has occurred because of the treatment of the total system of production in both biophysical and statistical models.

The knowledge gained and techniques developed are not complete. For additional gains in simulation the work of several disciplines needs to be integrated.

ACKNOWLEDGMENTS

Contribution from the Missouri Agricultural Experiment Station, Journal Series Number 8876.

This research has been partially sponsored through a grant from NOAA, U.S. Department of Commerce, Grant Number USDOC NA79AA-H-00104, which supports the program of USAID.

LITERATURE CITED

Baker, D. N., J. A. Landivar, and V. R. Reddy. 1979. *Plant responses to environmental conditions and modeling plant development.* Pages 69–109 in Proceedings Weather and Agriculture Symposium, 1979, W. L. Decker ed. Univ. Missouri Press.

Benci, J. F. 1974. *Soybean stomatal response to environmental conditions.* Ph.D. dissertation, University of Missouri-Columbia.

Bruce, R. R., and M. J. M. Romken, 1965. *Fruiting and growth characteristics of cotton in relation to soil moisture tension.* Agron. J. 51:135–40.

Curry, R. B. 1979. *Simulation of crop growth and production.* Proc. 10th Ann. Pittsburg Conf. on Modeling and Simulation.

Curry, R. B., G. E. Meyer, J. G. Streeter, and M. J. Mederski. 1980. *Simulation of the vegetative and reproductive growth of soybeans.* Pages 557–69 in *Proc. World Soybean Research Conf. 2.* Westview Press, Boulder, Co.

deWit, C. T. 1978. *Simulation of Assimilation, Respiration and Transportation of Crops.* John Wiley, New York.

Kanemasu, E. T., A. Feyerherm, J. Hanks, M. Keener, D. Lawlor, P. Rasmussen, H. Reetz, K. Saxton, and C. Wiegand. 1980. *Use of soil moisture information in crop yield models.* AGRISTARS pub. SM-MO-00462, NAS 9-14899.

Kreasuwun, Jiemjai. 1980. *Prediction of rice yields on the basis of climatic variables in Thailand.* Ph.D. dissertation, Universtiy of Missouri-Columbia.

LeDuc, S., C. Sakamoto, N. Stromment, and L. Steyaert. 1980. *Some problems associated with using climate-crop-yield models in an operational system: An overview.* Biometerology 7, Part 2, Pages 104–14 in Proc. of 8th Intern. Biometeor. Congr., Shefayin, Israel, Sept. 1979.

McCree, K. J. 1974. *Equations for the rate of dark respiration of white clover and grain sorghum as functions of dry weight, photosynthetic rate and temperature.* Crop Sci. 14:509–14.

Marquardt, D. W., and R. D. Snee. 1975. *Ridge regression in practice.* The American Statistician 29:3–20.

Nobel, P. S. 1975. *Introduction to Biophysical Plant Physiology.* W. F. Freeman & Co., San Francisco.

Richie, J. T. 1979. *Water management and water efficiencies for American Agriculture.* Pages 15–42 in Proc. Weather and Agric. Symp. 1979, W. L. Decker ed. Univ. Missouri Press.

Riesink, W. G. 1981. *Some insect pest management systems that utilize weather data.* Workshop on Applications of Weather Data to Agriculture and Forest Production, March 30–31, Anaheim, California.

Runge, E. C. A., and J. F. Benci. 1975. *Modeling corn production—estimating production under variable soil and climatic conditions*. Pages 194–214 in Proc. 30th Ann. Corn and Sorghum Res. Conf.

Steyaert, L., S. LeDuc, and J. McQuigg. 1978. *Atmospheric pressure and wheat yield modeling*. Agric. Meteorol. 19:23–34.

Strommen, N. D., C. M. Sakamoto, S. K. LeDuc, and D. E. Umberger. 1979. *Development of LACIE CCEA-L weather/wheat yield models*. Pages 99–108 in Proceedings of Technical Sessions, vol. I, the LACIE Symp. October 23–26, NASA-JSC July 1979.

Thompson, Louis M. 1962. *Evaluation of weather factors in the production of wheat*. J. Soil and Water Cons. 17:149–56.

———. 1963. *Evaluation of weather factors in the production of grain sorghum*. Agron. J. 55:182–85.

———. 1969. *Weather and technology in the production of wheat in the United States*. J. Soil and Water Cons. 24:219–24.

Wallace, H. A. 1920. *Mathematical inquiry into the effect of weather on corn yield in eight corn-belt states*. Mon. Wea. Rev. 48:439–46.

Walther, E. G., and P. R. Hayes. 1977. *Generating crop yield probabilities for policymakers*. Food Production/Climate Mission, Charles F. Kettering Foundation, Dayton, Ohio.

20] Radiation and Plant Response: A New View

by H. M. CATHEY*, L. E. CAMPBELL† and
R. W. THIMIJAN†

ABSTRACT

Traditionally, plants have been considered to respond to radiation in the 400–700 nm visible wavelength region and red, far-red 660, 730 nm phytochrome wavelength peaks. Research in growth chambers and greenhouses show that restricting consideration of this radiation in the 400–850 nm region is insufficient to explain the differences in plant response to various sources of radiation. Irradiance above 850 nm contributes and profoundly alters the growth responses. We propose a new theory of *binary classification* that divides plants into four groups depending on their thermal and spectral requirements. These are: (a) spectral-insensitive, thermal-insensitive; (b) spectral-insensitive, thermal-sensitive; (c) spectral-sensitive, thermal-insensitive; and (d) spectral-sensitive, thermal-sensitive. The theory is based on both plant response to radiation and the physical laws of emitted radiation, wherein all radiation upon absorption results in temperature increase in the absorbing medium. Thus, there is in theory a continuing spectrum from light to temperature. Our sources of radiation, including the sun, are not continuous over all wavelengths associated with the interval from light to temperature. Thermally sensitive plants will not tolerate prolonged exposure to nonvisible irradiance. Spectrally sensitive plants require broad or specific visible wavelengths of radiation—the ultraviolet, blue, and far-red wavelengths—to function efficiently.

NEW APPROACHES

Any appraisal as to how to reassemble the environmental elements for the culture of a wide range of plants must begin by establishing the plants' basic

*Director, National Arboretum, U.S. Department of Agriculture, Agricultural Research Service, Washington, D.C. 20002 and †U.S. Department of Agriculture, Agricultural Research Service, Beltsville Agricultural Research Center, Beltsville, Maryland 20705.

STRATEGIES OF PLANT REPRODUCTION (BARC Symposium number 6—Werner J. Meudt, ed.)
Allanheld, Osmun, Totowa

requirements for growth (Gaastra 1959; Evans 1963). Even though the first artificial lighting trial for plants was reported by Bailey in 1893, we still have not taken the steps, on a broad range of economic plants, to establish these baselines.

We saw from 1920 to the early 1960s the identification, manipulation, and commercial application of the photoperiodic responses of plants (Garner and Alland 1920; Downs et al. 1958; Lane, Cathey, and Evans 1965). A low-wattage incandescent lamp, mounted one to several meters above the plants, turned on for several hours continuously or intermittently each night, was sufficient to alter the flowering and the dormancy of a limited range of plants (Cathey et al. 1975). Chrysanthemums and carnations benefited particularly from their sensitivity to controlled daylength lighting. By lighting at night and by excluding the natural light they could be programmed to be salable for any specific date. Most commercial plants, however, were relatively insensitive to exposure to photoperiodically active lighting systems.

We saw beginning in the 1940s the development of fluorescent plant growth lamps that emit more red and blue wavelengths at the absorption peaks of chlorophyl than the conventional white lamps (Bickford and Dunn 1972; Cathey et al. 1975). These lamps were generally less efficient in radiant power output, and the red phosphors deteriorated more rapidly than conventional vision lamps. Their one distinctive advantage was enhancing colors of foliage and flowers. Other fluorescent wide-spectrum lamps were supposed to simulate the spectrum of the sun (Cathey et al. 1975). These lamps also improved color rendition but were lower in radiant power output than conventional lamps. Low-cost incandescent plant lights appealed to consumers because they were easy to install in conventional sockets, but these tinted incandescent lamps contributed no more to plant growth than conventional incandescent spotlights or floodlights. All incandescent lamps emit a beam of intense heat and because of their spectral composition promote stem elongation and paling of the foliage.

Recent plant research (Sevain 1964; Austin and Edrich 1974; Canham 1974; Duke et al. 1975) has identified that growth and flowering of a wide range of previously uncontrollable plants can be regulated when they were grown in greenhouses supplemented with the high-efficiency light sources. Their use has also required some highly complex analyses of both plant response and spectral radiation (Cathey and Campbell 1980). Based on these observations, we are proposing a new theory as to how plants utilize radiant energy.

We approached plant lighting on the basis of an efficiency of power conservation of sources to radiation rather than from simulation of the radiation from the sun (Campbell, Thimijan, and Cathey 1975). Plants were subjected to specific, selected, radiation environments. We used lamp emissions ratings of watts per square meter as the key to selecting lamps for plant growth studies (Cathey and Campbell 1974). Based on their radiation power distribution, we compared the various light sources with equal watts/m^2 from 400 to 850 nm. The traditional research on photosynthesis reported the radiation only from 400 nm to 700 nm (Balegh and Biddulph 1970; McCree 1971); where as we included the red–far-red absorbing regions from 700 to 850 nm to extend to the absorption range of photochrome (Jose and Vince-Prue 1978). When lamps of different spectral content were compared with equal irradiance W/m^2 (400 to 850 nm), we observed that various kinds of lamps were very similar in effectiveness for many plant growth responses (Cathey and Campbell 1979). For the detailed comparisons we selected cool white fluorescent (CWF) (blue, green, and yellow), high-pressure sodium (HPS) (peaks in yellow-orange with radiation into the green and red), and low-pressure sodium (LPS) (a narrow line in the yellow-orange) and exposed the plants growing in greenhouses to daily cycles of supplemental light during the day, day into the night, night, night into the day, and throughout the 24 hours. What emerged were several significant conclusions:

When compared at 24 W/m^2 (400 to 850nm) on 16 in 24 hr, or 6 W/m^2 for 24 hr, CWF, HPS and LPS were equally effective in regulating growth on a wide range of plants. We found that the spectral composition was not as criticial as equal irradiance in watts/meter square, in the region of 400 to 850 nm (Cathey and Campbell 1979).

Sunlight, even on the dimmest days in winter, was an essential part of the radiation system for plants in the greenhouse. Abnormalities observed with certain species grown under sodium lamps as a sole source in growth chambers were not evident on the plants grown in greenhouses with daylight and supplemental radiation from CWF, HPS, and LPS.

Lighting only during the day, or only at night during even the dimmest winter day, for 8 hr or less, was relatively ineffective for regulating growth. Lighting during the day and continuing into the night or lighting continuously throughout the 24 hr was the most-effective cycle to regulate growth in many species. Lighting for too many hours or with too great an intensity inhibited stem elongation, reduced leaf coloration and expansion, and in some species often severely crippled all growth processes. Plant species varied greatly in their tolerance or utilization of the supplemental light (Cathey and Campbell 1979).

The majority of annual and perennial plants developed (more accumulation of fresh and dry weight) and flowered much earlier than usually observed for the species grown only under natural day light. Regardless of their traditional classification as to their photoperiod responses, the majority of the species responded as if they were long-day plants. The growth responses were not as much related to daylength as to average total energy supplied over the 24 hr (Singh, Ogren, and Widholm 1974). Low-intensity incandescent light (0.9 W/m^2 400 to 850 nm), used as an extension to the day or as a night interruption, promoted the vegetative growth, and the plants were dark green in appearance. Only a few species exhibited this marked daylength dependence for growth. The obligate photoperiodic plants (short day for flowering or short day for dormancy) were the exception to general responses. These plants rapidly exhibited abnormalities of growth and flowering. They continued to require long nights for flowering or for the onset of dormancy. Supplemental lighting could be given to these plants only during the daylight hours.

These findings have two opposing concepts in them. First, we observed in growth chambers that many plants exhibited relatively nonspectral responses to lighting with CWF, HPS, and LPS lamps. When these sources were used to

Table 20.1 Radiation Power Distribution of Light Sources per 100 Watts of Total Radiation

Region	Spectrum limits	Light source (W)				
		DL	INC	CWF	HPS	LPS
Ultraviolet	Less than 400 nm	6	0.2	2	0.4	0.1
Light	400–850 nm	59	17	36	50	56
Infrared	850–2700 nm	33	74	1	12	3
Thermal	Beyond 2700 nm	2	9	61	38	41
Total		100	100	100	100	100

supplement the light available in the greenhouse in the winter, even spectral differences were not apparent on most plants. Second, we observed another opposing concept with certain other plants growing in greenhouses supplemented with CWF, HPS, and LPS. As we increased the intensity of radiation from 12 to 48 w/m^2 (400 to 850 nm) and extended the lighted period from 12 hrs, to 24 hrs, certain plants developed extremely chlorotic, crippled leaves and plants which actually weighed less than the unlighted ones. We believe that the explanation for this second type of plant growth can be traced to the distribution of energy between the visible (400 to 850 nm), infrared (850 to 2700 nm), and thermal (beyond 2700nm) (Tables 20.1 & 20.2). The radiant power of daylight is divided equally between the visible and the infrared (IR), with only traces in the thermal region. The artificial light sources present a very different ratio than daylight. All of the artificial light sources have significant irradiance in the thermal region. CWF and LPS have only trace amounts of IR, while HPS has a third as much IR as does daylight (Table 20.1). Some plants are apparently tolerant of intense and prolonged supplemental radiation, but others are intolerant of the radiation in the nonvisible region. In other words, it takes much more daylight to radiant heat plants than light from artificial light

Table 20.2 Radiation Power Distribution of Sources per 10,000 Lux

Region	Spectrum limits	Light source (W/m^2)				
		DL	INC	CWF	HPS	LPS
Ultraviolet	Less than 400 nm	5.6	0.8	1.3	0.3	0.04
Light	400–850 nm	55.1	90.0	30.0	33.8	21.8
Infrared	850–2700 nm	31.0	473.0	1.3	8.4	1.2
Thermal	Beyond 2700 nm	1.0	58.0	50.3	25.6	16.1
Total		93	622	83	68	39

sources. There are no filters or cooling systems that will remove all of the nonvisible irradiance. This nonvisible irradiance apparently has confounded experiments with artificial light sources from the very beginning experiments by Bailey in 1893.

In conclusion, we classified plants into four basic groups as follows. The spectral sensitivity is based on studies in growth chamber where the light sources are the sources of radiant energy. Thermal sensitivity is based on studies in greenhouses where natural light was supplemented with various energy-efficient light sources.

1. Spectral-insensitive, thermal-insensitive

Light	visible
Lamp	any source
Duration	24 hrs
w/m^2	12 to 48

We find that most bedding plants, foliage plants, small needle evergreens, and rose exhibit a spectral-thermal insensitive type of response when grown with supplemental light in greenhouses in the winter. Increasing the irradiance from 12 to 48 W/m^2 and from 12 to 24 hrs gives additional increments to the rate of fresh weight accumulation and reduces the number of days to flowering. Plants tolerate radiant heat absorbed from IR and thermal sources over a wide range of ambient temperatures. Higher temperatures may improve growth and development.

2. Spectral-insensitive, thermal-sensitive

Light	visible
Lamp	any source
Duration	to 16 hrs
W/m^2	12 to 24

We find that plants such as tomato and geranium exhibit spectral insensitive-thermal sensitive type of response when grown with supplemental lighting in greenhouses in the winter. Increasing the irradiance from 12 to 24 W/m^2 and from 12 to 16 hrs daily gives additional increments to the rate of fresh weight accumulation and reduces the number of days to flowering. Increasing the irradiance beyond 24 W/m^2 and the duration to 24 hrs daily did not give additional responses in growth. The plants actually appeared to shrink in size. We were observing a thermal overloading of the plants which inhibited leaf, stem, and lateral branch development. Heat from IR or thermal radiation, which raises the effective plant tissue temperature, is repressive for these plants.

3. Spectral-sensitive, Thermal-insensitive

Light	UV, blue, red, far-red
Lamp	any source & DL
Duration	24 hrs
W/m^2	12 to 48

We find that plants such as lettuce and strawberry exhibit this type of response when grown with supplemental lighting in greenhouses in the winter. Increasing the irradiance from 12 to 48 W/m^2 and from 12 to 24 hrs daily gives additional increments to the rate of fresh weight accumulation and reduces the number of days to flowering. It is essential that daylight, even traces of it for 5 to 7 hrs daily, be present in the greenhouse to correct spectral deficiencies of HPS and LPS.

Radiant heat from the sources is tolerated over a wide range of ambient temperatures. Higher temperatures may improve growth and development. Plant temperature within a fairly wide range is not critical.

4. Spectral-sensitive, thermal-sensitive

Light	UV, blue, red, far-red
Lamp	any source & DL
Duration	to 16 hrs
W/m^2	12 to 48

We find that plants such as chrysanthemum, carnation, poinsettia, and most deciduous trees exhibit spectral sensitive-thermal sensitive type of response when grown with supplemental lighting in greenhouses during the winter. Increasing the irradiance from 12 to 48 W/m^2 and from 12 to 16 hrs daily gives additional increments to fresh weight accumulation. Prolonging the daily exposure beyond 16 hrs. regardless of the type of artificial light sources used, created an environment where the plants developed chlorotic foliage.

Many of the deciduous woody plant species, regardless of the irradiance level and duration, ceased growth immediately when placed under the supplemental light systems when the experiments were conducted during the winter months. Other plants, of the species given supplemental lighting with incandescent lamps with an irradiance of 0.9 w/m^2 remained vegetative over the same period of time.

The timing of obligate short-day plants such as chrysanthemum and poinsettia was completely disrupted. Spectral composition, its irradiance level and duration, as well as ambient temperatures, were extremely critical for these plants. Green, rapidly elongating plants developed only when they were grown in an environment with fairly restricted spectral composition, irradiance level, and duration. Acceptable-appearing plants developed only when UV or blue, red, and far-red were supplied from natural daylight or incandescent-filament lamps in addition to supplemental HID lamps. Rapid flower initiation and development of acceptable plants occurred only when the daylength was restricted to 12 hrs or less.

A NEW VIEW

This summary of recent observations of plants grown in growth chambers and in greenhouses supplemented with radiation from HID lamps approaches plant growth from a new view. We have made no attempt to restrict or to ignore previously considered extraneous radiation. We believe that all of the radiation in the ultraviolet (less than 400 nm), visible (400 to 850 nm), infared (850 to 2700 nm), and thermal (beyond 2700 nm) affect plant growth directly on light-mediated systems or indirectly by altering the temperature of the plant tissue. Previous studies by other workers have emphasized the simulation of daylight with radiation between approximately 300 to 2700 nm (Aldrich and White 1969). The electric light sources, other than INC, emit nearly as much thermal as they do visible light. Simulation of daylight with these electric light sources is thus neither feasible nor possible. While in theory daylight simulation is possible, any combination of sources, including Xenon, gives a different spectrum from sunlight. When one wavelength region such as the visible region is matched, the IR or thermal region will be different. Recent studies with incandescent lamps, or their high intensity equivalents, also do not simulate daylight. Unlike daylight, light from incandescent lamps is high in infrared with less-visible radiation. Even with water filters and an efficient ventilation system, the radiation from incandescent lamps is unlike daylight. We believe these seemingly subtle differences have profound and dramatic effects on the growth systems of the plants and explain the often sharp disagreements between different workers. We have proposed a binary classification of plant growth based on spectral and thermal sensitivities to aid in creating the most productive and efficient environment. The classification permits many new approaches to plant growth. The research in growth chambers can no longer be considered as a separate situation from plants growing in greenhouses and out-of-doors (Krizek et al. 1968). Regardless of how the artificial light source is used—as a sole source (or in combinations) in warehouses or as a supplement in greenhouses, the same basic information can be utilized (ASHS Special Committee).

Plant scientist and engineers will have the opportunity to design and monitor an expanding area of growing facilities:

- Radiant environments for plants should be designed to utilize for growth the visible, the near—and far-infrareds, thermal, and the heat energy from the starting and ballast system.

- The growing space for greenhouses should be restricted to utilize the maximum potentials of solar heating and collection while reducing to a minimum the energy loss through walls, floors, and the ceiling (particularly at night).

· The structures should be decked and automated to handle several layers of plants and environmental requirements within the same space (White 1979).

In all cases, many unanswered questions confront plant scientists and engineers. We must develop experimental approaches to determine the baseline requirements of plants—the radiation required, the duration, the interactions with the environment, and the degree of tolerance to stress(es) that plants possess. We need the creative interactions of biological scientists, economists, and engineers to work toward energy-efficient crop production programs and facilities.

LITERATURE CITED

Aldrich, R. A., and J. W. White. 1969. *Solar radiation and plant growth in greenhouses.* Trans. Amer. Soc. Agric. Eng. 12:1.

ASHS Special Committee. *Growth chamber environments. 1977. Revised guidelines for reporting studies in controlled environment chambers.* Hort. Sci. 12:309–10.

Austin, R. B., and J. A. Edrich. 1974. *A companion of six sources of supplementary light for growing cereals in greenhouses during winter time.* J. Agric. Eng. Res. 19:339–45.

Bailey, L. H. 1893. *Greenhouse notes: Third report upon electro-horticulture.* Cornell Univ. Agric. Exp. Sta. Bull. 55:127–238.

Balegh, S. E., and O. Biddulph. 1970. *The photosynthetic action spectrum of the beanplant.* Plant Physiol. 46:1–5.

Bickford, E. D., and S. Dunn. 1972. *Lighting for Plant Growth.* Kent State Univ. Press, Kent, Ohio.

Brown, J. C., H. M. Cathey, J. H. Bennett, and R. W. Thimijan. 1979. *Effect of light quality and temperature on* Fe^{3+} reduction, and chlorophyll concentration in plants. Agron. J. 71:1015–21.

Campbell, L. E., R. W. Thimijan, and H. M. Cathey. 1975. *Spectral radiant power of lamps used in horticulture.* Trans. Amer. Soc. Agric. Eng. 19(5):952–56.

Canham, A. E. 1974. *Some recent developments in artificial lighting for protected crops.* Proc. 19th Intern. Hort. Congr. Warsaw, Sept. 11–18, 1974. Pages 267–76.

Cathey, H. M., and L. E. Campbell. 1974. *Lamps and lighting—a horticultural view.* Lighting Design and Application (IES) 4(11):41–51.

———. 1977. *Plant productivity: New approaches to efficient sources and environmental control.* Trans. Amer. Soc. Agric. Eng. 20(2):360–71.

———. 1979. *Relative efficiency of high- and low-pressure sodium and incandescent filament lamps used to supplement natural winter light in greenhouses.* J. Amer. Soc. Hort. Sci. 104:812–25.

———. 1980. *Light and lighting systems for horticultural plants.* J. Janick, ed. Pages 491–537 in *Horticultural Reviews*, vol. 2. AVI Publishing Co., Westport, Connecticut.

Cathey, H. M., L. E. Campbell, and R. W. Thimijan. 1975. *Comparative development of 11 plants grown under various fluorescent lamps and different duration of irradiation with and without additional incandescent lighting.* J. Amer. Soc. Hort. Sci. 103:781–91.

Cathey, H. M., F. F. Smith, L. E. Campbell, J. G. Hartsock, and J. U. McGuire. 1975. *Response of Acer rubrum L. to supplemental lighting, reflective aluminum soil mulch, and systemic soil insecticide.* J. Amer. Soc. Hort. Sci. 100:234–37.

Downs, R. J., H. A. Borthwick, and A. A. Piringer, Jr. 1958. *Comparison of incandescent and fluorescent lamps for lengthening photoperiods.* Proc. Amer. Soc. Hort. Sci. 71:568–78.

Duke, W. B., 1975. *Metal halide lamps for supplemental lighting in greenhouses. Crop response and spectral distribution.* Agron. J. 67:49–63.

Evans, L. T. 1963. *Extrapolation from controlled environments to the field.* Pages 421–37 in L. T. Evans, ed. *Environmental Control of Plant Growth.* Academic Press, New York.

Gaastra, P. 1959. *Phyotosynthesis of crop plants as influenced by light, carbon dioxide, temperature, and stomatal diffusion resistance.* Meded v.d. LBHS to Wageningen 59(13):11.

Garner, W. W., and H. A. Allard. 1920. *Flowering and fruiting of plants as controlled by the length of day.* Pages 377–400 in *USDA Yearbook 1920.* U.S. Dept. of Agric. Washington, D.C.

Jose, A. M., and D. Vince-Prue. 1978. *Phytochrome action: A reappraisal.* Photochem. & Photobiol. 27:209–16.

Krizek, D. T., W. A. Bailey, H. H. Klueter, and H. M. Cathey. 1968. *Controlled environments for seedling production.* Prod. Intern. Plant Prop. Soc. 18:273–80.

Lane, H. C., H. M. Cathey, and L. T. Evans. 1965. *The dependence of flowering in several long-day plants on the spectral composition of light extending the photoperiod.* Amer. J. Bot. 52:1006–14.

McCree, K. J. 1971. *The action spectrum absorptance and quantum yield of photosynthesis in crop plants.* Agr. Metrord. 9:191-216.

Singh, M., W. L. Ogren, and J. M. Widholm. 1974. *Photosynthetic characteristics of several C_3 and C_4 plant species grown under different light intensities.* Crop. Sci. 14:563–66.

Sevain, G. S. 1964. *The effect of supplementary illumination by mercury vapor lamps during periods of low natural light intensity on the production of Chrysanthemum cuttings.* Proc. Amer. Soc. Hort. Sci. 85:568–73.

White, J. W. 1979. *Energy efficient growing structures for controlled environment agriculture.* Pages 141–71 in J. Janick, ed. *Horticultural Reviews*, vol. 1. AVI Publishing Co., Westport, Connecticut.

21] Effects of Air Pollution on Crop Production

by WALTER W. HECK,*† UDO BLUM,†
RICHARD A. REINERT,*‡ and ALLEN S. HEAGLE*‡

ABSTRACT

This chapter presents a general overview of our current understanding of the effects of ozone, sulfur dioxide, and nitrogen dioxide on crop production. The importance of understanding pollutant sources and emission characteristics, meteorology, atmospheric chemistry, and atmospheric deposition is shown. Our present understanding of life stages that are sensitive to these pollutants is characterized and the effect on dry matter allocation is discussed as it affects biomass and yield. Plant leaves are most sensitive when they have just reached full expansion; foliar effects after flowering generally cause the greatest effect on seed yield. It is difficult to assess yield losses because fairly small losses occur in many crops over a large area. It is certain, however, that sizable annual losses occur throughout the United States; no area east of the Mississippi escapes some loss to important agricultural crops. Strategies to reduce crop losses to air pollutants are described, and the role of genetic variability is mentioned.

INTRODUCTION

The effects of air pollutants on crops have been studied for many years (Jacobson and Hill 1970; Mudd and Kozlowski 1975; Guderian 1977; Heck and Brandt 1977). Sulfur dioxide (SO_2) was identified from the smelting of ores, and its effects on vegetation were studied during the first half of the century (Heck and Brandt 1977). Little SO_2 effects research was done in the United States following this early work until SO_2 stress on cultivated and natural ecosystems

*Agricultural Research Service, U.S. Department of Agriculture, †Botany Department, North Carolina State University, and ‡Plant Pathology Department, North Carolina State University, Raleigh, NC 27650.

STRATEGIES OF PLANT REPRODUCTION (BARC Symposium number 6—Werner J. Meudt, ed.)
Allanheld, Osmun, Totowa

Table 21.1 List of Phytotoxic Air Pollutants in Order of Importance to Plant Systems

Pollutant[a]	Primary or secondary[b]	Major source(s)
O_3	Secondary	Atmospheric transformation associated with automotive emissions (NO_2, hydrocarbons)
SO_2	Primary	Power generation, smelter operations
NO_2	Primary and secondary	From direct release and atmospheric transformation (high temperature combustion, from NO); fertilizer production
HF	Primary	Super-phosphate, aluminum smelters
Ethylene	Primary	Combustion, natural
PAN-Oxid.	Secondary	Atmospheric transformation (automotive emissions, NO_2, hydrocarbons)
NO	Primary	Combustion, natural
Cl_2	Primary	Spills, manufacture
HCl	Primary	Burning of plastics
Toxic elements	Primary	Smelters, combustion processes
NH_3	Primary	Feedlots, natural
$SO_4^=$	Secondary	Atmospheric transformation (SO_2)
NO_3^-	Secondary	Atmospheric transformation (NO_2)
H_2S	Primary	Paper production, natural, geothermal
CO_2	Primary	Combustion, natural

[a]These are all gases except that F may occur in a particulate phase, the toxic elements occur as particulates or aerosols, and both $SO_4^=$ and NO_3^- occur as aerosols.
[b]Primary pollutants are emitted directly from a source, while secondary pollutants are formed through atmospheric transformation.

Note: This list is not meant to be complete but represents the most phytotoxic air pollutants endangering crop systems. The major component of acidic deposition ($SO_4^=$ and NO_3^-) and CO_2 have been poorly studied and may be more important than presently thought.

Source: Heck (1981) with modification.

increased with the growth (number and size) of coal-fired power plants. The photochemical oxidant complex was identified in the late 1940s as causing injury to plants (Middleton, Kendrick, and Schwalm 1950), but ozone (O_3) was not identified as the active phytotoxic component of this complex until 1958 (Richards, Middleton, and Hewitt 1958). Nitrogen dioxide (NO_2) is released in all high-temperature combusion processes and is an important chemical reactant in the photochemical formation on O_3. Generally, NO_2 alone is not directly associated with plant injury at most ambient concentrations (National Research Council 1977). In mixtures of gases (e.g., SO_2 and NO_2), however, NO_2 may cause injury to vegetation (Tingey et al. 1971).

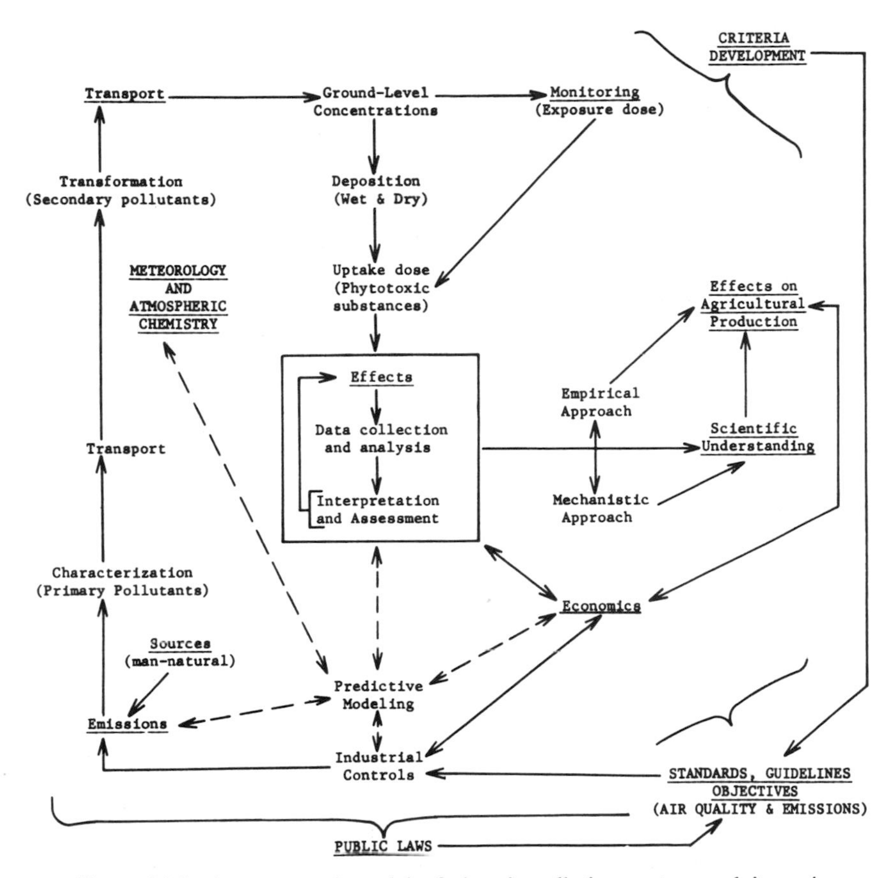

Figure 21.1. A conceptual model of the air pollution system and its major components. The solid lines suggest direct linkages; the dotted lines are areas of uncertainty associated with predictive concepts. The lines with an arrow for Criteria, Standards, and Public Laws show that all aspects of the system are included in developing these three pivotal parts. The central effects box refers to all effects, but this paper deals specifically with crops. (From Heck 1981.)

Ozone, SO_2, and NO_2 are the most important phytotoxic air pollutants in the United States (Table 21.1) because of the extensive area of their occurrence and their phytotoxicity at ambient concentrations, either alone or in mixtures. This chapter will focus on these three pollutants with emphasis on O_3. Ozone and NO_2 are ubiquitous across the US, while SO_2, mainly a local problem in the past, is becoming more ubiquitous because of the increase in electric generating plants that utilize coal and because of the increased use of tall stacks to dissipate pollutants. Other phytotoxic components in the atmosphere (e.g., gases or aerosols) may have regional or local importance (Table 21.1). These other pollutants are not covered in this chapter because their importance, on a national basis, is unclear.

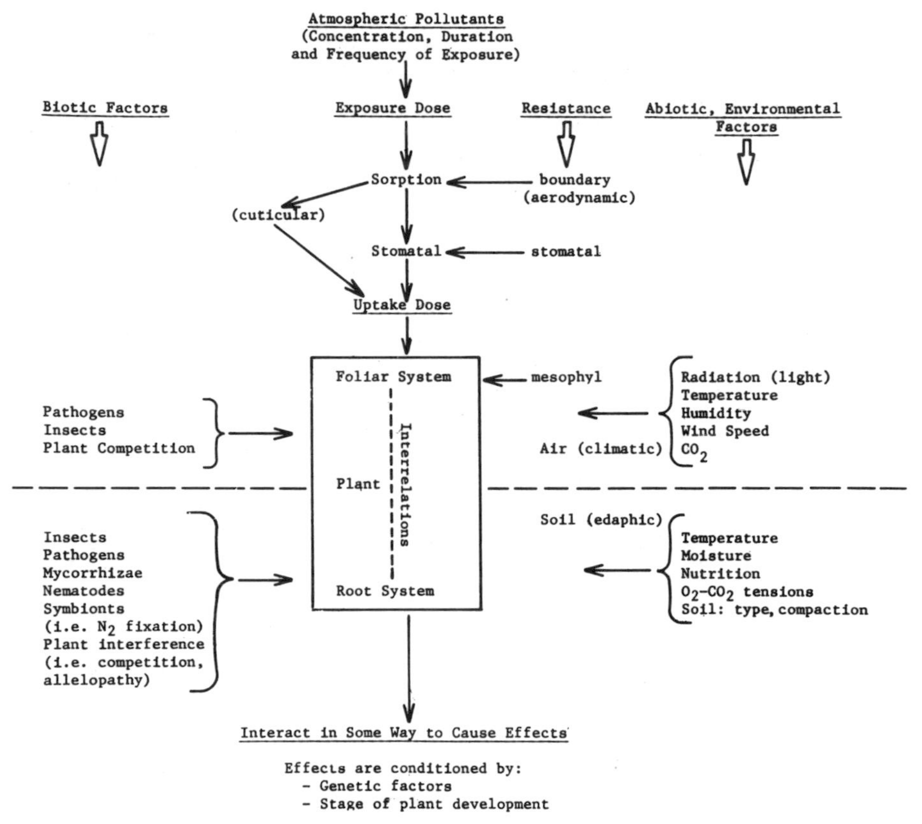

Figure 21.2. Biotic and abiotic factors that can affect the response of plants to an air pollutant exposure dose. (From Heck 1981.)

The magnitude of the air pollution problem suggests that crop scientists must develop some understanding of air pollution sciences. A conceptual model of the air pollution system is shown in Figure 21.1. It combines industrial, scientific, economic, and political concepts. For a more in-depth discussion of this model see Heck (1981).

In the United States there is increasing concern for the widespread occurrence of O_3, SO_2, and NO_2 in conjunction with each other at both chronic and acute concentrations. Several studies have shown that one pollutant may enhance, antagonize, or have no effect on the response of a given plant species to the other pollutants (Reinert, Heagle, and Heck 1975). Since these pollutants are part of the environmental complex surrounding crop species, their combined effects on crop production must be understood.

No discussion of pollutant effects on plants would be complete without mentioning that many biotic and abiotic stress factors affect the response of plants to gaseous pollutants. These biotic and abiotic factors have been

Table 21.2 The Relative Sensitivity of Crops to Air Pollutants at Different Life Stages

Stage of growth	Sensitivity[a]
Seeds	0
Germination	0
Seedling	5
Exponential growth	
Leaf (foliage)	3
Stem	1
Root[b]	4
Reproductive	
Pollen germination or growth	3
Floral development	2
Seed development	4-5

[a]Subjective estimates based on published data. They are primarily a reflection of foliar injury, but effects on biomass are reflected in these estimates. Sensitivity is graded on a scale of 0-5 with 0 being insensitive and 5 being very sensitive.
[b]The greater root sensitivity in relation to the foliage reflects preferential allocation of photosynthate to the foliage, when injury does occur.

Note: The sensitivity of a given life stage does not necessarily reflect its importance in crop production.

addressed in numerous review articles (Heck 1968; Heagle 1973; Heck, Mudd, and Miller 1977). The conceptual diagram shown in Figure 21.2 depicts the various factors that can affect the response of plants to O_3, SO_2, and NO_2.

In the following sections we discuss four concepts associated with an understanding of air pollution effects on crop production: sensitive life stages, dry weight partitioning in the plant, crop loss assessment, and strategies to reduce pollution effects.

SENSITIVE LIFE STAGES

Major goals in agricultural research are to increase yield and quality of crops. One of the first steps in achieving this goal is to understand how various stresses (e.g., air pollutants) affect crop production. Plants in nature are exposed to acute doses (relatively high concentrations of pollutants for short durations) and chronic doses (low concentrations of pollutants over a period of several or many days). Since plant sensitivity to a pollutant will vary over time to both acute and chronic exposures, knowledge of when a crop plant is most sensitive to a pollutant will aid in the development of models for estimating crop losses and may ultimately aid in the development of protective cultural measures.

Table 21.2 lists various plant life stages and our estimate of the relative sensitivity of each stage to O_3, SO_2, or NO_2. Seeds, germinating seeds, and developing seedlings prior to full expansion of the first leaf (leaves) are unaffected by O_3, SO_2, or NO_2 at current ambient concentrations. The most-sensitive life stage is the seedling stage, after the first true leaf reaches 75% of full expansion. An acute exposure at this stage will slow development and may delay yield or influence market quality of the crop (e.g., delayed harvest of tomato [Henderson and Reinert 1979]).

Both chronic and multiple acute exposures to O_3 (and probably SO_2 and NO_2) during a plant's exponential growth phase will reduce growth and may delay maturity (Oshima et al. 1979; Blum and Heck 1980). Nevertheless, chronic exposures will generally have a greater affect on plant yield than will a few acute exposures (Oshima et al. 1979, Figure 21.3). For leaf and root crops, the primary period of effect on these plant organs is during the exponential phase of growth. There is no evidence that O_3, SO_2, or NO_2 affect roots directly.

For seed and fruit crops, there is considerable evidence that both acute and chronic exposures, after the exponential phase of growth, seriously affect final seed yield because of photosynthate production (e.g., square loss in cotton; floral development in marigold [Reinert and Sanders 1981]). Reduction in fruit and seed numbers may also result from an unfavorable carbohydrate/nitrogen balance. Changes in seed quality have not been clearly shown (at least in relation to carbohydrate, protein, and fat ratios).

There is some evidence that air pollutants inhibit pollen germination and growth (Feder 1968).

DRY WEIGHT PARTITIONING IN PLANTS

Plant sensitivity is controlled partly by stomatal function and plant chemistry. Conditions that increase gas movement into leaves will increase plant sensitivity, and visa versa. Likewise, plants have variable capability for cellular repair or a variable ability to detoxify some level of the gases. Both the repair and detoxifying systems are affected by the growth stage and physiological health of the leaves. Thus, before we can understand the effects of an exposure of a given gas or gas mixture on crops, we must be able to determine gas uptake and know the physiological state of the plant tissue (Figure 21.2).

Immediately after acute or multiple low-level O_3 exposures, carbohydrate pools are modified within leaves of exposed plants. Both a depletion and an accumulation of various carbohydrate pools in the leaves (e.g., reducing sugars, sucrose, starch, total nonstructural carbohydrates) have been reported for a variety of species: soybean (Tingey, Fites, and Wickliff 1973), white pine (Wilkinson and Barnes 1973; Bennett, Heggestad, and McNulty 1977; Hanson and Stewart 1970), and unpublished data for clover.

Both the depletion of carbohydrate pools and the accumulation of carbohydrate pools in leaves suggest that sink-source relationships for plants are

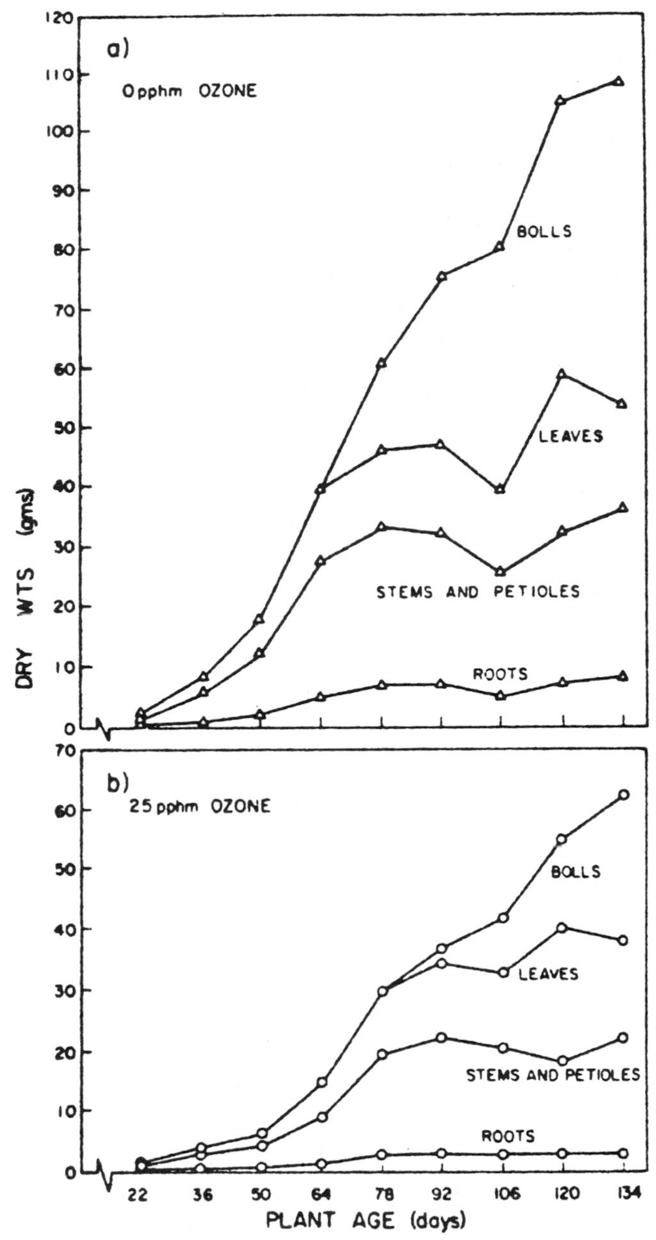

Figure 21.3. Plant (Acala SJ-2 cotton) and partitioned boll, leaf, stem, and petiole, and root growth of control plants and plants exposed to 0 or 0.25 ppm of 0_3 for 6 hr/day, 2 days/wk from 8 days of age to maturity. (From Oshima et al. 1979.)

Table 21.3 Effects of Simultaneous Exposure to NO_2, SO_2, and O_3 on Fresh Weight of Radish Roots

Gas concentration (ppm)		Root weight (g) per NO_2 concentration (ppm)		
SO_2	O_3	0.1	0.2	0.4
0.1	0.1	9.45	8.78	8.39
0.1	0.2	7.27	7.67	4.57
0.1	0.4	4.64	3.02	2.94
0.2	0.1	9.50	9.46	6.21
0.2	0.2	6.32	5.30	5.12
0.2	0.4	2.90	3.34	2.70
0.4	0.1	8.29	6.55	4.87
0.4	0.2	5.56	5.04	3.87
0.4	0.4	2.33	3.01	3.02

Note: Each value is the mean of 20 plants from four replications with an experimental unit of five plants. Least significant difference (LSD) value was 0.84 g, and the charcoal-filtered air bench control weighed 11.17 g. Radish plants were exposed one time for 3 hr at 16 days from seed and harvested at 23 days.

Source: Data from Reinert, Shriner, and Rawlings (1981).

modified by O_3. In the cases so far studied, this resulted in less carbon being allocated to roots (e.g., carrot—Bennett and Oshima 1976, Figure 21.4; soybean—Blum and Tingey 1977; radish—Reinert, Shriner, and Rawlings 1981, Table 21.3) and to developing flowers and fruits (e.g., begonia—Reinert and Nelson 1980; cotton—Oshima et al. 1979; Figure 21.3; pepper—Bennett, Oshima, and Lippert 1979; Figure 21.5).

For biennial and perennial crops the reduction of energy reserves in roots and other storage tissues may have an impact in subsequent years, since stored energy reserves for surviving periods of stress and overwintering are reduced (unpublished data for clover).

CROP LOSS ASSESSMENT

It is reasonble to say that between 75% and 90% of crop losses associated with air pollutants are due to chronic seasonal exposures to O_3 alone or in combination with NO_2 and SO_2. Thus, we must begin to develop meaningful estimates of crop losses due to O_3 dosages during the growing season. This information is needed by the Environmental Protection Agency (EPA) to develop criteria with which to promulgate standards. The use of cost/benefit analysis as a factor

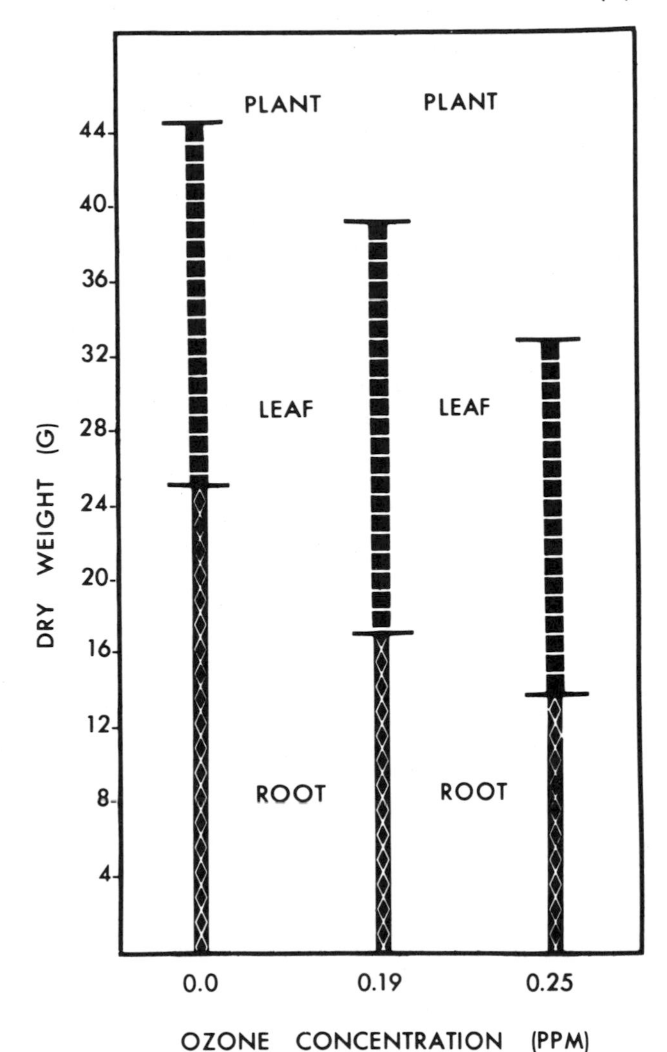

Figure 21.4. Dry weight partitioning to root and leaf of carrot exposed to different concentrations of 0_3 for 6 hr/day, 1.5 days/wk from first expanded leaf to maturity. (Adapted from data in Bennett and Oshima 1976.)

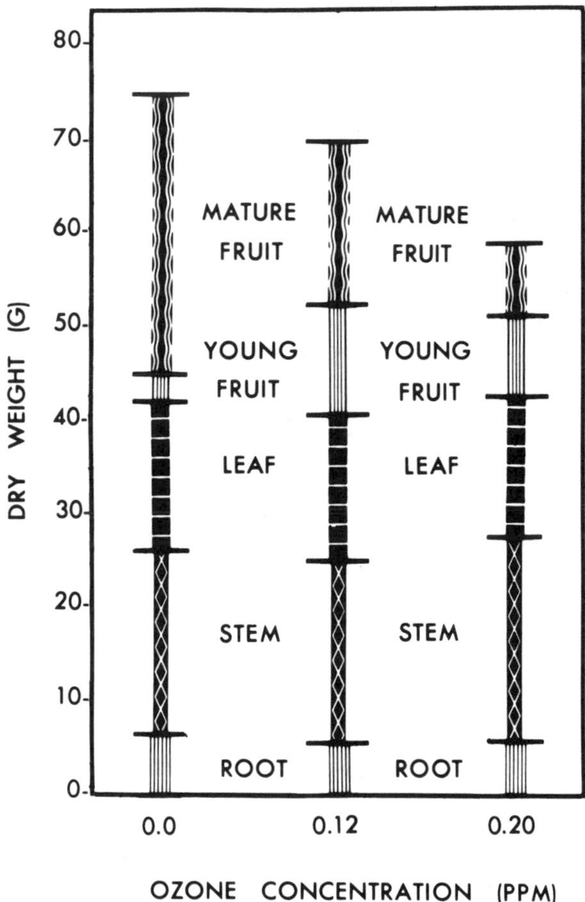

Figure 21.5 Dry weight allocation to different parts of pepper exposed to O_3 for 3 hr/day, 3 days/wk from 26 days of age to 103 days of age. (Adapted from data in Bennett et al. 1979.)

in pollution control strategy increases the need to more-accurately assess crop loss. Within the U.S. Department of Agriculture, we need to know the relative impact of air pollutants on different crop species in order to determine the amount of resources needed for the identification and development of resistant and tolerant cultivars and for research on ways to protect crops from air pollutants.

Measurement of crop loss caused by pollution stress is difficult. There is presently no universally accepted methodology for the assessment of air pollution effects. For episodic (acute) air pollution events in the field, epidemiological principles, similar to those used to assess crop loss caused by biotic

Table 21.4 Effects of O_3 on Ladino Clover Shoot Growth at Each of Three Harvests

Treatments[a]	Ozone concentration (ppm)[b]	Shoot growth at different harvests[c]			
		H_1	H_2	H_3	Season[d]
		mg dry wt/plant/wk			
Control	0.025	43.9	187.1	74.6	92.1
Ambient air		% of control			
Chamber	0.054	102_a	68_b	87_a	79_b
No chamber	0.054	116_a	55_c	99_a	77_b
Added O_3	0.081	130_a	24_d	32_b	51_c
LSD (%)		NS	10	23	

[a]Ozone was dispensed into chambers between 0930 and 1630 every day from June 19 to October 19, 1979.
[b]Ozone concentrations are shown as seasonal (June 19 to October 19) 7-hr/day averages. A small amount of O_3 was added to the chamber ambient air to bring the concentration up to the no-chamber treatment.
[c]Control values are given in mg/plant/wk, while treatment values are shown as percent of control. For the treatments the letter subscripts within a column (each harvest), when different, show treatment differences at the 5% level. Those with an "a" subscript are not different from the controls in a specific harvest.
[d]The seasonal values are adjusted for the varied lengths of each harvest.

Note: Harvest periods: H_1, April 29–July 26; H_2, July 27–September 13; H_3, September 14–October 19, for a total growth period of 175 days. Plant density was four clover and four fescue/pot.

Source: Data from Montes (1980).

pathogens, are probably valid. For chronic field exposures associated with O_3, however, this approach is not feasible for two reasons: (a) O_3 occurs throughout the eastern United States and the differences in average concentrations, over vast areas, are too slight to be called different; and (b) the concentrations of O_3 are near the threshold for measurable effects for many of the crop species. Thus, we generally would expect effects ranging from none or slight to possibly 10% to 15% reduction in productivity at any given location in any given year. The regional impacts of SO_2 and NO_2 alone are not known, but even at low concentrations we might expect an additive or greater than additive effect when either of these two gases is present with O_3 (Reinert, Shriner, and Rawlings 1981; Table 21.3).

Historically, crop loss assessment in the field has been most-successfully applied to severe air pollution problems from acute exposures associated with major nonmobile sources. In these studies, distance from the source and severity of injury can be used to develop dose-response functions.

Open-top field chambers (Heagle et al. 1979) have produced the most reliable dose-response data for chronic seasonal exposures, although a number

Table 21.5 Summary of Crop Yield at Different Seasonal Doses of O_3 (from studies in open-top field chambers)

Crop		Ozone concentration[a] (ppm)	Yield as % of the control for plants in[b]	
			Pots	Ground
Spinach		0.06	95	82
(4 cultivars)		0.10	73	62
		0.13	35	31
Winter wheat		0.06	102	97
(4 cultivars)		0.10	90	84
		0.13	73	67
Field corn		0.07	101	99
(Coker 16)		0.11	100	96
		0.15	87	85
Soybeans[c]	1977	0.07	102	—
(Davis)		0.12	64	—
		0.16	54	—
	1978	0.07	85	—
		0.10	70	—
		0.13	64	—

[a]The seasonal 7-hr/day (0930–1630 hrs) mean O_3 concentrations during the season when plants were exposed.
[b]Controls were plants in open-top chambers receiving charcoal-filtered air with seasonal 7-hr/day mean O_3 concentrations of 0.02–0.03 ppm.
[c]A two-year correlative study showed that the yield decreases for 'Davis' soybean at 0.10 ppm of O_3 was the same whether plants were grown in pots or in the ground.

Source: Heagle and Heck (1980), Table 2.

of other manipulative techniques have been tried or suggested. Open-top chambers permit the development of usable dose-response functions because they allow the control of pollutant doses in the air around the experimental plants. Thus, simulations of various pollution episodes can be accomplished. The open-top field chambers also cause the least possible change in the field environment.

The exposure regimes are based on the ambient O_3 concentration. Various constant increments of O_3 are added to the ambient concentration for a given number of hours per day to establish a series of step-wise doses (Heagle et al. 1979). Thus, crop yield can be determined at different doses, all with a constant relationship to ambient concentrations. Figure 21.6 shows an example of the seasonal 7-hr/day mean O_3 concentration for the charcoal filtered–air chamber (CF), the ambient air (AA–no chamber), the chamber with nonfiltered

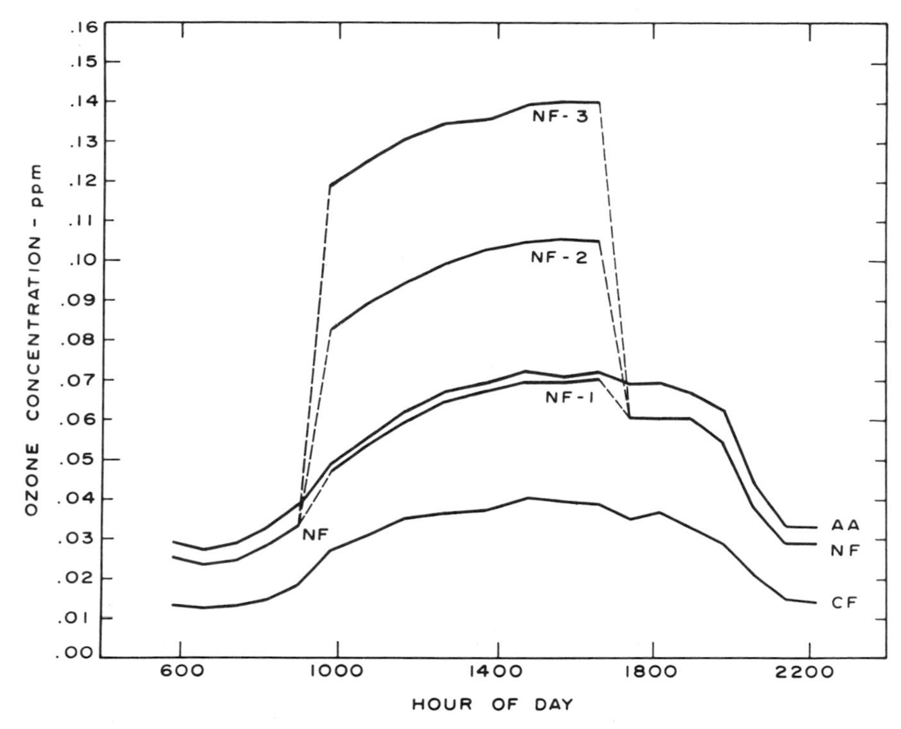

Figure 21.6. Seasonal mean 0_3 concentrations at different hours of the day in ambient air (AA), charcoal-filtered air chambers (CF), or in nonfiltered air chambers with different concentrations of 0_3 added (NF-1, NF-2, NF-3) for 7 hr (0930-1630 hr)/day. Concentrations shown are the means from 9 April through 31 May, 1977. (From Heagle, Spencer, and Letchworth 1979.)

air (NF), and the NF chamber treatments with 0.04 (NF-2) or 0.07 ppm (NF-3) of O_3 added to ambient concentrations for 7-hr/day (Heagle, Spencer, and Letchworth 1979). This design has been used to determine the effect of O_3 on the yield of several crop species (Tables 21.4 and 21.5; Figure 21.7). These yield data were used to estimate losses to several crops at different seasonal 7-hr/day mean O_3 concentrations (Table 21.6).

The use of the open-top chamber has been generally accepted by research workers within the air pollution field. It is the design of choice in a major EPA cooperative program (The National Crop Loss Assessment Network—NCLAN) to assess better the effects of pollutants on crop production. The NCLAN program will gather crop loss data from a number of regional sites. From these data, crop loss as a function of pollutant concentration will be used to develop crop loss models. As data become available, environmental parameters that are known or suspected of affecting plant response to the pollutants will be included in the crop loss assessment models.

Available data suggest that yield losses are occurring in soybean, wheat, and spinach at current ambient O_3 concentrations throughout the eastern United States (Table 21.6). There is considerable experimental and observational data

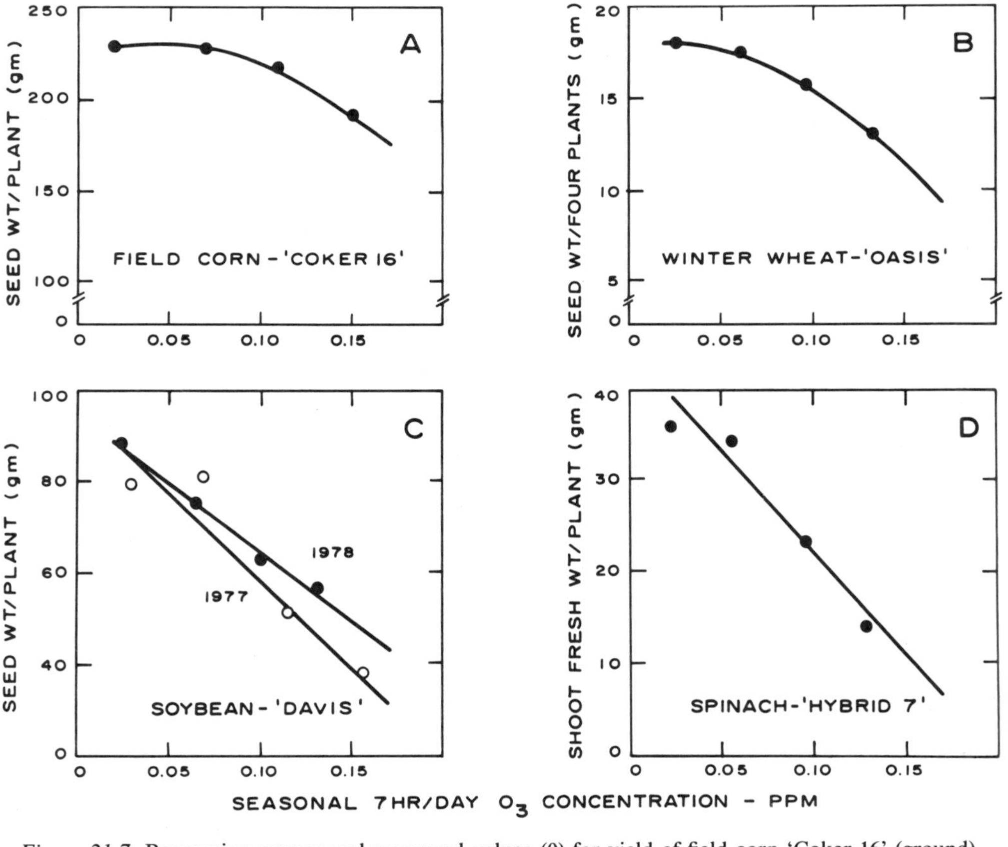

Figure 21.7. Regression curves and measured values (0) for yield of field corn 'Coker 16' (ground), winter wheat 'Oasis' (ground), soybean 'Davis' (pots), and spinach 'Hybrid 7' (ground) at different seasonal 7-hr/day mean 0_3 concentrations. (From Heagle and Heck 1980; Figure 4.)

to suggest that many other crop species are affected by O_3. The additional impact of effects from SO_2 and NO_2 in the presence of O_3 have not been addressed.

STRATEGIES TO REDUCE POLLUTION EFFECTS

The major strategy of the EPA is to reduce pollution effects in the United States by controlling emissions at the source (Figure 21.1). With the recent move to include cost/benefit analyses in setting standards, more emphasis must be placed on determining the economics of alternative strategies within the agricultural community.

Agricultural strategies considered to date include use of (a) cultural practices that reduce plant sensitivity (e.g., irrigation control), (b) protective sprays, (c) resistant cultivars, and (d) crop breeding programs. The last strategy is the one most accepted by plant scientists. There is evidence that this strategy will succeed, if current pollution levels do not significantly increase.

A number of crops have undergone at least some cultivar screening to determine the genetic variation in sensitivity to O_3, SO_2, and NO_2 (Table 21.7). These screens indicate that all crops tested have genetic resistance within their gene pool. Several attempts to select for resistance within crop species have been successful, but no major breeding program for pollutant resistance has been established.

Research concerning gene control of resistance has shown that the control mechanisms in crop species such as tobacco, petunia, sweet corn, and onion are different (Reinert, Heggestad, and Heck 1981). Tobacco shows an additive genetic variance with tolerance as a dominant factor; petunia is similar but shows only partial dominance; in sweet corn resistance is recessive; in onion, sensitivity to O_3 is dominant and involves a single gene control. Although these findings demonstrate a complexity of gene control for susceptibility to air pollutants, it may be relatively easy to develop breeding programs for certain species. In view of the options available, Lewis and Christiansen (1979) suggest that strategies of both indirect and direct breeding be used to improve plants for growth in air-polluted environments.

SUMMARY

Crop productivity is affected by the presence of O_3, SO_2, and NO_2 in ambient air. Losses probably vary from 0 to 15% depending on the crop species and cultivars used, the pollutant dose, and the environmental and biotic factors present during the growing season. The use of a series of open-top field chambers is currently the most widely accepted technique for assessing crop losses. Open-top chambers adequately simulate field conditions while permitting control of gas concentrations around the test species. Plants have a number of sensitive life stages, but the effects of the pollutants on carbohydrate production (photosynthesis) is critical whatever the growth stage. There

Table 21.6 Estimated National Annual Crop Losses at Different Seasonal Seven-hour per Day Mean O_3 Concentrations

Crop	National value[a] 1978 ($ billions)	Estimated percentage loss[b]/dollar loss ($ billions) per seasonal 7-hr/day mean O_3 concentration (ppm)			
		.06	.07	.08	.10
Field corn	14.9	+0.2/0.0[c]	0.4/0.1	1.2/0.2	3.8/0.6
Winter wheat	3.7	4.7/0.2	7.3/0.3	10.3/0.4	17.8/0.6
Soybeans	12.2	13.3/1.6	17.3/2.1	21.3/2.6	29.2/3.6
Spinach	<.1	20.0/0.0	26.6/0.0	32.3/0.0	46.6/0.0
All other[d]	23.0	5.4/1.2	7.6/1.7	9.8/2.3	15.0/3.5
Totals	53.9	5.6/3.0	7.8/4.2	10.1/5.5	15.4/8.3

[a]Crop production values were obtained from references 24–27 in Heagle and Heck (1980). Values for "all other" crops include those for field and vegetable crops but not those for fruit, nut, ornamental, or forest species.
[b]Regression analysis estimates for field corn (one cultivar), winter wheat (four cultivars) and spinach (four cultivars) grown in the ground and for soybeans (one cultivar) grown in 15-liter pots (means for 2 years). Estimated percentage losses at each concentration were calculated using estimated values at 0.03 ppm of O_3 as the controls (0.0265 for soybean).
[c]The + value indicates an estimated percentage gain in yield.
[d]Percentage losses for all other crops were estimated as the mean predicted losses for field corn, winter wheat, soybeans and spinach, with values for spinach arbitrarily weighted at 20%.

Note: The seasonal 7-hr/day mean O_3 concentration is defined as the mean concentration from 0930 to 1630 hours during the period of exposure (early vegetative to mature stages).

Source: Heagle and Heck (1980), Table 4.

is sufficient genetic variability to make breeding for resistant cultivars an acceptable agricultural strategy to alleviate pollution effects.

ACKNOWLEDGMENTS

Cooperative investigations of the U.S. Department of Agriculture and North Carolina State University. Paper No. 8040 of the Journal Series of the North Carolina Agricultural Research Service, Raleigh, NC.

LITERATURE CITED

Bennett, J. H., H. E. Heggestad, and I. B. McNulty. 1977. *Ozone and leaf physiology.* Pages 323–30 in Proc. 4th Ann. Plant Growth Regulator Working Group.

Table 21.7 Number of Crop Cultivars Tested for Sensitivity to O_3, SO_2, or NO_2

Species	Number of cultivars for each pollutant			
	O_3	SO_2	NO_2	Ambient air
Alfalfa	23	–	–	2
Corn, field	11	4	–	–
Forage legumes	10	–	–	–
Safflower	12	–	–	–
Soybean	>40	21	–	4
Tobacco	>73	12	–	–
Turfgrass	28	28	–	>48
Bean	>80	33	–	>387
Grape	22	–	–	143
Potato	>37	–	–	>75
Spinach	11	–	–	–
Tomato	338	–	–	–
Winter wheat	14	7	–	–

Source: From Reinert, Heggestad, and Heck (1981); this is not a complete listing.

Bennett, J. P., and R. J. Oshima. 1976. *Carrot injury and yield response to ozone*. J. Amer. Soc. Hort. Sci. 101:638–39.

Bennett, J. P., R. J. Oshima, and L. F. Lippert. 1979. *Effects of ozone on injury and dry matter partitioning in pepper plants*. Environ. and Exp. Bot. 19:33–39.

Blum, U., and W. W. Heck. 1980. *Effects of acute ozone exposures on bush snap bean at various stages in its life cycle*. Environ. and Exp. Bot. 20:73–85.

Blum, U., and D. T. Tingey. 1977. *A study of the potential ways in which ozone could reduce root growth and nodulation of soybeans*. Atmos. Environ. 11:737–39.

Feder, W. A. 1968. *Reduction in tobacco pollen germination and tube elongation, induced by low levels of ozone*. Science 160:1122.

Guderian, R. 1977. *Air Pollution: Phytotoxicity of Acid Gases and Its Significance in Air Pollution Control*. Ecological Studies 22. Springer-Verlag, Berlin, Heidelberg, New York.

Hanson, G. P., and W. S. Stewart. 1970. *Photochemical oxidant effects on starch hydrolysis in leaves*. Science 168:1223–24.

Heagle, A. S. 1973. *Interactions between air pollutants and plant parasites*. Ann. Rev. Phytopath. 11:365–88.

Heagle, A. S., and W. W. Heck. 1980. *Field methods to assess crop losses due to oxidant air pollutants*. Pages 296–305 in *Proc. E. C. Stakman Comm. Symp. Crop Loss Assessment*. Misc. Publ. #7. Agric. Exp. Sta., Univ. of Minn.

Heagle, A. S., R. B. Philbeck, H. H. Rogers, and M. B. Letchworth. 1979. *Dispensing and monitoring ozone in open-top field chambers for plant effects studies*. Phytopathology 69:15–20.

Heagle, A. S., Suzanne Spencer, and M. B. Letchworth. 1979. *Yield response of winter wheat to chronic doses of ozone*. Can. J. Bot. 57:1999–2005.

Heck, W. W. 1968. *Factors influencing expression of oxidant damage to plants*. Ann. Rev. Phytopath. 6:165–88.

———. 1982. *Future directions in air pollution research*. In Proc. of the 32nd Univ. of Nottingham School in Agricultural Science: *Effects of Air Pollution on Agriculture and Horticulture*. Univ. of Nottingham, Nottingham, England. (forthcoming)

Heck, W. W., and C. S. Brandt. 1977. *Effects on vegetation: native, crops, forests*. Pages 157–229 in *Air Pollution*, A. C. Stern, ed. 3rd ed., vol. 2b. Academic Press, New York.

Heck, W. W., J. B. Mudd, and P. R. Miller. 1977. *Plants and microorganisms*. Pages 437-585 in *Ozone and Other Photochemical Oxidants*. National Acad. Sci., Washington, D.C.

Henderson, W. R., and R. A. Reinert. 1979. *Yield response of four fresh market tomato cultivars after acute ozone exposure in the seedling stage*. J. Amer. Soc. Hort. Sci. 104:754–59.

Jacobson, J. S., and A. C. Hill, eds. 1970. *Recognition of Air Pollution Injury to Vegetation: A Pictorial Atlas*. Air Poll. Control Assoc., Pittsburgh, PA.

Lewis, C. F., and M. N. Christiansen. 1979. *Breeding plants for stress environments*. In *Plant Breeding Symposium 2*. Iowa State Univ., Ames.

Middleton, J. T., J. B. Kendrick, Jr., and H. W. Schwalm. 1950. *Injury to herbaceous plants by smog or air pollution*. Plant Dis. Rep. 34:245–52.

Montes, R. A. 1980. *Effects of O_3 and nitrogen fertilizer on forage production and nitrogen fixation of a clover-fescue system*. Ph.D. dissertation, North Carolina State Univ., Raleigh.

Mudd, J. B., and T. T. Kozlowski, eds. 1975. *Response of Plants to Air Pollutants*. Academic Press, New York.

National Research Council. 1977. *Nitrogen Oxides*. Committee on Medical and Biologic Effects of Environmental Pollutants. Nat. Acad. Sci., Washington, D.C.

Oshima, R. J., P. K. Braegelmann, R. B. Flagler, and R. R. Teso. 1979. *The effects of ozone on the growth, yield and partitioning of dry matter in cotton*. J. Environ. Qual. 8:474–79.

Reinert, R. A., and P. V. Nelson. 1980. *Sensitivity and growth of five elatior begonia cultivars to SO_2 and O_3, alone and in combination*. J. Amer. Soc. Hort. Sci. 105:721–23.

Reinert, R. A., and J. S. Sanders. 1982. *Growth of radish and marigold following repeated exposure to NO_2, SO_2, and O_3 alone and in mixture*. Plant Disease 66:122–24.

Reinert, R. A., A. S. Heagle, and W. W. Heck. 1975. *Plant response to pollutant combinations*. Pages 122–39 in *Response of Plants to Air Pollutants*, J. B. Mudd and T. T. Kozlowski, eds. Academic Press, New York.

Reinert, R. A., H. E. Heggestad, and W. W. Heck. 1981. *Response and genetic modification of plants for tolerance to air pollutants*. Pages 381–458 in *Breeding Plants for Marginal Environments*, C. F. Lewis and M. N. Christiansen, eds. John Wiley & Sons, New York.

Reinert, R. A., D. S. Shriner, and J. O. Rawlings. 1981. *Response of radish to NO_2, SO_2, and O_3 in all combinations of three concentrations*. J. Environ. Qual. 11:52–57.

Richards, B. L., J. T. Middleton, and W. B. Hewitt. 1958. *Air pollution with relation to agronomic crops: V. Oxidant stipple of grape*. Agron. J. 50:559–61.

Sanders, J. S. 1980. Effects of ozone, sulfur dioxide, and nitrogen dioxide alone and in mixtures on azaleas. Ph.D. dissertation, North Carolina State Univ., Raleigh.

Tingey, D. T., R. C. Fites, and C. Wickliff. 1973. *Foliar sensitivity of soybeans to ozone as related to several leaf parameters*. Environ. Pollut. 4:183–92.

Tingey, D. T., R. A. Reinert, J. A. Dunning, and W. W. Heck. 1971. *Vegetation injury from the interaction of nitrogen dioxide and sulfur dioxide*. Phytopathology 61:1506–11.

Wilkinson, T. G., and R. L. Barnes. 1973. *Effects of ozone on $^{14}CO_2$ fixation pattern in pine*. Can. J. Bot. 51:1573–78.

22] Climate Simulations

by ROBERT J. DOWNS*

ABSTRACT

Simulation of environmental events and climate models in plant growth chambers is easily accomplished from an engineering point of view. Problems in climate simulations, however, are found in the development of the climate model to be simulated and in adequately describing the plant growth and development expected to result from the simulation.

INTRODUCTION

Controlled-environment chambers for plant growth are rarely equipped to simulate the vagaries of the natural environment. They are seldom intended to do so, because the original concept of the plant growth chamber was to keep constant as much of the environment as possible in order to study the physiological consequences of varying one climatic factor at a time. Thus the pioneering controlled-environment rooms used by Borthwick, Hendricks, and Parker at Beltsville were designed to maintain a constant temperature and radiant flux density so the physiological action of what was then called the photoperiod pigment (i.e., phytochrome) could be studied (Downs 1980). More-recent controlled-environment rooms may provide a higher radiant flux density, keep temperature variations smaller, and perhaps allow a greater degree of humidity control, but the methods of using them have changed little. This does not mean that environmental simulation chambers cannot be constructed, or present plant growth chambers modified to enable simulation. From an engineering point of view, the problems with programming environmental events simply cannot be blamed on an inability to design and construct the necessary facilities. Any sort of environmental condition, and its daily or seasonal course, can be simulated without much difficulty; although some, such as the daily progress and magnitude of solar radiation, can be very costly.

*North Carolina State University, Southern Plant Environment Laboratories, 2003 Gardner Hall, Raleigh, North Carolina 27650.

STRATEGIES OF PLANT REPRODUCTION (BARC Symposium number 6—Werner J. Meudt, ed.)
Allanheld, Osmun, Totowa

Table 22.1 Growth of Ransom Soybean and Tokyo Cross Turnip after 24 and 35 Days, Respectively, in Different Substrates

	Soybean		Turnip	
Substrate	Leaf area (cm^2)	Fresh weight (g)	Leaf weight (g)	Root weight (g)
Peat-lite	830.94_a	25.46_a	89.70_a	110.91_a
Vermiculite	400.90_c	15.24_{bc}	44.76_b	54.52_b
Pine bark	569.79_b	17.98_b	20.65_c	17.58_c
Calcined clay[a]	305.62_d	11.59_c	7.22_d	2.14_d

[a] Baking temperatures too low for complete calcining.

Note: Subscripts indicate that individual readings followed by the same letter are not significantly different at the 95% level.

The fact is that very few chamber users are taking advantage of the simulation capabilities their facilities now have. The apparent lack of effort in using the methods available is due principally to the belief that the entire complex, but inadequately defined, microclimate must be simulated and to an indistinct concept of the objectives of the simulation.

The expenditure of time and effort in developing climate models is largely futile, however, unless satisfactory plant growth can be attained. Etiolation, nutritional problems, and generally poor growth are common traits of plants grown in controlled environments. Irrespective of the frequent occurrence of undesirable plant growth, most plant scientists assume from the outset that they know how to grow plants in controlled environments. Consequently, each scientist bases cultural practices and selects environmental conditions chiefly from experience in the field and greenhouse, or sometimes because of a report in the literature based on some other person's concept of the proper conditions. Tanner and Hume (1976), for example, noted that no two of thirty soybean investigators interviewed used the same potting mixture. The reasoning behind the choice of substrate usually is not clear because the selection rarely is based on a comparative plant growth study. Yet the comparative test is essential because, while some species react strongly to an undesirable substrate, or to support media inappropriately labeled inert (Table 22.1), the response of other plants is more subtle and reduced growth would not be obvious if only a single substrate was used. In addition to an unscientific choice of substrate, frequency and amount of water and nutrients often are arbitrary decisions, based, I have observed, as much on the convenience of the investigator as on the needs of the plant. Whatever the cause, when unsatisfactory plant growth makes it obvious that the researcher does not know how to grow plants in controlled environments, the causal agent is most often assumed to be some imagined deficiency of the facility—typically, poor humidity control, low radiant flux density, or the lack of climate simulation capability.

Table 22.2 Effect of a Weather Record-Based, Continuously Variable Temperature Program on Leaf Area and Pod Development of Ransom Soybean and Bush Blue Lake Bean

Temperature regime	Soybean		Snap bean	
	Leaf area[a] (cm^2)	Pod weight[b] (mg)	Leaf area[a] (cm^2)	Pod weight[b] (mg)
Programmed	720_c	330_b	658_c	0_c
Day/night 29°/23°C	974_b	520_a	837_b	450_b
Constant 26°C	1136_a	500_a	1233_a	1520_a

[a]Total leaf area per plant at 5 weeks.
[b]Average weight of pods more than 2 cm long at 7 weeks.

Note: Individual readings followed by the same letter indicate no significant difference at the 95% level.

After the investigator has developed the cultural practices necessary to grow plants reasonably well in controlled-environment rooms, the real problems of climate simulation can be attacked. The first difficulty encountered is defining the objectives of the program. This may seem deceptively simple to do if the objective is to develop a frost hardiness testing model or one to measure resistance to air pollution. In the past few years, however, considerable emphasis has been placed on growing "normal" plants. In some cases this emphasis is undoubtedly due to generally unsatisfactory plant growth caused by some unsuspected influence of the environment, misconceived selection of environmental conditions, or simply to poor choice of cultural practices. Cauliflower plants grown in controlled-environment rooms, for example, frequently suffer from leaf tip burn and the curds may become glassy if the humidity is constant and relatively high (Wiebe and Krug 1974). *Nicotiana thyrsiflora,* which is found at altitudes of 3500 meters, has a history of producing fasciated stems when grown indoors, because the temperatures used are based on experience with commercial tobacco rather than on the natural habitat of the species (Downs and Burk 1980).

Nevertheless, plants resembling the field phenotype are beneficial to most research programs and are essential if the results of experiments in controlled environments, or in the greenhouse, are to be extrapolated to the real world and if plant performance in the field and natural habitat are to be predicted. Stress research requires, in addition, that the different stages of plant development proceed at relative rates similar to those in nature so environmental stress can be applied to plants in the same physiological status as the stress would find them under natural conditions.

Unfortunately, reference to the "normal" plant as the field phenotype is not very enlightening because field phenotypes rarely are defined in sufficient detail to allow duplication. Field crops, for instance, usually are described in terms of yield per hectare. The individual researcher may have a mental image of the proper appearance of the plants, but factual information on branching, stem length, and leaf area at different times during the growing season, or even at maturity, are difficult to obtain. Lack of a definitive field phenotype description is partly due to the large variation that can occur from year to year with the same variety grown in the same field with identical cultural practices. Average seed size of soybean, for example, can vary as much as 60%. Ecophenes and physiological ecotypes make description of a normal plant growing in a natural habitat even more difficult than describing a normal crop plant. Seed from Potentilla plants growing in coastal areas, for example, consistently produce smaller plants than when the seeds are derived from Potentilla growing inland at higher altitudes. Therefore, the first task of the crop scientist, ecologist, or stress physiologist is to describe, in detail, the plant he is trying to reproduce in order to have a basis for evaluating climate simulations.

TEMPERATURE

Climate models to date have stressed temperature. Went (1957) approximated the average day temperature from weather records as

$$T_{day} = T_{max} - 1/4(T_{max} - T_{min}$$

and the average night temperature as

$$T_{night} = T_{min} + 1/4(T_{max} - T_{min}).$$

This procedure is useful for selecting a day/night temperature alternation that is likely to satisfy the thermoperiod requirement of a species from a restricted geographic area, but it does not imply that the resulting day/night temperature is optimum or that normal plants will result.

Some investigators have programmed a continuously variable diurnal temperature program from standard meteorological data. The assumption is that these programs represent natural conditions more closely than day/night alternations of temperature and, *ipso facto,* induce a more-natural plant response. To improve this natural state, 12-hr. days often are used with the temperature program, in spite of the fact that meteorological data clearly show that 12-hr. days are not customary during the growing season in the temperate zone. The results of comparing plant growth under a diurnal program to that obtained with the average, or average day and average night, temperature of the program, however, are not impressive. Bean plants produce the same number of nodes under the three temperature regimes, but leaf area and pod production are significantly less under the diurnal one (Table 22.2). A single diurnal temperature program derived from weather records simply does not insure that the plants will grow better, or resemble the field phenotype more closely, than a constant temperature.

Diurnal temperature programs, however, can be essential in some kinds of programs, such as frost hardiness testing. Schultz (1978), for example, developed a program for testing Eucalyptus species for adaptability to the south-

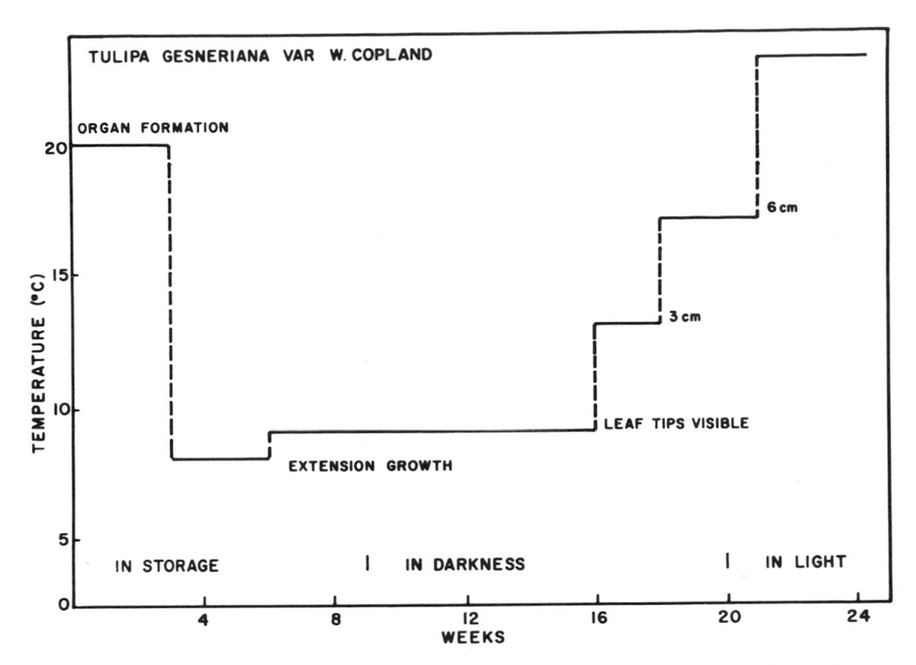

Figure 22.1. Optimum temperature during each stage of development for early flowering of tulip (Hartsema 1961).

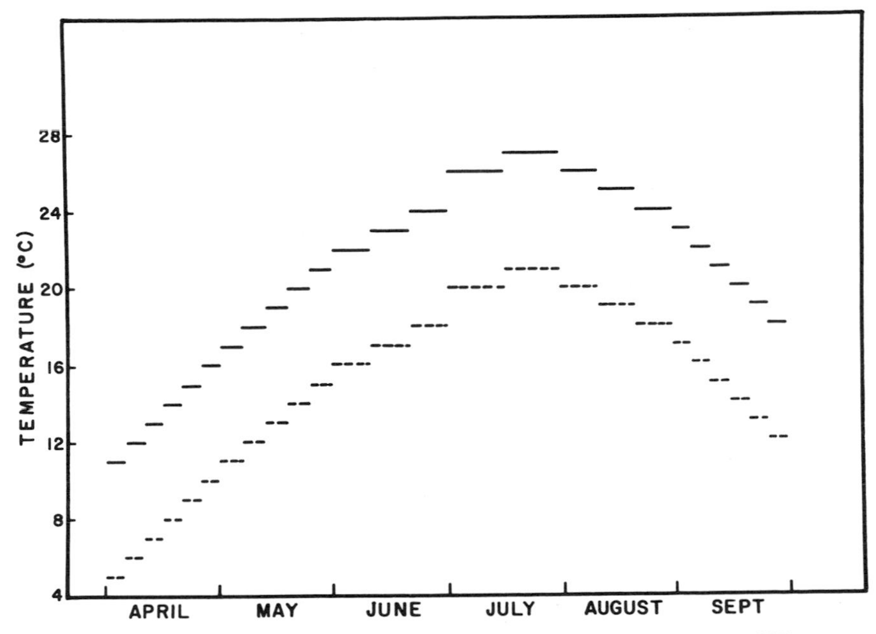

Figure 22.2. Seasonal model of day and night temperatures (Romanov 1980).

eastern United States where test winters occur only every eight to ten years. Seedlings grown at 26/22°C day/night temperatures and long days were hardened off during a two-week period prior to the test by gradually lowering temperatures and short days. The test period consisted of 3 cycles of a temperature program, based on conditions occuring during test winters in the planting habitat. Each dark period included 10 hrs below freezing with a minimum of −5°C and a warming cycle during the light period that reached a maximum of 21°C. Simulating the natural, cold-damaging cycle allowed the screening program to confidently eliminate 50% of the Eucalyptus sources before field planting.

Diurnal simulations, thus, are valuable research tools if developed and used wisely, but it seems obvious that a diurnal program rarely will result in normal plant growth unless seasonal changes also are considered. Seasonal thermoperiodicity is known to be important for plants that require a cold period for breaking dormancy (Coville 1920), or for flowering (Stout 1946), and the familiar studies of Blaauw and coworkers (Hartesma 1961) have made common knowledge of the fact that each stage of plant growth has its own optimum temperature (Fig. 22.1). Nevertheless, the implications of this information seem largely ignored and many users of controlled-environment chambers appear to assume, quite erroneously, that a seasonal progression of temperature applies only to plants that require a cold period.

Several seasonal temperature models have been developed from meteorological data (Bliss 1958; Bretschneider-Herrmann 1969, 1974; Raper 1971; Pletzer 1973; Romanov 1980). Maximum and minimum temperatures are rarely used directly but are transposed to effective temperatures by methods ranging from a rather complicated Fourier series to simple root-mean-square values. Most of these models are concerned primarily with simulating the natural temperature progression, so they use a relatively large number of temperature steps (Fig. 22.2). The only correlation between these models and plant development, however, seems to be coordinating planting, or the beginning of the experiment, with the proper point in the temperature program. The multistep models thus assume that the plant will develop at the same rate in controlled environments as it does in the field. Plants, however, may, and usually do, grow more rapidly in plant growth chambers—if for no other reason than the lack of water stress—so a better approach may be to correlate temperature changes with the developmental stage of the plant. An effective seasonal temperature stimulation could then be achieved with fewer temperature changes, each correlated with a stage of plant development (Fig. 22.3).

Most authors of temperature models agree that a minimum of 15 years of meteorological data is needed to give the model a sound base. Data over a 40–50-year period, however, is more likely to result in a model consistent with real climate conditions. In Hungary, for example, the years 1970–1979 were the warmest winters on record (Rajki 1980). If frost resistance of small grains breeding stock had been left to natural selection in the field or by testing under a controlled-environment simulation based on weather records mainly from this decade, the nurseries would contain many lines and varieties that lack adequate frost resistance.

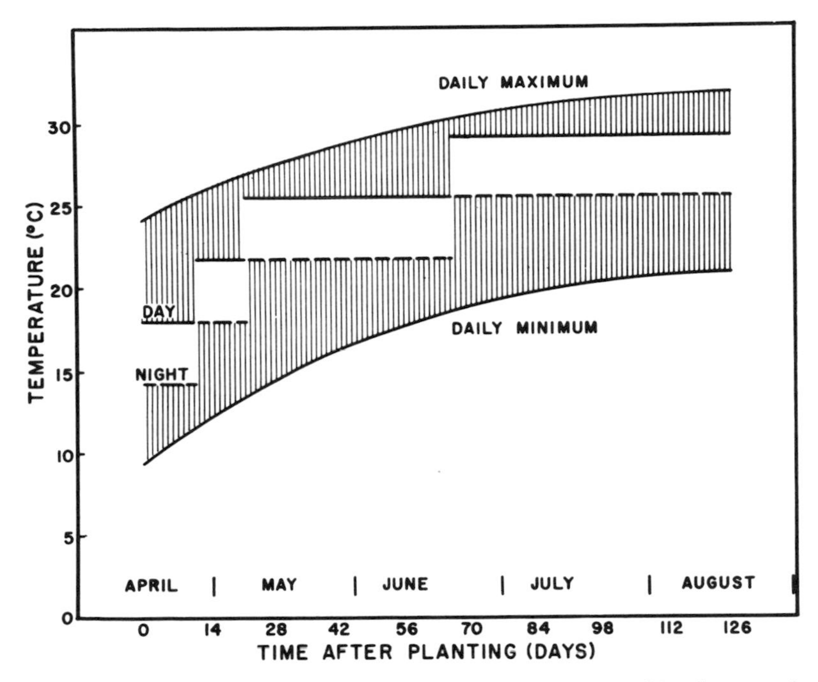

Figure 22.3. Seasonal temperature model adjusted to the stage of development of tobacco plants (Raper 1971).

PHOTOPERIOD

Although a seasonal temperature model may suffice for some frost resistance testing program, simulation of a single environmental factor is rarely enough to insure normal plant development. A multidimensional program is necessary, although all environmental factors cannot, and probably need not, be included. Some climate models include temperature and daylength (Fig. 22.4), but if the species or variety being studied is not photoperiodically sensitive, as with tobacco (Raper and Downs 1976), the inclusion of a daylength simulation can be superfluous. Many so-called day-neutral varieties, however, do respond to photoperiod, although not in terms of flower initiation. Fiskeby soybean, for example, is day-neutral with respect to flowering (Criswell and Hume 1972), but daylength nevertheless exerts considerable influence on plant development. Moreover, the daylength effect is evident when the plants are exposed to only a two-week period of long days prior to transfer to short days, a procedure frequently practiced with short-day varieties (Table 22.3).

RADIANT FLUX DENSITY

The abrupt light-dark transition and the fact that photosynthetically active radiation rarely exceeds 750 μE m^{-2} s^{-1} in controlled-environment rooms does

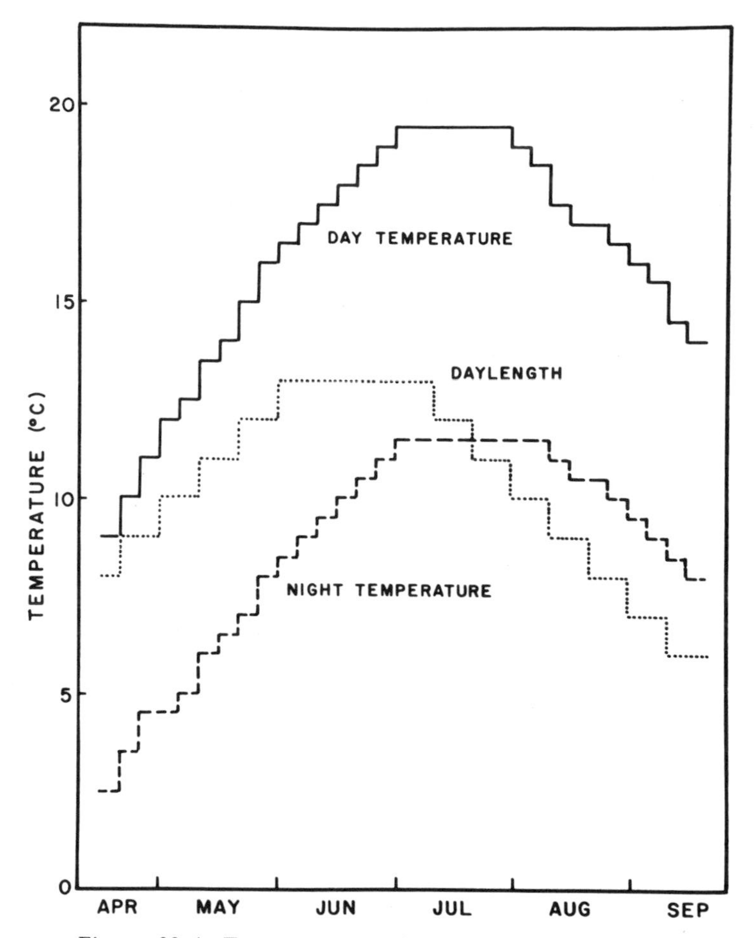

Figure 22.4. Temperature and daylength climate model (Bretschneider-Herrmann 1969).

not prohibit achievement of the objectives of a simulation (Raper and Downs 1976). In practice the theoretically desirable PAR levels of 1500 $\mu E\ m^{-2}\ s^{-1}$ or more seem to cause more plant growth problems than the solve, since plants grown under high-irradiance conditions are generally abnormal and not typical of either greenhouse or field-grown material (Warrington, Edge, and Green 1978). It can be argued that the problems with using very high light levels are due to improper cultural practices because similar problems were encountered when irradiance levels were increased threefold by the development of the 1500 ma fluorescent lamp. It can also be argued that the plant growth difficulties arise because of the unnatural radiation environment created by using very high levels of PAR throughout the light period instead of simulating the diurnal course of PAR, where photon flux density is less than the maximum

Table 22.3 Effect of Daylength on Growth of Fiskeby and Ransom Soybean

Daylength: weeks 0–2	2–7	Internode length (mm)		Mean pod weight (mg)		Stem length (cm)		Plant weight (g)	
		Fiskeby	Ransom	Fiskeby	Ransom	Fiskeby	Ransom	Fiskeby	Ransom
LD	LD	85_a	99_a	364_c	0_c	111_a	155_a	76.1_a	165.9_a
LD	LD	68_b	73_b	592_b	447_b	80_b	91_b	45.1_b	78.7_b
SD	SD	54_c	50_c	689_a	625_a	53_c	37_c	26.0_c	46.7_c

Note: SD = 9 hr high-intensity light; LD = 9 hr high-intensity light plus a 3-hr interruption of the dark period with low-intensity incandescent lamps.

Individual readings followed by the same letter indicate no significant difference at the 95% level.

Data pooled from 4 day/night temperatures.

for most of the day. The fact remains, however, that at present few investigators know how to use high photon flux densities effectively.

Seasonal changes in light intensity are probably not essential to most climate models. Little factual information is available, however, and few conclusions can be drawn from Rajki's (1980) model, which simulated the natural decrease in temperature, photoperiod, and illumination, because the relative biological effectiveness of each factor simulated was not reported.

LIGHT QUALITY

When plants are grown in controlled-environment rooms, the investigator must, irrespective of his wishes in the matter, deal with light quality. During the early development of plant growth chambers, Parker and Borthwick (1949) showed that substantial increases in dry weight and carbohydrate production were obtained by supplementing the radiation from a carbon arc lamp with a low level of light from incandescent lamps. Borthwick also added incandescent lamps to one of the first fluorescent-lighted rooms, and later Went (1957) reported that the fluorescent-incandescent combination resulted in greater weight production than fluorescent alone. Primarily as a result of these reports, it has been assumed that incandescent lamps are essential for good plant growth and that adding 10% of the light by incandescent lamps in optimum. Neither assumption is necessarily valid. The increased plant weight due to the incandescent lamps often results from increased stem length, and the increased length seems proportional to the amount of light provided by the incandescent lamps (Summerfield and Huxley 1972; Hassan and Newton 1975). The longer stems induced by the incandescent light undoubtedly result in the more-normal appearance of some plants, but Tanner and Hume (1976) have suggested that the incandescent radiation is an important preventative to achieving a replica

Table 22.4 **Effect of Radiation from Incandescent Lamps during the Main Light Period on Growth of Ransom Soybean after 32 Days**

Light source	Nodes (number)	Main axis: Length (cm)	Main axis: Leaf area (cm^2)	Branches: Number	Branches: Leaf area (cm^2)
Fluorescent	11.3_a	76.9_b	1235_b	7.8_b	1993_a
Fluorescent plus incandescent	11.5_a	92.7_a	1410_a	5.4_a	1622_b

Note: Individual readings followed by the same letter indicate no significant difference at the 95% level.

Table 22.5 **Effect of the Souce of Photoperiod Control Lighting on Growth of Ransom Soybean in Temperature-Controlled Greenhouses**

Light source	Stem length (cm)	Leaf area (cm^2)	Fresh weight (g)	Pod[a] number	Pod weight (g)
Incandescent	68_a	4859_a	178.9_a	77_a	0.926_b
Fluorescent	42_b	2926_b	88.6_b	66_b	1.112_a

[a] Pods larger than 2 cm.

Note: Individual readings followed by the same letter indicate no significant difference at the 95% level.

Table 22.6 **Effect of Light Quality for a 30-minute Period after the Close of the High-Intensity Light Period on Growth of Tobacco Seedlings**

Variety	Light source	Stem length (cm)	Fifth leaf: Length (cm)	Fifth leaf: Width (cm)
Coker 319	Fluorescent	6.3	9.7	16.2
	Incandescent	13.7	10.0	19.7
NC 2326	Fluorescent	5.7	9.0	15.8
	Incandescent	10.0	10.2	19.5

Source: Downs and Hellmers (1975).

of the soybean field phenotype in controlled environments. Experimental evidence shows that incandescent lamps equal to 27% of the fluorescent watts do produce significant elongation in soybean and also reduce the number of branches and total leaf area (Table 22.4).

Some plant growth chambers are designed to allow the fluorescent and incandescent lamps to be manipulated independently in order to have more flexibility in separating photoperiodic effects from photosynthetic ones when different daylengths are used. Controlled-environment users know that the spectral distribution just prior to the onset of the dark period has, through its action on phytochrome, considerable influence on the way plants grow. Thus it comes as no surprise that the light source used to obtain a daylength simulation by extending the high-intensity light period with low-intensity light is equally as important as photoperiod and temperature for producing plants of reasonable size (Table 22.5). Less commonly accepted is the fact that failure to turn off the light sources of a plant growth chamber simultaneously can inadvertently cause a significant difference in plant growth between two chambers ostensibly providing the same environmental conditions (Table 22.6).

CARBON DIOXIDE

Maintenance of normal CO_2 levels, the order of 350 ppm, is absolutely essential for developing plants that resemble the field phenotype in controlled environments. Most controlled-environment rooms are equipped with a make-up air system, commonly advertised as sufficient to prevent depletion of CO_2 in the chamber. This is patently nonsense, because simple calculations, such as those performed by Morse (1963), show that 75% or more of the chamber air must be exchanged each minute to avoid significant CO_2 depletion. Very few make-up air systems are large enough to provide the required air exchange rate, and very few chamber cooling systems are designed to operate with a once-through air flow. With a typical make-up air system operating, corn plants will reduce the CO_2 level in the chamber from a building ambient of 350 to 80 ppm within an hour after the onset of the light period. Tobacco and soybean, with higher CO_2 compensation points, reduce the CO_2 level below 200 ppm (Fig. 22.5).

If C_3 plants such as bean or cucumber are placed in a chamber already partially occupied by C_4 plants such as corn, the effect of the low CO_2 level caused by the C_4 plants will be quickly revealed by yellowing and reduced growth of the C_3 species (Fig. 22.6). The effect of the plant-induced CO_2 stress may not be obvious on the plants that produce the stress, however. Soybean leaf area, for example, is reduced by the order of 50% if the plants are allowed to deplete the chamber CO_2 (H. D. Gross, personal communication), but the plants appear healthy; without a comparison with soybeans growing with normal CO_2 levels, the grower would likely assume the plants are typical of what could be produced in the chamber. All species, of course, do not respond to CO_2 depletion in the same way. Tobacco leaf area and leaf specific weight, for example, are not greatly altered but, without CO_2 control, tobacco inter-

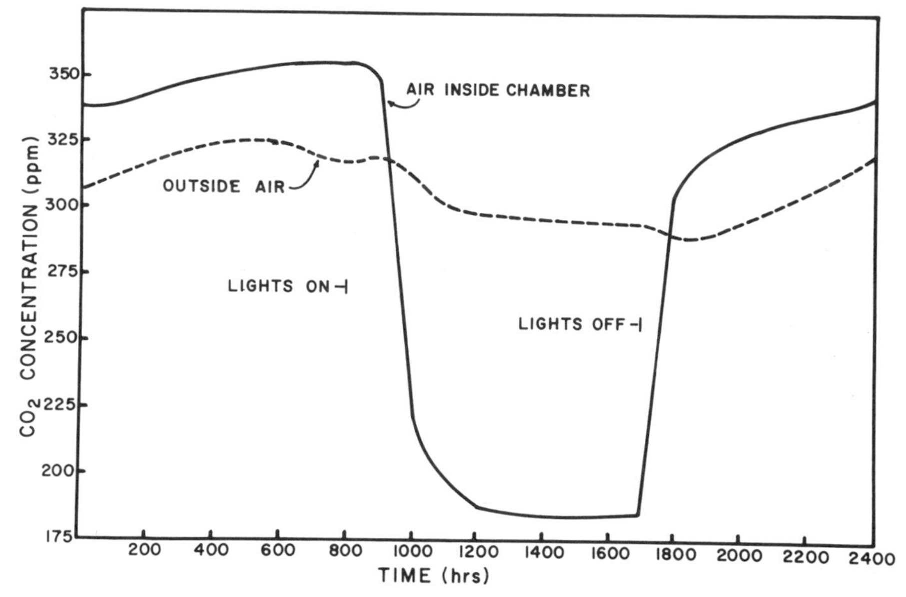

Figure 22.5. Time course of CO_2 concentration in a controlled-environment chamber filled with 4-wk-old bean plants growing at 26°/22° C day/night temperature and a photon flux density of PAR of 670 uE m^{-2} s^{-1} (Downs 1980).

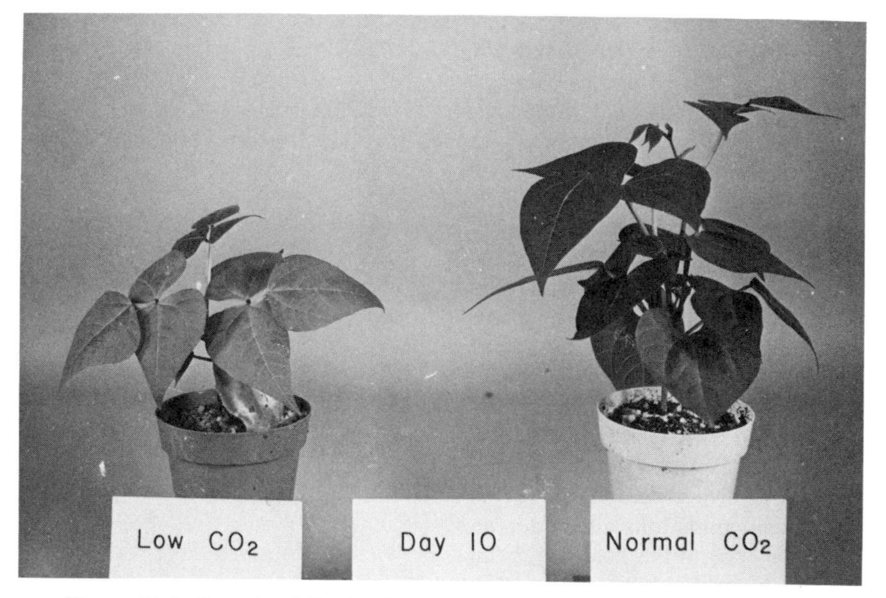

Figure 22.6. Growth of Bush Blue Lake beans 5 days after being placed in controlled-environment rooms half-filled with 1 m corn plants in which Co_2 was not controlled (*left*) or controlled at 350 ppm (*right*) (Downs 1980).

nodes are only 70% of the length customarily produced by field-grown plants and the leaves exhibit a pronounced epinastic curvature that is rarely observed under field conditions (Raper and Downs 1976).

NUTRITION

Temperature, photoperiod, and light quality are obvious environmental factors to include in a climate model. Less-obvious, but equally important, elements of the plants' environment, such as nutrition and water, are likely to be overlooked in the development of climate simulations. Nutrient availability for plants grown in plant growth chambers usually is kept high throughout the growth period, irrespective of the fact that production of most field crops is based on high nutrient levels during early growth and diminishing levels as the growing season proceeds. In controlled environments, continuously high nutrient levels can lengthen the fruit-ripening stage considerably and may, with tobacco, for example, result in very abnormal plants. Yet, in spite of the obvious need to consider nutrition as a function of stage of plant development, Raper's program for development of the field phenotype of tobacco (Raper and Downs 1976) seems to be the only environmental simulation that includes the seasonal decrease in nitrogen availability typical of field conditions.

WATER

Contrary to many ecology and plant physiology texts, water, not CO_2, is usually the factor limiting plant growth under natural conditions. Climate simulations, therefore, must consider the fact that most plants grow in water-limited environments. Leaf area and yield of soybeans grown in the field, for example, can be increased 40% by irrigation. Water availability in controlled environments, however, is routinely kept high, most often by watering the substrate copiously two or more times each day. It is easy to demonstrate, however, that the typical watering procedure does not guarantee that the plants are under optimum water conditions (Table 22.7). Moreover, watering procedures commonly used on plants growing in containers can create a stress condition caused by watering to substrate capacity, alternating with a rapid decrease in water availability. Tomatoes, for example, watered twice during an 8- or 10-hr period of high radiant flux density will not exhibit signs of wilting, but the leaves will curl and the fruit will usually develop blossom end rot. If the cyclic decrease in water availability is minimized by automatic watering or avoided by drip irrigation, the leaves and fruits will develop normally.

Climate simulations aimed at water stress must consider that the development of stress in the field is a gradual process because, since root density is low, all of the root system is not deprived of water simultaneously. Consequently the plant in the field is capable of making a certain amount of osmotic adjustment and may retain a positive turgor pressure over a wide range of leaf water content (Begg 1979; Turner 1979). In containers, however, root density

Table 22.7 Corn Growth in 15-cm Pots after 65 Days under Three Watering Regimes in a Greenhouse at 26°/22°C Day/Night Temperatures

Watering regime	Fresh weight (g)
Drip irrigation (5 ml per minute)	1137.67
Automatic watering (4 X daily)	957.91
Hand watering (2 X daily)	438.18

is high and the entire root system is simultaneously subjected to the water deficit. Obviously, the practice of growing plants under near optimum water conditions, then suddenly depriving them of water, is not likely to yield results that are readily extrapolated to natural conditions.

High levels of water availability may not influence the climate model designed to simulate the field phenotype of some species, but the lack of any water stress is probably the major reason for the atypical lush growth of many species, such as soybeans, in controlled environments.

THE ROOT ZONE

Few investigators make any attempt to control the environment of the root. Moreover, critical features of the root environment such as temperature and pH are unknown factors in most controlled-environment research. Substrate temperatures usually are assumed to approximate the temperature of the top of the plant, and substrate pH is presumed to be close to that of the applied water or nutrient solution. In practice, however, substrate temperatures are lower in clay pots than in plastic ones, and the root system of plants grown in soil beds almost always is cooler than that of plants grown in containers. Substrate temperature also changes when water is applied, sometimes drastically, since city water during winter can be as low as 10°C. Moreover, under high radiant flux conditions, the temperature of a substrate at seed level can be much higher than the air temperature (Fig. 22.7). Considerable change also can take place in the pH of the water or nutrient solution as it passes through the substrate. Composted pine bark and calcined clays baked at suboptimal temperatures, for example, alter the pH of a 6.5 nutrient solution to 4.2 as it passes through the substrate. Low pH has several effects on plant development other than the familiar iron chlorosis, including reduced nitrogen uptake and dry weight production, and often the root system is affected more than the aerial portion of the plant (T. W. Rufty Personal Communication).

Root environment can, of course, have a considerable effect on plant development. Root temperature, for example, has as great an effect on growth of tobacco as aerial temperature or nitrate concentration between 3.5 and 14.6 mmol, especially if the temperature is low, 16°C, rather than warm, 24° or 32° C. The reduction in dry matter production and nitrogen accumulation asso-

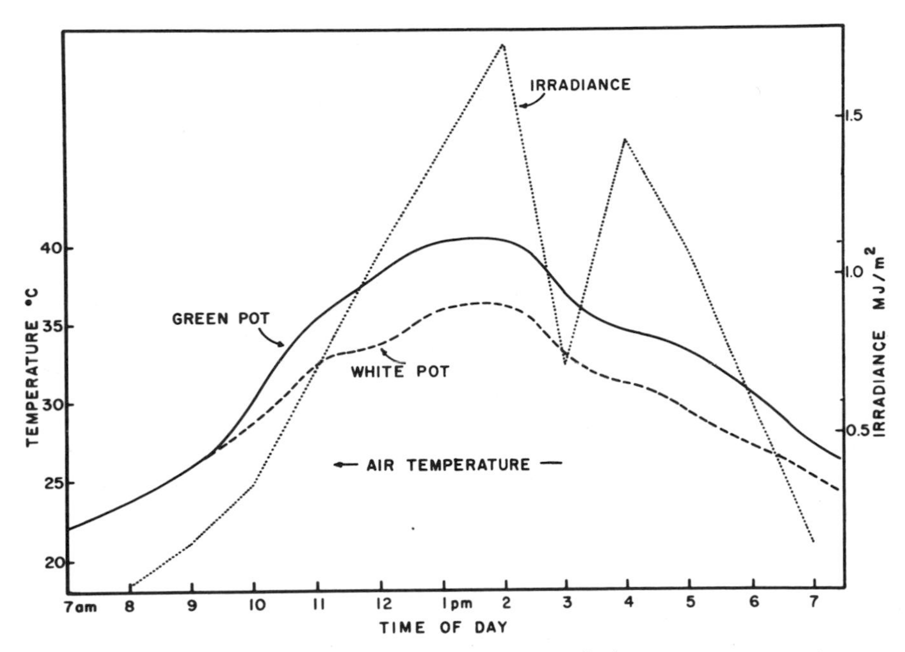

Figure 22.7. Effect of pot color on the daily course of substrate temperature at a depth of 2.5 cm.

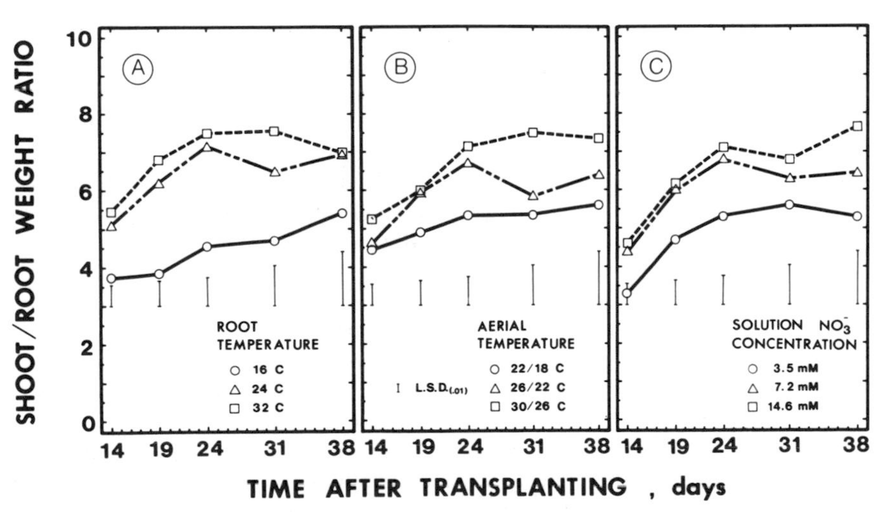

Figure 22.8. Mean effects of root and aerial temperatures and solution nitrate concentration on shoot: root dry weight ratios of tobacco plants (Osmond and Raper 1981).

ciated with the reduction in root temperature is enhanced as aerial temperatures increase (Osmond and Raper 1981). Similar results were obtained with soybean plants (Rufty, Raper and Jackson 1981) where the first response to low root temperature is a decrease in emergence of new leaves and later decreases in leaf area and dry weight.

Although root zone control probably is not imperative to the success of many climate models, root zone environments can be controlled and various conditions simulated when necessary. Several root zone control systems have been described (Fölster 1974; Matsui, Eguchi, and Hamakoga 1972). Merva and Kilic (1972), for example, modified a commercial growth chamber to provide substrate temperature control, and Osmond, York, and Raper (1981) described a root zone control system designed to be placed within walk-in controlled-environment rooms. The degree, uniformity, and range of root temperature, pH, nutrient, and water availability control obviously depends on the methods used. Simply placing the container in one controlled area and the top part of the plant in another will not provide the degree of control, or response rate to a change of conditions that can be afforded by a liquid culture system. The liquid culture system reported by Osmond, York, and Raper (1981), for instance, can provide precise nutrient and pH control, while the root temperature can be set anywhere within the range of 5° to 30° C over the same range of air temperatures.

CONCLUSIONS

Paul Kramer (personal communication) noted that over the past thousand years rather negligible changes in climate have taken place, whereas short-term climate or weather has varied significantly. His point is that climate simulation is much too broad a subject to work with effectively. As a result most research deals with how one, two, or perhaps three environmental factors affect some biochemical event, physiological process, or stage of plant development. Climate models, therefore, do not really simulate climate but serve only to mimic the average diurnal or seasonal progression of a few environmental factors. Such climate models, nevertheless, can be developed to reach any clearly defined objective and to produce any kind of plant that is sufficiently described. Moreover, the majority of the models can be simulated in existing plant growth chambers with only minor modifications. The simulations, nevertheless must be compatible with agrestal situations, prevent depletion of CO_2, and manipulate light quality wisely. The simulation also must refrain from mixing the invariably suboptimal climate model with optimum nutrition levels and water availability.

LITERATURE CITED

Begg, J. E. 1979. *Differences in response of plants grown in controlled conditions and in the field.* Phytotronic Newsletter 18:81–84.

Bliss, C. I. 1958. *Periodic regression in biology and climatology.* Conn. Agric. Exp. Sta. Bull. 615. New Haven, Conn.

Bretschneider-Herrmann, B. 1969. *Design of climatic programs in phytotrons*. Pages 30–31 in P. Chouard and N. de Bilderling ed. *Phytotronique I. Centre Natl. de la Recherche Sci.* Paris.

———. 1974. *Effect of some simple temperature programs on spring wheat*. Acta Hort. 39:117–26.

Coville, F. 1920. *The influence of cold in stimulating the growth of plants*. J. Agric. Res. 20:151–60.

Criswell, J. G., and D. J. Hume. 1972. *Variation in sensitivity to photoperiod among early maturing soybean varieties*. Crop Sci. 12:657–60.

Downs, R. J. 1980. *Phytotrons*. The Bot. Rev. 46:447–89.

Downs, R. J., and L. G. Burk. 1980. *Environmental control of fasciation in Nicotiana thyrsiflora*. Tob. Sci. 24:12–14.

Downs, R. J., and H. Hellmers. 1975. *Environment and the Experimental Control of Plant Growth*. Academic Press, London.

Fölster, E. 1974. *The influence of the root space temperature on the growth of young cucumbers*. Acta Hort. 39:153–60.

Hartsema, A. M. 1961. *Influence of temperatures on flower formation and flowering of bulbous and tuberous plants*. Handbuch der Pflanzenphysiologie 14:123–67.

Hassan, M. R. A., and P. Newton. 1975. *Growth of Chrysanthemum morifolium cultivar Pollyanne in natural and artificial light*. Pages 154–64 in *P. Chouard and* N. de Bilderling, eds. *Phytotronics in Agricultural and Horticultural Research*. Gauthier-Villars, Paris.

Matsui, T., H. Eguchi, and M. Hamakoga. 1972. *Interference of environmental factors with temperature effects on plant growth. II. Soil temperature*. Environ. Control in Biol. 10:58–62.

Merva, G. E., and N. Kilic. 1972. *A multichannel facility for studying water potential distibution in the root zone under simulated field conditions*. Pages 159–64 *in* R. W. Brown and B. P. Van Haversen, eds. *Psychrometry in Water Relations Research*. Utah Agri. Exp. Sta.

Morse, R. N. 1963. *Phytotron design criteria: engineering considerations*. Pages 20–61 *in Engineering Aspects of Environment Control for Plant Growth*. CSIRO, Melbourne, Australia.

Osmond, D. L., E. K. York, and C. D. Raper, Jr. 1981. *A system for independently controlled root and shoot temperatures for nutrient uptake studies*. Tob. Sci. 25:11–12.

Osmond, Deanna L., and C. D. Raper, Jr. 1981. *Growth and nitrogen accululation in tobacco plants as affected by nitrate concentration, root temperature, and aerial temperature*. Agron. J. (in press).

Parker, M. W., and H. A. Borthwick. 1949. *Growth and composition of Biloxi soybean grown in controlled environment with radiation from different carbon arc sources*. Plant Physiol. 24:345–58.

Pletzer, J. 1973. *Climate model for phytotron studies*. Acta Agron. Hung. 22:67–80.

Rajki. E. 1980. *Winter hardiness—frost resistance*. Acta Agron. Hung. 29:451–68.

Raper, C. D., Jr. 1971. *Factors affecting the development of flue-cured tobacco grown in artificial environments. III. Morphological behavior of leaves in simulated temperature, light duration and nutrition progressions during growth*. Agron. J. 63:848–52.

Raper, C. D., Jr., and R. J. Downs. 1976. *Field phenotype in phytotron culture: a case study of tobacco*. Bot. Rev. 42:317–43.

Romanov, V. B. 1980. *Climate modelization in phytotronic installations*. Phytotronic Newsletter 21:52–60.

Rufty, T. W., C. D. Raper, Jr., and W. A. Jackson. 1981. *Nitrogen assimilation, root growth and whole plant responses of soybean to root temperature, and to CO_2 and light in the aerial environment*. New Phytologist. (in press).

Schultz, Emily, B. 1978. *Cold hardiness testing of Eucalyptus*. Pages 137–40 *in* R. J. Downs, ed. *NCSU Phytotron Annual Report.*, Raleigh, North Carolina.

Stout, M. 1946. *Relation of temperature to reproduction in sugar beets*. J. Agric. Res. 72:49–68.

Summerfield, R. J., and P. A. Huxley. 1972. *Management and husbandry problems of growing soybean and cowpea cultivars under artificial conditions*. Rept. # 3. Internatl. Inst. Trop. Agric. Trop. Grain Legume Physiol. Proj. Reading Univ., Berkshire, England.

Tanner, J. W., and D. J. Hume. 1976. *The use of growth chambers in soybean research*. Pages 342–51 *in* L. D. Hill, ed. *World Soybean Research*. The Interstate Printers and Publishers, Danville, Ill.

Turner, N. C. 1979. *Drought resistance and adaptations to water deficits in crop plants*. Pages 343–72 in H. Mussell and R. C. Staples, ed. *Stress Physiology in Crop Plants*. Wiley Interscience, New York.

Warrington, I. J., E. A. Edge, and L. M. Green. 1978. *Plant growth under high radiant energy fluxes*. Ann. Bot. 42:1305–13.

Went, F. W. 1957. *Environmental control of Plant Growth*. Chronica Botanica, Ronald Press, New York.

Wiebe, H. J., and H. Krug. 1974. *Physiological problems of experiments with cauliflower in growth chambers*. Acta Hort. 39:105–13.

Index of Authors*

Aaronson, S. A., 62
Aasheim, T., 219
Abe, S., 79
Abeles, F. B., 149, 150
Adamski, T., 55
Adedipe, N. O., 224
Adler, J., 153
Adorjan, J., 254
Ajami, A, 162, 167
Al-Charachafchi, F., 137
Aldrich, R. A., 329
Alexander, D. E., 56
Alexander, J., 158
Al-Hadithi, T. R. A. M., 209
Allard, H. A., 12, 100, 324
Allard, R. W., 28
Altman, P. L., 103
Altschul, A. M., 244
Alvim, R., 159
Ammirato, P. V., 292
Ananiev, E. V., 62
Anderson, E., 58
Anderson, G. R., 123
Anderson, I. C., 78
Anderson, M. K., 56
Andrews, F. W. 212
Appleyard, M., 79
Armstrong, J. A., 39
Asahira, T., 187
Ascher, P. D., 41, 43, 44
Ashford, A. E., 42
Anhford, M., 76
Aspinall, O., 78
Atsmon, D., 172
Auderset, G., 267
Aubertin, M., 42
Augsten, H., 267
Austin, R. B., 79, 167, 249, 258, 324
Axmann, H., 256
Ayala, F. J., 63

Bacchi, O., 291
Baer, A. S., 53
Bailey, L. H., 327
Bailey, W. A., 324
Baker, D. N., 311, 312
Baker, H. G., 39, 40
Baker, I., 39, 40
Baker, L. R., 171
Baldev, B., 164
Balegh, S. E., 324
Ballard, L. A. T., 90
Balls, W. L., 204
Barber, J., 43
Barber, J. T., 43
Barnard, C., 281
Barnes, L. W., 252
Barnes, R. L., 338
Barros, R. S., 159
Batjer, L. P., 278
Bayev, A. A., Jr., 62
Bazin, M., 173
Beauchamp, E. G., 249
Beckett, J. B., 56
Beemster, A. B. R., 100
Beever, J. E., 215, 219, 221
Beevers, L., 253
Begg, J. E., 363
Belliard, G., 297
Bellman, S. K., 172
Benci, J. F., 308, 309
Bennett, J. P., 338, 340, 341, 342
BEN-TAL, Y., 9, 157-179
Berg, M. A., 248, 249
Berg, R., 55
Berger, B., 138
Bergquist, R. R., 60
Berko, N. F., 215
Bernard, R. L., 223
Bernier, G., 90, 159, 165, 166, 267
Berrie, A. M. M., 225
Beverley, S. M., 63
Bhattacharjya, S. S., 43
Biain, M. M., 190
Bickford, E. D., 324
Biddulph, O., 324
Bidney, D., 296, 297
Billingsley, H. D., 275

*Senior authors of literature cited are listed. Authors' names in capital letters are contributors to chapters in these proceedings.

Bingham, E. T., 295
Bingham, J., 249, 258
Biran, I., 186, 187, 189
Bismuth, F., 268
Biswas, P. K., 159
Blaas, J., 43, 44
Blaauw, O. H., 139
Blaauw-Jensen, G., 139
Black, C. A., 249
Blackwell, R. D., 249, 258
Blake, J., 89, 100
Bledsoe, C. S., 159
Bliss, C. I., 356
BLUM, U., 333, 350
Boatwright, G. O., 249
Bodson, M. 90, 160, 165, 264, 265
Boisard, J., 129
Boland, G., 255
Bonner, D., 161
Bonny, 36
Booth, A., 182, 183, 186, 188, 217
Borah, M. H., 185
Bornman, C. H., 291
Borst, H. L., 247, 249
Borthwick, H. A., 81, 100, 101, 111, 112, 119, 123, 170, 324, 351, 359
Bouwkamp, J. C., 219
Bowen, M. R., 161
Braegelmann, P. K., 338, 339, 340, 341
Brandt, C. S., 333
Brede, J. M., 267
Bredemeijer, G. M. M., 43, 44
Brenchley, W. E., 215
Bretschneider-Herrmann, B., 356, 358
Brewbaker, J. L., 40
Briant, R. E., 219
Briggs, W. R., 119, 120, 122, 123, 125, 126, 128, 129, 141, 163, 164, 170
Brill, W. J., 252
Bris, B., 44
Brouwer, R., 215
Borwn, N. A. C., 185
Brown, W. V., 13
Browning, G., 227
Bruce, R. R., 311
Brun, W. A., 252
Brunori, A., 256
Buban, T., 283
Bui-Dang-Ha, D., 294
Bukovac, M. J., 281
Bullwinkel, B., 219
Bunning, E., 82
Bunt, A. C., 90
Burg, E. A., 166, 283
Burg, S. P., 283
Bruk, L. G., 353
Burns, J. A., 55
Burton, G. W., 54
Busch, R. H., 253, 254
Bush, G. L., 64
Bushnell, W. R., 42
Butler, W. L., 100, 120, 122, 125, 126, 128, 129
Button, J., 291
Byers, R. E., 171
Cagampage, G. B., 255, 256
Cain, J. C., 275, 280, 281
Caldwell, B. E., 252
Callahan, R., 62
Calos, M. P., 63
Cameron, D. F., 78
Campbell, J. H., 62
CAMPBELL, L. E., 323, 331
Canaani, E., 62
Canham, A. E., 324
Caporello, D. C., 26
Carlquist, S., 27
Carlson, P. S., 289
Carpenter, B. H., 158
Carr, D. J., 161, 164
Carroll, T. W., 5
Cartwright, P. M., 79
Carty, C. E., 138
Case, S. M., 64
CATHEY, H. M., 76, 81, 92, 100, 323-31
Cella, R., 57
Cetas, C. B., 267
Chabin, A., 173
Chailakhyan, M. K., 158, 159, 161, 163, 173, 219
Chambers, G. K., 63
Champault, A., 173
Chan, B. G., 275, 280, 281
Chanter, D. O., 90
Chang, Z., 55
Chapman, H. W., 186
Charles-Edwards, D. A., 264, 266
Charnay, D., 187
Chaturvedi, H. C., 291
Chelsky, D., 153
Chen, S., 61
Chevalier, P., 251
Chmeliauskaite, V. G., 62
Chorney, W., 125
Chouard, P., 163
Chowhury, S., 54
CHRISTIANSEN, M. N., 237-42, 347
Ciferri, O., 57
Clark, R. B., 240
Cleland, C. E., 113
CLELAND, C. F., 157-179
Cleland, R. E., 154
Clements, H. F., 84
Clinch, P. E. M., 186
Clouston, T. W., 212
Cockshull, K. E., 76, 77, 91, 264, 266
Coe, E. H., Jr., 53
Cohen, D., 202
Cohen, R. J., 152
Colbert, K. A., 221
Cole, L. C., 8, 202
Coleman, R. A., 125
Colin, L. J., 40
Collin, H. A., 292
Collins, G. B., 297
Conde, M. R., 297
Constabel, F., 294
Cook, M. G., 264
Cooke, R. J., 112
Cooper, A. J., 212

Cordonnier, M. M., 119, 130
Covich, A., 38
Correns, C., 43
Courduroux, J. C., 186, 187
Coustaut, D., 44
Coville, F., 356
Cowan, T. R., 269
Cox, T. S., 248
Crafts, C. B., 158
Cranvin, D. T., 248
CREGAN, P. B., 243, 262
Criswell, J. G., 357
Crouch, M. L., 198
Crowley, M. P., 186
Croy, L. I., 253, 255
Cruz, L. J., 255, 256
Cumming, B. G., 75, 76
Cummin, W. R., 154
Cundiff, S. C., 123, 128
Cunningham, A., 158, 161, 164
Curry, R. B., 310, 311
Cutter, E. G., 188

Dahlquist, F. W., 153
Dai, L., 61
Dale, R. M. K., 198
Dalling, M. J., 254, 255
Dalton, P. J., 164
Dane, F., 44
Daniels, S. M., 119, 128, 130
Darlington, C. D., 3, 4, 9, 12, 14, 15, 31
Darwin, C., 7, 8, 14, 26, 43, 52
Dauphin, B., 173
Dauphin-Guerin, B., 173
Davey, J. E., 221
David, J., 63
Davies, C. R., 217, 225, 227
Davies, P. J., 148, 159, 164, 225, 227
Davis, L. D., 281
Davis, S. D., 225
Davis, W. H., 185
Davy, J. E., 183
Dawson, A. L., 267
Dearing, R. D., 289
Deckard, E. L., 249, 253, 254
DECKER, W. L., 239, 307, 321
DeCock, A. W. A. M., 38
DeFranco, A. L., 153
Degman, E. S., 275, 276, 281, 282
DEITZER, G. F., 99, 115, 126
Delaigue, M., 173
DeLarco, J., 62
Delbart, C., 44
Delbruck, M., 152
DeLint, P. J. A. L., 100, 112
deMooy, C. J., 252
De Nettancourt, D., 43
Dennis, F. G., Jr., 274, 275
Dennis, D. T., 164
Dempsey, E., 56
Deputy, J., 264
Derman, B. D., 223
de Stigter, H. C. M., 212, 215
De Turk, E. E., 249, 250
Deurenberg, J. J. M., 44
Devine, T. E., 252, 253
deWet, J. M. J., 56, 58, 59, 60, 62
De Wit, C. T., 223, 225, 309
DeZeeuw, D., 268
Dhillon, B. S., 251, 254
Dickinson, H. G., 6, 40, 43
Dickson, R. E. 268
Dilley, D. R., 171
Dimalla, G. C., 189
Dittmer, D. S., 103
Dobzhansky, T., 14
Doebley, J. F., 228
Doflein, F., 203, 204
Dohrmann, U., 150
Dooner, H. K., 61
Dooskin, R. H., 125
DOWNS, R. J., 81, 100, 170, 324, 351-68
Drewlow, L. W., 41
Duan, X., 61
Ducker, S., 38
Dudits, D., 294
Dudley, J. W., 250, 253, 258
Dugue, N., 165
Duke, W. B., 324
Dumas de Vaulx, R., 56
Dunlop, J., 169
Dunn, S., 324
Dunning, J. A., 334
Durand, B., 173
Durand, R., 173
Dure, L., 198
Dyson, P. W., 189

Earley, E. B., 249
East, E. M., 43
Eaton, F. M., 212
Eberhard, D., 251, 254
Eck, H. V., 253, 254
Edge, E. A., 358
Edgerton, L. J., 277
Edrich, J. A., 324
Egli, D. B., 248
Eguchi, H., 366
Eichhorn, M., 267
El-Antably, H. M. M., 166, 187
El-din, A., 185
Elgin, S. C. R., 62
Elridge, N., 64
Emery, W. H. P., 13
Engelbrecht, L., 218, 223
Engels, W. R., 55
Epel, B. L., 129
Ergle, D. R., 212
Erickson, R. O., 58
Errera, L., 206
Esan, E. B., 290, 291
Esashi, Y., 186
Esmama, B. V., 255
Evans, A., 100
EVANS, D. A., 287, 301
Evans, J. O., 159
Evans, L. T., 74, 75, 77, 78, 81, 84, 87, 92, 100, 112, 113, 158, 159, 166, 223, 264, 324

Evenson, W. E., 32, 40
Ewing, E. C., 204
Ewing, E. E., 191

Faegri, K., 36, 37, 39
Farquhar, G. D., 269
Faust, M., 283
Feder, W. A., 338
Feldman, L. J., 167
Ferguson, D. B., 253
Ferrari, T. E., 43
Feyerherm, A., 314
Figuoroa, A., 256
Finch, R. A., 56
Fischer, R. A., 23, 223
Fishbein, L., 53
Fisher, J. E., 100
Fites, R. C., 338
Flagler, R. B., 338-341
Flavell, R. B., 61, 62, 63
Flesher, D., 244
Fletcher, L. A., 275, 276, 281, 282
Fletcher, R. A., 224
FLICK, C. E., 287-301
Folster, E., 366
Fontaine, D., 165, 266
Ford, M. A., 249, 258
Fork, D. C., 120
Frankel, R., 40
Frankenberger, E. A., 295
Frankland, B., 117
Fredericq, H., 81
Frey, K. J., 54, 248
Friedlander, M., 172
Friend, D. F. G., 85, 92, 264
Friend, D. J. C., 100, 160
Frolova, I. A., 159, 160
Frydman, V. M., 227
Fryxell, P., 28
Fryxell, P. A., 12
Furuya, M., 123, 126

Gaastra, P., 324
Gahan, P. B., 267
Galau, G. A., 198
Galinat, W. C., 60
Gallori, E., 152
Galston, A. W., 148
Galun, E., 40, 172
Gamborg, O. L., 293, 294
Ganders, F. R., 26
Gardella, C., 42
Gardner, G., 119, 152
Garner, W. W., 12, 100, 324
Gaskin, R., 227
Gasser, J. K. R., 249
Gausman, H. W., 283
Georgiev, G. P., 62
Gerstel, D. U., 55
Ghiselin, M., 22, 31, 32
Gibson, A. H., 78
Gifford, R. M., 189
Gilissen, L. J. W., 44
Gilmore, E. C., 253
Glass, A. D. M., 169
Godley, E. J., 28
Goldberg, B., 101
Goldschmidt, M. H. M., 283
Gong, Z., 61
Goodman, P. J., 251
Goodwin, P. B., 266
Gordon, S. A., 125
Goren, R., 281
Gorski, J., 152
Gotoh, T., 54
Gottschall, V., 225
Gould, S. J., 64
Goy, M. F., 153
Graebe, J. E., 227
Grant, V., 3, 14, 31
Green, L. M., 358
Green, M. M., 55, 62
Greenway, S., 198
Gregory, L. E., 182, 186
Greppin, H., 267
Grochowska, M. J., 274
Gruen, H. E., 158
Grun, P., 42
Guderian, R., 333
Guinn, G., 239
Gupta, S. C., 56, 165
Gur, I., 186, 187, 189
Gustafsson, A., 9, 10
Guttridge, C. G., 79, 159, 164, 185
Guzowska, I., 297

Haas, H. J., 249
Haberlandt, G., 206
HACKETT, W. P., 160, 164, 263-72
Hadley, H. H., 249, 254
Hageman, R. H., 244, 249, 251, 253, 254, 256, 257, 258
Hagerup, O., 36
Hahn, T. R., 129
Haldane, J. B. S., 9
Halevy, A. H., 171, 186, 187, 189, 281
Halloran, B., 152
Halloran, G. M., 254, 259
Ham, G. E., 252, 253
Hamakoga, M., 366
Hammond, L. C., 247, 249
Hamner, K. C., 78, 80, 81, 84, 100, 158, 161, 164
Hamner, P. A., 79
Hanks, J., 314
Hanna, W. W., 54
Hansen, D. J., 172
Hanson, A. D., 149
Hanson, G. P., 338
Hanway, J. J., 246, 247, 248, 249
Haour, F., 151, 153
Harada, H., 219, 265
Hardwick, R. C., 220, 223
Hardy, R. W. F., 238, 244, 252
HARLAN, J. R., 51-69
Harley, C. P., 275, 276, 281, 282
Harmey, M. A., 186
Harper, J. L., 202-204
Harris, G. P., 76, 89, 90, 217

Hartmann, K. M., 101
Hartsema, A. M., 355, 356
Hartsock, J. G., 324
Hasegawa, P. M., 292
Hashimoto, T., 166
Hassan, M. R. A., 359
Havelange, A., 165, 166, 267
Havelka, U. D., 244, 252
Hay, R. E., 249
Hayes, P. R., 316
Hayes, R., 112
Head, G. C., 212
HEAGLE, A. S., 333-50
Healy, W. E., 90
Hebert, T. T., 254
HECK, W. W., 333-50
Hedden, P., 227
Heftmann, E., 164
Heggestad, H. E., 338, 347
Heim, B., 127
Heins, R. D., 90
Heinz, D. J., 289, 296
Hellmers, H., 360
Helson, V. A., 100
Henderson, J. H. M., 159
Henderson, W. R., 338
Hendricks, S. B., 81, 82, 100, 111, 112, 120, 123, 137, 240, 351
Henson, I. E., 166, 185, 221, 267, 268
Herbert, T. T., 256
Hermsen, J. G., 42
Herner, R. C., 171
Herrero, M., 43
Herrmann, B., 356
Hertel, R., 150
HESLOP-HARRISON, J., 3-18, 39, 40, 43, 44, 171, 173
Heslop-Harrison, Y., 40, 43
Hesse, M., 37, 38
Hevely, R. H., 39
Hewitt, W. B., 334
Higgins, T. J. V., 154, 266
Hildebrandt, A. C., 289, 292
Hill, A. C., 333
Hilldebrand, F., 201
Hillman, J., 166, 187
Hillman, W. S., 85, 126, 168
Hinata, K., 44, 292
Ho, M. W., 64
Hoad, G. V., 161, 221, 274
Hodson, H. K., 161, 164
Hoener, I. R., 250
Hoffman, M., 43
Holland, R. W. K., 100
Holmes, D. P., 251
Holmes, M. G., 84, 101, 103, 112, 122
Holton, F. S., 204
Hong-Yuan, Y., 55
Honma, S., 219
Hood, L., 62
Hoogenboom, N. G., 41, 42
Hooker, A. L., 60
Hooymans, J. J. M., 251
Hopkins, D. W., 123
Hopkins, W. G., 125
Horavka, B., 166
Horgan, R., 185
Horridge, J. S., 264, 266
Hoving, E., 42
Howell, D. J., 39
Hristov, K., 54
Hsu, F., 198
Hudson, J. P., 212
Huet, J., 275
Hughes, A. P., 76
Hume, D. J., 352, 357, 359
Humphries, E. C., 189
Hunt, L. A., 224
Hunt, R. E., 119, 126, 127, 128, 130, 139
Hunter, R. B., 249
Hurley, L. V., 295
Huston, H. A., 249
Huxley, P. A., 359

Ikeda, K., 84, 85
Ikeda, F., 54
Ihle, J. N., 198
Ihrke, C. A., 55
Iltis, H. H., 228
Ilyin, Y. V., 62
Ingram, L. O., 138
Ingram, T. J., 227
Inoue, Y., 119
Isebrands, J. G., 268
Islam, A. K. M. R., 55
Ito, H., 282
Iverson, T. H., 219

Jabben, M., 112, 126, 127
Jackson, M. B., 149
Jackson, V. G., 215
Jackson, W. A., 366
Jacobs, W. P., 159, 219
Jacobsen, E., 295
Jacobson, H. J., 150
Jacobson, J. S., 333
Jacobson, J. V., 154
Jacqmard, A., 159, 165, 166, 267
Jacques, 163
Jacques, M., 89
Jacques, R., 89
Jain, S. K., 28, 29
Jaiswal, V. S., 172-173
Jalani, B. S., 54
Jan, N. Y., 152
Janick, J., 223, 292
Jans, P. J. H. TH., 42
Jansen, P., 42
Jeffcoat, B., 217
Jelaska, S., 292
Jenner, E. L., 123
Jennings, A. M. V., 187, 188
Jensen, C. J., 284, 300
Jenson, E. V., 150
Jeppson, R. G., 249, 254
Johnson, C. B., 125
Johnson, R. R., 249, 254
Johnson, V. A., 254

Jones, R. L., 152
Jones, C. E., 40
Jones, H., 217
Jones, M. G., 163
Jones, R. L., 152, 227
Jones, W. J., Jr., 249
Jose, A. M., 101, 325
Jost, L., 43
Jugenheimer, R. W., 250
Juliano, B. O., 255, 256

Kaatz, D. M., 295
Kahanak, G. M., 225
Kahlem, G., 173
Kamienska, A., 165
Kan, Y. W., 63
Kandeler, R., 167, 168
Kanemasu, E. T., 314
Kannenberg, L. W., 249
Kao, K. N., 292, 293, 294
Karaszewska, A., 274
Kartha, K. K., 294
Kasche, V., 119
Kasha, K. J., 297
Kasperbauer, M. J., 297
Kato, H., 290
Kato, T., 282
Katsumata, H., 79
Katzenellenbogen, B. S., 152
Kays, S. E., 150
Kedar, N., 171
Keener, M., 314
KENDE, H., 147-56, 164, 221
Kendrick, J. B., Jr., 334
Kendrick, R. E., 112, 120, 122
Kermicle, J. L., 56
Khatoon, S., 162
Khryanin, V. N., 173
Khurana, J. P., 167
Kidd, G. H., 125, 126
Kilic, N., 366
Kim, I. S., 129
Kinet, J. M., 165, 166, 267
King, M. C., 62
King, R. W., 77, 78, 81, 82, 84, 86, 162, 165, 188, 198, 266, 267
Kirby, E. J. M., 79
Klebs, G., 99, 264
Klein, D., 251, 254
Klein, W. H., 101
Klepper, L. A., 253
Klueter, H. H., 329
Knowles, F., 246, 249
Knox, R. B., 12, 13, 38, 42, 43, 184
Kochba, J., 291
Kofranek, A. M., 265
Kolar, J. J., 252
Konar, R. N., 40, 290
Konishi, T., 54
Koon, G., 89
Koshland, D. E., Jr., 153
Koster, R. 42
Kozlowski, T. T., 333
Krauss, A., 187, 192
Kreasuwun, J., 308
Kreber, R. A., 55
Krekule, J., 74, 162, 163, 166, 268, 281
Kribben, F. J., 171
Krishanmurthi, M., 296
Krishnamoorthy, H. N., 163
Krizek, D. T., 329
Kroes, H. W., 43
Kronsted, W. E., 55
Krug, H., 353
Krul, W. R., 292
Kubota, J., 265
Kulguskin, V. V., 62
Kumar, D., 182, 183, 186, 188
Kumar, S., 167
Kunishi, A., 284
Kunzel, G., 55
Kuraishi, S., 282
Kuykendall, L. D., 252
Kvien, C., 253

Labib, G., 289
Lagarias, J. C., 139
Laibach, F., 171
Lal, K., 163
Lal, P., 246, 249, 255
Lewontin, R. C., 14, 202-203
Lieberman, M., 284
Lincoln, R. G., 158, 161, 164
Lindemuth, H., 219
Linder, R., 44
Lindoo, S. J., 223, 224
Ling, E., 149
Linschitz, H., 119
LINSKENS, H. F., 35-49
Lippert, L. F., 340
Liverman, J. L., 100, 281
Lloyd, D. G., 21, 23, 24, 32
Lockhard, J. A., 149, 225
Loomis, W. E., 212
Looney, N. E., 241
Louis, J. P., 173
Loussaert, D., 253
Lovell, P. H., 186
Loveys, B. R., 221
Luckwill, L. C., 221, 274, 275, 281
Lyubomirskaya, N. V., 62

McCartney, H. A., 84
McClintock, B., 55, 61
McCree, K. J., 225, 310, 324
McDaniel, C. N., 268
McDavid, C. R., 219
McDonald, J. F., 63
McGuire, J. U., 324
Macior, L. W., 39
McKee, H. S., 244
Mackenzie, I. A., 5, 6, 294
Mackenzie, J. M., Jr., 125, 126, 128
McMillan, C., 75
Macmillan, J., 227, 274
McNeal, F. H., 248, 249
McNulty, I. B., 338
McQuigg, J., 314

Madec, P., 186
Magness, J. R., 282
Maguire, M. P., 60
Maheshwari, P., 5
Maheshwari, S. C., 165, 166, 167
Major, D. J., 172
Makinen, Y. L. A., 43
Malik, N. S. A., 225
Maloof, C., 265
Malthus, T. R., 51, 52, 237
Manabe, K., 123
Mancinelli, A. L., 101, 107, 120, 125, 129, 141
Mandoli, D. F., 141
Mannetje, L. T., 78
Maretzki, A., 296
Marme, D., 81, 129, 139
Marquardt, D. W., 314
Marschner, H., 187, 192
Marshall, C., 219
Martin, F. W., 40
Martin, S. L., 63
Martinez, T., 252, 254
Marx, G. A., 164, 225, 227
Marx, J. L., 62
Maskell, E. J., 263
Mason, T. G., 204, 263
Masure, M. P., 275, 276, 281, 282
Mather, K., 3, 8, 9, 14, 41
Mathis, P., 119
Matricon, J., 152
Matsui, T., 366
Matsuoka, H., 292
Mattern, P. J., 254
Mattirolo, O., 203, 204
Mattsson, O., 43
Mauk, C. S., 186, 187
Maxwell, F. G., 161
Mayfield, D. L., 158
Mayhew, D. E., 5
Maynard Smith, Jr., 22, 23, 24, 25, 32
Mayr, E., 64
Medawar, P. B., 16
Mederski, M. J., 310, 311
Melchers, G., 289
Melton, B., 44
Menge, I., 186
Menzel, C. M., 187, 189, 191
Merlo, D. J., 64
Merva, G. E., 366
Mes, M. G., 186
Metzger, J. D., 163, 267
Meyer, G. E., 310, 311
Michayluk, M. R., 292, 293, 294
Michniewicz, M., 165
Micke, A., 54
Middleton, G. K., 254, 256
Middleton, J. T., 334
Miginiac, E., 165, 266, 268
Miller, C. O., 128, 158, 289
Miller, E. C., 254
Miller, J. C., Jr., 252
Miller, J. H., 63
Miller, P. R., 337
Milthorpe, F., 185
Misiura, E., 297
Mitra, G. C., 291
Modi, M. S., 249, 255
Mohan Ram, H. Y., 172-173
Mohr, H., 121, 124
Mok, D. W. S., 56
Molisch, H., 203, 207
Monselise, S. P., 281
Monserrate, R. V., 255
Montes, R. A., 343
Moorby, J., 189, 217
Moore, J. N., 173
Morgan, C. L., 249, 258
Morgan, D. C., 101, 102, 112, 120
Morgan, D. G., 78, 170
Morgan, P., 113
Morgan, P. W., 283
Morris, D. A., 219
Morse, R. N., 361
Moschetto, Y., 44
Moss, G. I., 171
Mothes, K., 218, 221, 223
Mozer, T. J., 154
Mudd, J. B., 333, 337
Muir, R. M., 282
Mukade, K., 55
Mulcahy, D. L., 26
Mullins, M. G., 266
Mumford, F. E., 123
Murashige, T., 186, 187, 190, 241, 288, 290, 291, 292, 295
Murfet, I. C., 159, 164, 225
Murneek, A. C., 204, 205, 206, 212
Murty, B. R., 42
Musgrave, A., 149, 150

Nadar, H. M., 289
Nagao, M., 186
Nagata, T., 294
Nagayoshi, H., 79
Nagy, A. H., 56
Nakashima, N., 85, 86
Nakayama, A., 81
Nakayama, S., 166
Nanda, K. K., 163, 167
Nasrallah, M. E., 43
Natajara, K., 290
Naylor, F. L., 170, 171
Negbi, M., 123
Negi, S. S., 173
Nelson, O. E., 58, 61
Nelson, P. V., 340
Neuffer, M. G., 53, 54
Newton, P., 359
Nickell, L. G., 241, 296
Nicoloff, H., 55
Niedergang-Kamien, E., 223
Nielson, C. S., 171
Nightingale, G. T., 215
Niklas, K. J., 64
Ninnemann, H., 164
Nishio, T., 43
Nitsch, C., 265
Nitsch, J. P., 187, 190, 265, 274, 275

Nobel, P. S., 309
Nooden, L. D., 218, 223, 224, 225, 227
Norris, K. H., 100
Norman, A. G., 249
Nyberg, D., 22, 23

O'Brian, T., 101, 112
Oelze-Harow, H., 124
Ogawa, Y., 86, 162, 165, 266
Ogren, W. L., 325
Ohlrogge, A. J., 254
Okagami, N., 186
Okazawa, Y., 186, 187, 189, 191
Oktan, Y., 225
Okuno, K., 54
Olmo, H. P., 173
Olsder, Th. J., 42
Oota, Y., 85, 86, 159
Opler, P. A., 39, 40
ORNDUFF, R., 21-33
Ornston, L. N., 63
Osborne, D. J., 227
Oshima, R. J., 338, 339, 340, 341
Osmond, D. L., 365, 366
Owens, L. D., 284

Painter, R. H., 161
Palmer, C. E., 186, 187
Pandey, K. K., 42, 43
Parker, M. W., 81, 100, 119, 351, 359
Parisi, B., 57
Pasztor, K., 54
Patrick, J. W., 266
Patton, J. L., 64
Paul, K. B., 159
Pecket, R. C., 137
Pedersen, K., 219
Pelletier, G., 297
Peloquin, S. J., 44, 56
Pence, V. C., 292
Perennec, P., 186
Perez, C. M., 255
Pesek, J., 252
Peterson, D. M., 253
Peterson, P. A., 55
Peterson, R. L., 163
Petinov, N. S., 215
Pettitt, J., 38
Pharis, R. P., 163, 172, 266
Philbeck, R. B., 343, 344
Phillips, D. A., 161, 166
Phillips, I. D. J., 148, 264
Phinney, B. O., 227
Pierard, D., 159
Pieterse, A. H., 167
Pike, C. S., 119, 122, 123, 126
Piringer, A. A., 100, 170, 324
Pittendrigh, C. S., 85
Platz, B. B., 128
Pletzer, J., 356
Pohjakallio, O., 79
Pohl, F., 37
Pollmer, W. G., 251, 254
Post, K., 76
Powell, J. B., 54
Powell, M. C., 90
PRATT, L. H., 117, 133, 139, 140, 141
Price, E. G., 62
Price, H. J., 292
Priestly, C. A., 282
Probst, A. H., 219
Proebsting, W. M., 164, 225, 227
Pryor, L. D., 42
Pulham, H. R., 203
Purvis, A. C., 253

Quail, P. H., 84, 129
Quatrano, R. S., 198
Quebedeaux, B., 198, 244, 252
Quedado, R., 160, 264

Rabie, H., 54
Rabino, I., 101, 107, 120, 129, 141
Raghavan, V., 159, 292, 293
Railton, I. D., 189, 191
Rains, D. W., 251
Rajki, E., 356, 359
Raju, M. V. S., 165
Ramina, A., 160, 265
Rangaswamy, N. S., 291
Rao, B. G. S., 60
Rao, S. C., 251, 255
Raper, C. D., Jr., 356, 357, 358, 363, 365, 366
Rapoport, H., 139
Rapp, U., 62
Rappaport, L., 163, 187
Raschke, K., 154
Rasmussen, P., 314
Raup, D. M., 64
Rawlings, J. O., 340, 343
Ray, P. M., 150
Reanney, D., 57
Reddy, G. G., 249, 255
Reddy, V. R., 311, 312
Reed, S. M., 294
Reetz, H., 314
Reichart, C., 203
Reid, J. B., 159, 164, 225
Reinbergs, E., 297
Reinert, J., 292
REINERT, R. A., 333-350
Reynolds, J. R., 292
Rhoades, M. M., 56
Rice, H. V., 119, 128
Richards, B. L., 334
Richie, J. T., 312
Riesink, W. G., 319
Rimpau, J., 63
Ringe, F., 265
Roberts, R. H., 208, 215
Robichaud, C. S., 197
Rogan, P. G., 185
Rogers, H. H., 343, 344
Rohmeder, E., 204
Romanov, V. B., 355, 356
Rombach, J., 87
Romken, M. J. M., 311
Rood, S. B., 172

Ropers, H. J., 227
Ross, C., 159
Ross, C. W., 148, 159
Rossini, C., 265
Rottenburg, T., 167
Roux, S. J., 128
Rowell, J. C., 198
Rubinstein, B., 150
Ruddat, M., 164
Rudich, J., 171
Rufty, T. W., 366
Runge, E. C. A., 308
Rupp, D. C., 223
Russo, V. E. A., 152
Ryugo, K., 283

Saad, F. S., 78
Sacher, R. M., 172
SACHS, R. M., 160, 164, 219, 263-272, 282, 283
Sachs, T., 218, 282
Saez, J. M., 151, 153
Safran, Y., 291
Sagar, G. R., 219
Saito, T., 172
SAKAMOTO, C. M., 307-321
Sala, F., 57
Salama, A. M. S., 185
Salisbury, F. B., 85, 148, 162
Salmon, J., 159
Sanders, J. S., 338
Sandhoff, K., 44
Sattelmacher, B., 192
Satter, R. L., 148
Saunders, J. W., 295
Saunders, P. F., 112, 159
Saunders, P. T., 64
Sawhney, N., 163
Sawhney, S., 163
Saxton, K., 314
Sayre, J. D., 249
Schafer, E., 121, 125, 126
Scharfetter, E., 167
Schmidt, J. W., 254
Schmidt, W., 121
Schneider, D., 152
Schneider, M. J., 100, 111, 112
Schooler, A. B., 56
Schopf, T. J. M., 64
Schrader, L. E., 251, 253
Schultz, E. B., 354
Schutte, H. R., 218
Schwabe, W. W., 76, 166
Schwalm, H. W., 334
Schwemmle, J., 5
Scott, T. K., 219
Scudo, F. M., 23
Scully, N. J., 100
Sears, E. R., 55
Seidlova, F., 162, 163, 165, 281
Sell, H. M., 171
Semenoff, S., 292
Sengupta, C., 292
Seth, A. K., 217, 218
Seth, J., 254, 256
Seth, P. N., 165, 166
Sevain, G. S., 324
Shahin, E., 296, 297
Sharma, K. C., 246, 249
SHARP, W. R., 287-301
Shepard, J. F., 293, 294, 296, 297
Shepherd, K. W., 55
Sheridan, W. F., 54
Sherwood, S. B., 159
Shibles, R., 78
Shichi, H., 153
Shifriss, O., 173
Shimazaki, Y., 119, 123
Shivanna, K. R., 39, 40
Shriner, D. S., 340, 343
Shropshire, W., Jr., 101
Sibi, M., 296
Siegelman, H. W., 100, 120, 125, 128
Sigurbjornsson, R., 54
Silva, J. M., 283
Simberloff, D., 64
Simone, G. W., 60
Simpson, E., 56
Sinclair, T. R., 223, 225
Singer, D. S., 63
Singh, M., 325
Skene, K. G. M., 221
Skoog, F., 186, 190, 289
Sloger, C., 252
Small, J. G. C., 139
Smith, A. R., 189
Smith, C. A., 40
Smith, D. B., 61-63
Smith, D. L., 185
Smith, F. F., 324
Smith, H., 101, 103, 112, 119, 122, 128, 129
Smith, H. H., 289
Smith, J. M. B., 27
Smith, O., 212
Smith, O. E., 186, 187
Smith, R. H., 292
Smith, W. O., Jr., 120, 129, 130
Snape, J. W., 56
Snee, R. D., 314
Solbrig, O. T., 14
Somers, R. L., 153
SOMMER, H. E., 287-301
Sondahl, M. R., 292
Song, P. S., 129
Song, X., 61
Sood, V., 167
Sopory, S. K., 295
Sotta, B., 165
Spahlinger, D. A., 292
Spaldon, E., 252
Spanjers, A. W., 44
Sparrow, D. H. B., 55
Spencer, R., 89, 100
Spencer, S., 345
Spiegel-Roy, P., 291
Spratt, E. D., 249
Springer, M. S., 153
Spruit, C. J. P., 120, 122, 139
Srinivasan, C., 266

Stadel, J. M., 151
Stahl, C. A., 152
Stahly, E. A., 283
Stalker, H. T., 56, 58-60
Stanley, R. G., 39, 40
Stebbins, G. L., 3, 10, 14, 40, 52
Steenackers, V., 42
STEFFENS, G. L., 237-242
Stettler, R. F., 42
Stewart, W. S., 338
Steyaert, L., 314
Stiles, J. I., Jr., 159
Stoddart, J. L., 163, 228
Stojianov, M. J., 219
Stolwijk, J. A. J., 100
Stone, H. J., 125, 126
Stout, M., 356
Strasburger, E., 43
Straub, J., 43
Street, H. E., 219
Street, Y. E., 290, 291
Streeter, J. G., 310, 311
Strijbosch, T., 212
Strommen, N. D., 316
Stromment, N., 315, 316
Struckmeyer, B. E., 208, 215
Stuart, N. W., 164
Stuckey, I. H., 215
Suge, H., 163, 166, 172
Sukhapinda, K., 55
Summerfield, R. J., 77, 78, 359
Sunderland, N., 293
SUSSEX, I. M., 197-199
Swaminathan, M. S., 42
Sweetser, P. B., 198
Szilagyi, L., 56

Tabbak, C., 172
Tafazoli, E., 76, 164
Takahashi, H., 167, 172
Takebe, I., 289, 294
Takeuchi, M., 290
Takimoto, A., 79, 81, 82, 84, 85, 87, 100, 167
Tanaka, O., 167, 168
Tanner, J. W., 352, 359
Tasker, R., 126
Taylor, M., 249, 258
Taylorson, R. B., 101, 137
Tazawa, M., 292
Tcha, K. H., 89
Tell, G. P., 153
Teller, G., 173
Templeton, A. R., 63, 64
Teso, R. R., 338-341
Thatcher, L. E., 247, 249
Thimann, K. V., 219, 225, 282
Thimijan, R. W., 323-331
Thomas, H., 228
Thompson, L. M., 313
Thompson, P. A., 185
Thomson, P. G., 219
Thornley, J. H. M., 264, 266
Thorpe, T. A., 294, 300
Thurman, D. A., 219
Tigchelaar, E. C., 295
Tingey, D. T., 334, 338, 340
Tisserat, B., 290, 291
Tizio, R., 186, 189, 190
Todaro, G., 62
Tokuyasu, K. T., 129
Toole, E. H., 123
Toole, V. K., 100, 101, 123, 135
Torrey, J. G., 219
Totten, R. E., 293, 294
Tournois, J., 99
Triplett, B. A., 198
Tromp, P. J., 281
T'se, A., 266, 268
T'se, A. T. Y., 165
Turgeon, R., 268
Turner, N. C., 363

Udebo, A. E., 187
Umberger, D. E., 316
Umemura, K., 159
Upper, C. D., 164

Vaadia, Y., 148
Van Bavel, C. H. M., 225
Van den Honert, T. H., 251
Van der Donk, J. A. W. M., 43
Vanderhoef, L. N., 152
Van der Pijl, L., 36, 37, 39
Van der Post, C. J., 215
VAN DER WOUDE, W. J., 135-143
Van Dobben, W. H., 212, 215
Van Herpen, M. M. A., 44
Van Overbeek, J., 219
Van Staden, J., 183, 185, 189, 221, 267
Varner, J. E., 152
Vasil, I. K., 289, 292-294
Vasil, V., 292-294
Vedel, F., 297
Venis, M. A., 150
Venkataraman, R., 165, 166
Vest, G., 252
Vince, D., 89, 100
Vincent, T. L., 203
VINCE-PRUE, D., 73-97, 101, 117, 123, 124, 159, 163, 164, 169, 325

Wacek, T. J., 252
Wada, K., 165
Wagenaar, E. G., 56
Wagner, E., 267
Walbot, V., 199
Wall, P. C., 79
Wallace, D. H., 43, 52
Wallace, H. A., 313
Walther, E. G., 315
Wang, T. L., 185
Wang, Z., 61
Wardell, W. C., 264, 268
Wardlaw, I. F., 223
WAREING, P. F., 148, 159, 166, 181-195, 217, 218, 221, 264, 267, 268
Warner, R. W., 253
Warren-Wilson, Jr., 264

Warrington, I. J., 358
Wassink, E. C., 100
Watanabe, K., 167
Watkin, J. E., 246, 249
Watson, C. A., 249
Watson, P. J., 150
Weaver, P., 274
Weaver, R. J., 241
Weaver, R. W., 252
Webber, H. J., 291
Weber, C. R., 247, 248, 249, 252, 253
Weber, D. F., 252, 253
Weismann, A., 22, 23
Weiss, C., 148
Weller, S. G., 30
Went, F. W., 354, 359
Wessels, M. R., 151
West, C. A., 164
Westermann, D. T., 252
Wetter, L. R., 293, 294
Whalley, D. N., 79
White, D. W. R., 294
White, J. W., 329, 330
Wickliff, C., 338
Widholm, J. M., 325
Wiebe, G. A., 100
Wiebe, H. J., 353
Wiegand, C., 314
Wien, H. C., 77, 78
Wienhues, A., 55
Wilkins, H. F., 90
Wilkinson, T. G., 338
Willemse, M. T. M., 39, 40
Williams, E. A., 113
Williams, G. C., 14, 22, 23, 25, 31, 32
Williams, L., 292
WILLIAMS, M. W., 241, 273-286
Williams, R. F., 254
Willing, R. R., 42
Wilson, A. C., 62, 64
Wilson, E. O., 22, 23
Wilson, G. C., 254
Wilson, J. H., 253-255
Wilton, O. C., 208
Winter, F. L., 250
Withrow, R. B., 111
Wittwer, S. H., 241
Woittiez, R. D., 39, 40
Wong, J., 197
Woodworth, C. M., 250
WOOLHOUSE, H. W., 201-233, 264
Woolley, D. J., 183, 184
Worley, J. F., 292
Wright, S., 9, 14, 15
Wright, S. T. S., 149

Yampolsky, C., 8, 9
Yampolsky, H., 8, 9
Yamamoto, K. T., 120, 129
Yang, W., 61
Yasuda, S., 43
Yeoman, M. M., 290
Yeung, E. C., 163, 294, 300
York, E. K., 366

Zary, K. W., 252
Zeevaart, J. A. D., 100, 113, 149, 158, 160-164, 166, 167, 188, 267, 282
Zehni, M. S., 78, 170
Zeng, Y., 61
Zenkteller, M., 297
Zhadanova, L. P., 219
Zhou, G., 61
Zieserl, J. F., 253
Zimmer, E. A., 63
Zimmerman, R. H., 281
Zollinger, W. D., 128
Zu, D., 61
Zwar, J. A., 154

Index of Subjects

Abiotic vectors, 37
Abscisic acid (ABA)
closure of stomates, 154
dormancy, 159
flowering, 166, 281-83
in fruit, 224
geotropism, 219
induction of mRNAs by, 154
inhibition of hydrolases by, 154
inhibition of mRNA translation by, 198
in leaves, 188
seed development and, 198
seed dormancy and, 198
seed embryos, accumulation of, 198
and senescence, 224
sex expression, 172
transport of, 188
tuberization by, 181, 187-89
Abscission zone, 149, 150
Abutilon Thomponii, 219
Accelerator theory, 43
Acetylene reduction activity (ARA), 252
Acridine orange, 53
Action spectrum, for inhibition, 81
Adaptation, 152
Adenylate
cyclase, 151
cyclase-coupled systems, 151
energy charge, 267
kinase (AK), 267
Adventitious embryony, 10
Aflotoxin, 53
Agamospermy, 12, 31
Agave americana, 203
Ageratum, 208
Agglutini, 198
Agropyron, 182
Agropyron repens, 158
Air pollutants
acute doses of pollutants, 337, 342, 343
air pollution system model, 335
chronic doses of pollutants, 337, 343
corn yield loss, 346, 348
cost/benefit analysis, 340-42, 347
crop breeding programs, 347
crop loss assessment, 340-47
cultural practices, 347
dose-response data, 343-45
dry weight partitioning in plants, 338-40
genetic variability, 347-48
the national crop loss assessment network (NCLAN), 345
nitrogen dioxide (NO_2), 334-38, 340, 343, 347, 349
open-top field chambers, 343-45
ozone (O_3), 334-38, 340-47, 349
photochemical oxidant complex, 334
primary pollutants, 334
protective sprays, 347
secondary pollutants, 334
sensitivity at different life stages, 337-38
sink-source relationships, 338-40
soybean yield loss, 345, 346, 348
spinach yield loss, 345, 346, 348
strategies to reduce pollution effects, 347
sulfur dioxide (SO_2), 333-38, 340, 343, 347, 349
wheat yield loss, 345, 346, 348
Alfalfa, 292, 294-95
Allium, 79
Allogamy, 36, 297
Allotetraploid, 201
α-amylase, 147, 152-54
Alternation of generations, 5, 10
Aminoethoxyvinyl glycine (AVG), 171-72, 282-84
AMO-1618, 225
cAMP, see Cyclic AMP
Anagallis, 266
Anagallis arvensis L., 165, 268
Androdioecy, 8
Androgenesis, 288, 293-95
Andropogoneae, 13, 63
Anemophily, 37
Animal hormones, 151
Anther culture, 294
Anthesins, 163
Anthesis, 245, 248, 265
Antibody theory, 43
Antipodals, 5
Anti-repressor theory, 43
Aphids, 161
Apical dominance, 163, 181-82, 282-83
Apical meristem, 265-66
Apomictic reproduction, 11-12
Apomixis, 10, 21, 29, 31, 32, 56, 287
Apospory, 10, 11
Apples, 274-86
biennial bearing in, 274-79

Golden Delicious, 274-84
Spur Delicious, 278
Starkrimson Delicious, 278-80
Winesap, 276
Yellow Newton, 275-76
Aquatic plants, 149-50
Asexual reproduction, 9, 23, 287
Asparagus, 294, 297
Aster, 202
Attractants, 39
Autogamy, 14, 36
Auxins
apical dominance by, 163, 282, 283
benzoxazolinones, 150
diffusible, 221
floral inhibition by, 162-63
floral initiation by, 281-83
flowering, 282
indole-3-acetic acid, 183
indole-3-acetic acid (IAA), production by fungi, 158
proton excretion, 153, 154
receptor sites, 150
root growth and, 219
senescence and, 217, 219
sex expression, 171-73
sink activity and, 218, 266
stimulation of ethylene production by, 150, 166
stolon production by, 181, 183
supernatant factor, 150
tissue sensitivity to, 149
transport of, 184
Axillary bud propagation, 181, 288, 295

Bacteria, 63, 152-53, 252-53
Barley, 78, 99, 100, 152-54, 254, 294-95, 297
Base substitution, 53
Beans, 252, 353-54, 361
Benzoic acid, 167
Benzoxazolinones, 150
Benzyladenine, *see* Cytokinin
Biloxi soybean, 100
Bioregulants
aminoethoxyvinylglycine (AVG) 282-84
biennial bearing, in apples, 277-79
(2-chloroethyl)-phosphonic acid (ethephon), 278-83
4,6-dinitro-ortho-cresol (DNOC), 277-80
ICI pp 333, 284
1-naphthaleneacetamide (NAAm), 278, 279, 284
naphthaleneacetic acid (NAA), 150, 278-80, 284
1-naphthyl-N-methylcarbamate (carbaryl), 278-80, 284
seed abortion, 279-80
succinic acid-2,2-dimethylhydrazide (diaminozioe), 278-80, 283
to control flowering, 277-79
to control fruit set, 274-75, 277-80
2,3,5-triiodobenzoic acid (TIBA), 283
Binary classification theory, 323, 327-29
Biophysical models, 308
Biotic pollination, 35, 38
Bipartite theory, 43
Blue light, 86, 101
Bothriochloa, 63
Bougainvillea, 264-68
Brassica, 264
Brassica campestris L., 92, 162-63
Brassica napus, 198, 297
Breeding populations, 14
Bromeliads, 167
5-Bromouracil, 53
Bronze mutant locus, 61
Bryophyllum daigremontianum, 164
Bunsen-Roscoe law of reciprocity, 101

C_3 plants, 361
C_4 plants, 202, 245, 361
Cacao, 292
Calamintha, 89
Callistephus chinensis, 76
Callitriche platycarpa, 149
Callose, 208, 212
Cambial activity, 208-9
Cannabis, 173
Cannabis sativa L., 171-72
Carbaryl, 278
Carbohydrates, 338
Carbon dioxide (CO_2), 361-63
Carbon fixation, 238-39
Carnation, 76, 88, 90, 324, 328
Carraway, 292
Carrot, 292, 294
Caryopses, 13
Caryopteris x *clandonensis* A. Simmonds, 170
Cauliflower, 353
Celery, 292
Cereals, 78, 289, 294
Chasmogamy, 12, 13
Cheiranthus cheiri, 203
Chenopodium spp., 75, 76, 81, 101, 161-67
Chicory, 265
Chlormequat (CCC), 189
(2-Chloroethyl)-phosphonic acid (ethephon), 189, 278-83
Chromosome, 55-56
Chrysanthemum spp., 76, 77, 88, 90, 92, 93, 208, 264, 266, 324
Circadian rhythm, 78, 82, 85, 101-15
Citrus, 290-91
Cleistogamy, 12, 13
Climate simulation, 351-68
carbon arc lamp, 359
carbon dioxide (CO_2), 361-63
controlled environment chambers, 351-68
diurnal temperature programs, 354
fluorescent lamp, 359-61
incandescent lamp, 359-61
pH, 364-66
temperature, 354
water, 363-66
Clonal-selection theory, 43
Cloning, 295-96
Clover, 294
Coffee, 290, 292
Colchicine, 56, 297
Coleus, 219

Collembola, 39
Common rust, 60
Consumption theory, 43
Corn, 172, 294, 297, 308, 346, 348, 361
Correlation analysis, 314
Cosmos spp., 208
Cost/benefit analysis, 340-42, 347
Cotton, 204, 219, 238, 263, 292, 295, 311-12
Cowpea, 252
Critical daylength, 70-97
Critical nightlength, 83
Crop loss assessment, 340-47
Crop moisture ratio, 317-18
Cross-pollination, 43, 44
Cryogenic spectrophotometry, 119
Cucumber spp., 171-72, 215, 361
Cucumis sativus L., 171, 212
Cyclic AMP (cAMP), 51
Cysteine, 128
Cytokinin, 147-49, 289
  α-amino butyric acid movement and, 218
  assimilate transport, 165
  benzyladenine, 183
  conversion of stolon to orthotropic shoots by, 181
  distribution of, 221
  flowering, 165-66, 266
  in fruit, 221
  in leaves, 224
  metabolism of, 221
  production by fungi, 158
  promotion of tuber initiation by, 181, 186-87
  roots as source of, 183-85
  senescence and, 217, 224
  sex expression, 172, 173
  sink demand, 266
  stolon development, 183
  tissue sensitivity to, 149
  transport of, 184
  zeatin, 183
  zeatin riboside, 184

Dahlia spp., 182
Daminozide (B9), 189, 278
Dark phase, *see* Skoto-phase
Dark respiration, 310
Date palm, 292
Datura, 294
Daylength, 99, 357-58
Day-neutral plants, 75
Delphinium, 208
Demography, 24, 27-28
Desensitization, 151
Dianthus caryophyllus, 89
Dichanthium spp., 63
2, 4-Dichlorophenoxyacetic acid, 289, 292
Dieldrin, 53
Dimer hypothesis, 43
Dimethyl sulfide, 55
4, 6-Dinitro-ortho-cresol (DNOC), 277-80
Dioecious plants, 25-27, 171-72
Dioecy, 8
Dioscorea, 182
Diplospory, 10, 11
Disease resistance, 60
Dithionite, 123
DNA, 293, 296-97
  base substitution, 53
  chlorinated hydrocarbons, effect of, 53
  coding triplet, changing sense of the, 53
  frame shift, 53
Dormancy, 13, 159
Double fertilization, 5
Douglas fir, 295
Driselase, 293
Drosophila mercatorum, 62-63

Eggplant, 292
Embryo culture, 294-95, 297-300
Embryogenesis, 197-99, 287-303
Embryo sac, 4, 5, 15
Endive, 292
End-of-day exposure, 85
Endopeptidases, 118-19
Entomophily, 37
Environmental factors, 12, 307-321
Enzyme theory, 43
Ephyrophily, 38
Epilobium montanum, 218
Epinephrine, 151
Erschopfungstod, 204
Escherichia coli, 138
Estrogen, 151
Ethanol, 137-40
Ethephon (2-chloroethyl)-phosphonic acid, 189, 278-83
Ethylene
  aquatic plants and, 149
  auxin effect on ethylene synthesis, 150
  effect on sporangiophore of *Phycomyces,* 152
  flowering, 149, 166-67
  and leaf abscission, 149
  potentiation of GA response, 149
  product of ethephon breakdown, 283
  and senescence, 227
  sex expression, 171-73
  suppressants in fruit trees, 284
  tissue sensitivity to, 149
Ethylenediaminetetraacetic acid, 55
Eucalyptus, 354-56
Evapotranspiration, 313-14
Evocation, 158
Exhaustion death, 204

Fagus sylvatica, 204
Far-red energy, 99, 100
Far-red light, 100, 117-39
Fatty acids, 44
Feed-back-loop, 220, 221
Fertilization barrier, 41
Flash activation analysis, 119
Flowering
  apical dome diameter and, 264-66
  assimilate supply, 160
  bioregulation of, 282-84
  CO_2 effect on, 265
  critical time, 274, 280, 281
  cytokinin-induced promotion of, 266

daylength and, 265
evocation, 265
development, 170
flower-inhibitory substances, 158-60, 162, 165
in fruit trees, 273-86
hormonal control of, 161-79, 241, 266
inducing substances for, 157-79
induction of, 99-117, 163-64
inhibition of, 162-63, 265
initiation of, 76-77, 268, 273-74, 280-83
meristeme d'attente, 266
nutrient diversion hypothesis, 160, 266, 268-69
nutritional hypothesis and, 266
parthenocarpy and, 280
phloem loading, 265
photosynthates and, 265, 282
photosynthesis, 264-66
resource allocation and, 265
root growth and inhibition of, 268
seed development and, 273, 275
short-day plants, 160
stimulus, 266, 282
stimulus diversion hypothesis, 266, 269
sugar concentrations and, 265
temperature and, 265
vegetative growth and, 281
Fragaria × *ananassa* Duch., 76, 164, 185
Frame shift, 53
Fraxinus excelsior, 204
Frost hardiness, 354-56
Fruiting, 238, 277-80, 284
Fuchsia, 89

Gametophyte, 5, 7, 287-89
Geitonogamy, 29, 37
Gene mutation, 52
Gene regulation, 52, 58-65
Genetic
crossover, 61
drift, 65
fixation, 56
recall, 62-3
recombination, 52, 56-7
revolution, 64
variability, 245-59, 296, 347
Geotropism, 219
Geranium, 208, 327
Germination, 117, 135-43, 159, 198, 238
Gibberellin (GA)
aleurone layer, 152
α-amylase, 152
auxin production stimulated by, 282
axillary bud promotion, 181
binding sites, 150
dose-response, 152
floral initiation, 282-83
and flowering, 149, 161, 163-65, 282
inhibition of flowering by, 268
inhibition of tuber initiation by 181, 183, 186
and photoperiod, 226
production by fungi, 158
rhizome development by, 185
seed development and, 274-76
senescence and, 217, 225, 227
sex expression, 171-73
sink demand and, 266
stolon production by, 181, 183
synthesis, 149
tissue sensitivity to, 149
Glucagon binding sites, 151
Glucose-6-phosphate dehydrogenase (GDH), 267-68
Gluteraldehyde, 128
Glycine Max (L.) Merrill, 78, 208, 222, 245-47
Glycolipids, 44
Glycosphingolipids, 44
Gonadotropin, 151
Gossypium hirsutum, 212
Grain legumes, 78
Gramineae, 63, 215, 289, 294
Grand Rapids lettuce, 135-43
Grape, 292, 294
Grasses, 123, 289
Group selection, 14
Growth regulators, 147-56, 241
Gynodioecy, 8
Gynogenesis, 288, 293

Haploid plant production, 294, 297
Harvest index, *see* Nitrogen
Hedalgo, 208
Helianthus tuberosus, 182, 190, 207
Hermaphrodite, 8, 24-25
Heterostyly, 9, 25-26
Heterozygosity, 8
High energy response (HER), 101
High irradiance response (HIR), 101, 112, 120, 135, 141-42, 358-59
Histidine, 128
Homeostasis, 151
Homozygosity, 36
Hordelymus, 295
Hordeum, 295
Hordeum vulgare L., 99-115, 251
Hormonal control of flowering, 157-79
Hormonal control of tuberization, 185-90
Hormone sensitivity, 149-51
Hormones: (*see also* Abscisic acid, Auxins, Cytokinin, Ethylene, Gibberellin)
adaptation to, 152-53
in crop production, 241
dose-response relationship, 152-53
flowering, 157-79, 241
mode of action, 147-56
modulation, 149-51
negative cooperativity, 153
plant function, alteration of, 241
receptor, 153
Scatchard plots, 153
senescence and, 215-17
sex expression, 171-73, 241
status in fruit, 218
and stolon development, 183
stomatal control, 154
Hycanthone, 53
Hydrophily, 38
Hyoscyamus, 100, 294
Hyoscyamus niger L., 160

IAA, *see* Auxin
ICI pp333, 284
Immunity theory, 43
Impatiens balsamina L., 167
Impomoea tricolor, 149
Inbreeding, 8, 12, 28
Incompatibility, 29, 41-4
Indole-3-acetic acid (IAA), *see* Auxin
Induced-embryogenic determined cells (IEDC), 289-93
Induced-organogenic determined cells (IODC), 288-90
Infrared radiation, 326, 329
Intermediate-day plants, 75
In vitro propagation (IVP), 287-303
Ipomoea batatas, 182

Jerusalem artichoke, 207
Jute, 295

Kalanchoe, 81
Karyotype, 55
Kinetin, *see* Cytokin
Kranz anatomy, 202

Lactuca sativa L., 136
Large phytochrome, see Phytochrome
Law of universal regression, 313
Least squares analyses, 308
Lectin, 198
Legumes, 252-53, 294
Lemna spp., 85, 161, 164, 167-70
Lettuce, 296-97, 328
Leydig cells, 151
Light-off signal, 84
Linestuff theory, 43
Lipids, 44
Lolium spp., 92, 100, 150
Long-day plants, 75, 160, 163-64
Luffa cylindrica Roem, 172
Lycopersicon asculentum, 204, 212, 294
Lysosomal enzymes, 44
Lysosomes, 44

Make-up air systems, 361
Maize, 63, 245-47
Maize – *Tripsacum,* 58-65
Mangroves, 197
Manihot, 182
Master-gene theory, 43
Mathematical models of plant growth, 307-21
Megachromosomes, 55
Meiosis, 14
Membrane hyperfluidity, 136-42
Membrane permeability, 81
2-Mercaptoethanol, 125-26
Mercurialis annua L., 173
Metabolic death, 203
Metal ions, 123
6-Methoxy-2-benzoxazolinone, 150
Mineral nutrition, 239-40
Mirex, 53
Mobile dispersed genetic elements, 62
Modiola caroliniana, 219

Monocarpism, 201
Monoecious plants, 171
Monomorphic, 26
mRNAs, 154, 198
Multigene families, 62
Mutagenic agents, 296
Mycotoxins, 53

1-Naphthaleneacetamide (NAAm), 278-79, 284
1-Naphthaleneacetic acid, 150, 278, 289, 291
1-Napthyl N-methylcarbamate (carbaryl), 278-80, 284
Natural selection, 8, 14, 57
Nectar, 39
Nicotiana, 294
N. glauca, 289
N. langsdorfii, 289
N. sylvestris L., 160
N. tabacum, 289
N. tabacum L. var. Maryland mammouth, 160-61
N. tabacum L. var. Trapezond, 160
N. thyrsiflora, 353
Night-break, 81
Nitrate reductase, 253, 58
Nitrogen
  balance, 309
  dioxide, 332-50
  fixation, 239, 251-53
  flux, 245
  harvest indices, 248-49, 253-54
  loss, 255
  metabolism, 243-62
  nutrition, 192, 244
  oxides, 53
  remobilization, 248-49, 254-57
  Remobilization Efficiency (NRE), 254
  storage, 205
  translocation, 251
  uptake 251-53, 366
Nitrous acid, 53
Nucellus, 10
Null method, 84
Nutrient diversion hypothesis, 160, 264
Nutrition, 363-66

Oats, 125
Olfactants, 152
Open-top field chambers, 343-45
Organogenesis, 287-89, 293-95
Orthotropic habit, 182
Outbreeding mechanism, 8, 12, 28
Outcrossing, 29-30
Ovarial-substance theory, 43
Ozone, 53, 332-50

Panicoideae, 289
Panicum, 63
Parthenogenesis, 11, 279-80, 288
P-chloromercuribenzoate, 128
Pea, 219
Pears, 275
Pearl millet, 63, 292, 294
Pectinase, 293

Pentose phosphate pathway (PPP), 267-68
Pepper, 294
Perianth, 38
Perigon, 38
Perilla crispa (Thunb) Tanaka, 162
P. frutescens L. Britt., 208-9, 212, 219
Peroxidase-isoenzyme theory, 43
Petunia, 294
Pfr phytochrome, 81-2, 100, 118-34, 136-42
Pharbitis, 162-66, 265-66
Pharbitis nil, 79, 80, 87-88, 159
Pharbitis nil cv. Violet, 79
Phaseoleae, 78
Phaseolin, 198
Phaseolus vulgaris L., 170, 198, 218, 252
Phenology, 27
Phenotypic plasticity, 12
Phenylmethylsulfonyl fluoride, 126
Phloem, 161, 208-9
Phloem transport, 265
Phosphodiesterase, 151
Photochemical oxidant, 333-34
Photocontrol, 75
Photon fluence rate, 120, 135, 138-40
Photoperiod
  adenylate energy charge and, 276
  adenylate kinase activity and, 267
  continuous light, 76
  crop yield, 77
  daylength, 74-76, 324-25, 327-29
  dry weight, 105, 107
  flowering, 74
  and gibberellins, 225
  glucose-6-phosphate dehydrogenase activity and, 276
  induction, 144, 157-60, 169-70
  sensitivity, 76, 157-79
  and storage organs, 79
  temperature, 78, 79, 81
  time-measuring, 82
  tuberization, 79
  water stress, 74, 78
Photoperiodic induction, 149, 157-60, 169-70
Photoperiodic sensitivity, 76, 157-79
Photoperiodism, 100, 117, 135, 141-42
Photo-phase, 82, 85, 88
Photoreceptor, 80, 112, 118
Photoreversible chromopeptide, 119
Photosynthesis, 264-65, 282
Phycomyces blakesleeanus, 152
Phytochrome
  absorption spectrum, 118
  binding, 129
  bound, 142
  and chlorophyll absorbance, 122
  in cotyledons, 86
  cryogenic spectrophotometry, 119
  destruction in light grown plants, 126
  destruction of protein moiety, 125
  ethylene accumulation, 125
  flash activation, 119
  floral induction in LDP, 81
  floral induction in SDP, 81
  large phytochrome (LP), 119
  membrane property, 81, 136-37, 142
  molecular properties, 117-34, 139
  night break, 81, 100
  nonphotochemical reactions, 122
  pelletability, 124, 139, 141
  photostationary equilibrium, 120
  photothermal interaction, 135
  phototransformation pathway, 119-39, 141
  proteolytic degradation, 119
  radioimmunoassay for, 126-27
  reversion, 81, 122-28
  seed germination, 135-43
  sequestering, 126, 128-29, 140-41
  small phytochrome, 120
  thermal adaptation, 137
  transformation, 83, 118-22, 135-43
Phytotoxic air pollutants, 333-50
Pistia stratiotes L., 167
Pisum sativum L., 164-65, 222, 225
Plagiotropic shoots, 182
Plant hormones, 147-56
Poa annua var. reptans (Hauskins) Timm., 201
Poacoideae, 298
Poinsettia, 328
Pollen tube, 338
Pollination, 8, 13, 15, 297-300
Pollinators, 27, 29, 35, 38
Polycarpism, 201
Polychlorinated biphenyl (PCB), 53
Polyembryony, 13, 291
Polygonatum, 202
Potato, 182-83, 294-97
Potential evapotranspiration (PET), 313-18
Potentilla, 354
Prechilling, 136-37
Preembryogenic determined cells (PEDC), 290-92
Principle of non-functioning, 42
Propagules, 10
Protandry, 9
Protein production, 250, 256
Protein-synthesis, 159, 198, 255-57, 283
Proto-angiosperms, 15
Protogyny, 9
Protoplast fusion, 293-94, 297
Pseudobulbils, 291
Pseudocopulation, 15, 39
Pseudogamy, 12
Pteridium, 182
Pumpkin, 292
Pyridine nucleotides, 292

Quack grass, 185

Radiant flux density, 351, 357-59
Ranunculus sceleratus, 290
Rapeseed, 294
Raphamus sativus, 215
Red-light, 100–101, 117-34
Refractoriness, 151
Regulator theory, 43
Regression statistics, 313-14
Reproductive versatility, 12
Reseda odorata, 203

Resource allocation, 265
Rhizobitoxine, 283
Rhizobium, 252-53
Rhizome, 182
Rhodopsin, 153
Rhozyme, 293
Rice, 61, 78, 255-57, 290, 295, 297, 308
Ricinus communis L., 173, 208
RNA-synthesis, 257
Root
  growth, 214-15, 219, 268, 289
  meristem, 148
  temperature, 339
  zone environment, 364-66
Rosaceae, 327
Rottboellia exaltata, 13-14
Rubber, 294
Runners, 182
Rye, 61, 294

Saccharininae, 63
Salicylic acid, 162, 167-69
Sambucus, 202
Sandoz 9789, 126
Scrofularia, 266
Scrophularia arguta Sol., 165
Secale, 295
Secondary xylem, 209
Seed
  abortion, 279-80
  development, 198, 274
  dormancy, 197-98
  germination, 117, 135-43, 159, 198, 238
Selection screens, 57
Selective pressure, 52
Self-fertilization, 56
Self-incompatibility, 8, 41, 297
Selfing, 28
Self-pollinating, 12, 28, 44
Senescence, 201-33
Sex expression, 171-73
Sexual-affinity theory, 43
Sexual polymorphism, 8
S-genes, 41-4
Shamouti orange, 291
Shoot regeneration, 148
Short-day, 75, 160, 182, 209, 215
Sib-mating, 56
Silene armeria L., 164
Sinapis alba L., 102, 124, 159, 166, 265-68
Sink activation, 263-64, 267
Sink-source, 207, 263-72, 311, 338-40
Skoto-phase, 82
Soil-water relationships, 309, 312
Solanum spp. 181-95, 294
Somatic embryogenesis, 290-93
Sorghum spp., 61, 63, 78, 253-54
Soya bean, 219
Soybean spp., 152, 245-47, 310-12, 345-48, 352, 354, 357, 360-64, 366
Spectral sensitivity, 323-31
Spinach, 149, 173, 267, 345-48
Sporophyte-gametophyte transition, 5-6
Statistical models, 308-21
Stem apex, 148
Steroids, 159
Stigma, 37-9
Stigmatic exudate, 39
Stoffwechseltod, 203
Stolon development, 181-95
Stomata, 154, 309-10, 338
Storage protein, 198
Strawberry, 79, 294, 328
Stress, 239, 313-14, 353
Stylosanthes, 77
Sub-sexual, 12
Succinic acid-2, 2-dimethylhydrazide (daminozide), 279-83
Sugarcane, 84, 289, 296
Sulfur dioxide, 53, 333-50
Sulfur oxides, 53
Sunflower, 185
Sweet potato, 182, 294
Synergids, 5, 12
Syringa, 202

Temperature, 240, 265, 351-66
Teosinte, 58
Themeda australis, 74-5
Thermal radiation, 326
Thermal sensitivity, 323-31
Thermoperiodism, 354-56
Tillering, 13
Tobacco, 56, 160, 221, 268, 294, 297, 354, 360-64
Tomato, 56, 204, 268, 295, 327, 363
Totipotency, 289
2, 3, 5-Triiodobenzoic acid (TIBA), 283
Tri-partite theory, 43
Tripsacoid, 58
Tripsacum spp., 58-60, 295
Triticum spp. 245-47, 295
Tuberization, 79, 181-95, 207
Turnip, 352
Twilight, 85, 101
Tyrosine, 128

Ultraviolent radiation, 328

Vicia faba, 203
Vinyl chlorides, 53
Visable radiation, 323-31

Water, 363-66
Water stress, 240, 314, 356, 363-64
Wheat, 61-2, 78, 149, 245-46, 294, 313-14, 345-48

Xanthium spp., 75, 80, 85, 100, 159-71, 185, 208, 221, 265
Xenogamy, 37
Xenopus, 62

Yams, 182
Yeast, 291

Zea spp., 58, 63, 171, 212, 245-47, 295
Zeatin, 183
^{14}C-Zeatin riboside (ZR), 184

	DATE DUE		